DESIGN-BUILD

DESIGN-BUILD ESSENTIALS

THE DESIGN-BUILD LIBRARY

Volume I: Design-Build Essentials
Volume II: Design-Build: The Integrated Project Team
Volume III: Design-Build Project Management
Volume IV: Design-Build: An Owner's Reference

DESIGN-BUILD
DESIGN-BUILD ESSENTIALS

Barbara J. Jackson, Ph.D., DBIA
Director, California Center for Construction Education,
  Construction Management Department
College of Architecture and Environmental Design
California Polytechnic State University, San Luis Obispo

DELMAR
CENGAGE Learning

Australia • Brazil • Japan • Korea • Mexico • Singapore • Spain • United Kingdom • United States

Design-Build Essentials
Barbara J. Jackson

Vice President, Career and Professional Editorial: Dave Garza

Director of Learning Solutions: Sandy Clark

Senior Acquisitions Editor: James DeVoe

Managing Editor: Larry Main

Product Manager: Mary Clyne

Development: Barbara Clemens, Ann Shaffer, Mary Pat Shaffer

Editorial Assistant: Cris Savino

Vice President, Career and Professional Marketing: Jennifer McAvey

Marketing Director: Deborah Yarnell

Associate Marketing Manager: Mark Pierro

Production Director: Wendy Troeger

Production Manager: Mark Bernard

Content Project Manager: Mike Tubbert

Art Director: Casey Kirchmayer

Chapter 1
Macintosh is a registered trademark of Apple Inc.
The Austin Method is a registered trademark of The Austin Company.
Designated Design-Build Professional is a trademark of the Design-Build Institute of America.

Chapter 3
MasterFormat is a trademark of the Construction Specifications Institute

Chapter 7
ADePT, ADePT Design Builder, and ADePT Design Manager are trademarks of Adept Management Ltd.

Chapter 8
LEED is a registered trademark of the U.S. Green Building Council

Chapter 9
OPS, BIMstorm, and BIMBomb are trademarks of ONUMA, Inc.
buildingSMART alliance is a trademark of the National Institute of Building Sciences
FileMaker is a registered trademark of FileMaker, Inc.
ArchiCAD is a registered trademark of Graphisoft R&D Zrt
Certified Design-Build Professional is a trademark of the Design-Build Institute of America
Expedia is a registered trademark of Expedia, Inc

© 2011 Delmar, Cengage Learning

ALL RIGHTS RESERVED. No part of this work covered by the copyright herein may be reproduced, transmitted, stored, or used in any form or by any means graphic, electronic, or mechanical, including but not limited to photocopying, recording, scanning, digitizing, taping, Web distribution, information networks, or information storage and retrieval systems, except as permitted under Section 107 or 108 of the 1976 United States Copyright Act, without the prior written permission of the publisher.

> For product information and technology assistance, contact us at
> **Professional & Career Group Customer Support, 1-800-648-7450**
>
> For permission to use material from this text or product, submit all requests online at **cengage.com/permissions**.
> Further permissions questions can be emailed to **permissionrequest@cengage.com**.

Library of Congress Control Number: 2009942094

ISBN-13: 978-1-4283-5303-9
ISBN-10: 1-4283-5303-8

Delmar
5 Maxwell Drive
Clifton Park, NY 12065-2919
USA

Cengage Learning products are represented in Canada by Nelson Education, Ltd.

For your lifelong learning solutions, visit **delmar.cengage.com**

Visit our corporate website at **cengage.com**.

Notice to the Reader
Publisher does not warrant or guarantee any of the products described herein or perform any independent analysis in connection with any of the product information contained herein. Publisher does not assume, and expressly disclaims, any obligation to obtain and include information other than that provided to it by the manufacturer. The reader is expressly warned to consider and adopt all safety precautions that might be indicated by the activities described herein and to avoid all potential hazards. By following the instructions contained herein, the reader willingly assumes all risks in connection with such instructions. The publisher makes no representations or warranties of any kind, including but not limited to, the warranties of fitness for particular purpose or merchantability, nor are any such representations implied with respect to the material set forth herein, and the publisher takes no responsibility with respect to such material. The publisher shall not be liable for any special, consequential, or exemplary damages resulting, in whole or part, from the readers' use of, or reliance upon, this material.

Printed in the United States of America
5 6 7 8 9 10 11 22 21 20 19 18

DEDICATION

I dedicate this book to the next generation of design-builders—the young constructors, architects, and engineers who will lead this magnificent industry through its most significant transformation.

Barbara Jackson

CONTENTS

PREFACE XI

ABOUT THE AUTHOR XV

ACKNOWLEDGMENTS XVI

CHAPTER 1   **TRANSFORMATION OF AN INDUSTRY**   3

From Master Builder to Segregated Services   4
Severing the Ties That Bind   4   *The Slippery Slope*   6

Low-Bid Mentality   6

Economic Conditions & Outside Influences   8
Decade of Decadence   8   *Change on the Horizon*   9
More Bad News   10

In Search of a Better Way   11

Constructor as a Professional   13

Re-Emergence of Design-Build   15
Design-Build Gains Prominence   16   *The Design-Build Institute of America*   18
Legislative Action   20   *Federal Acquisition Regulations Modification*   21

Case Study: The Power of a Promise   22
The Lead-In   22   *The Story*   22   *The Close*   26

Increased Collaboration and Integration   28
The Collaborative Process Institute White Paper   28   *CURT White Papers*   29

The Role of Technology   31
Building Information Modeling   31   *Architects Getting on Board*   32

New Initiatives, New Strategies   33
Integrated Project Delivery   33   *Lean Construction*   34
Relational Contracting   35   *Leadership in Energy and Environmental Design—LEED*   36

Back to the Future   37
The New Master Builder   37

CHAPTER 2   **PLANNING, DESIGNING, AND CONSTRUCTING PROJECTS**   41

Project Delivery Methods   42
Design-Bid-Build   42   *CM-at-Risk*   47   *Design-Build*   51

Comparing the Project Delivery Methods   57

Delivering Beyond Expectations   59
Total Facility Solutions   64

Case Study: The Sweet Smell of Success   60
  The Lead-In   60   *The Story*   60
  The Close   63
Managing Design-Build Projects   65
Making Smart Project Delivery Decisions   67
  Doing Your Homework   68
Different in Every Way   69

CHAPTER 3

FUNDAMENTAL CHARACTERISTICS OF DESIGN-BUILD   73

Solicitation and Basis of Award   74
  Requests for Proposals (RFP) Versus Invitation for Bids (IFB)   74
  Contractual Configuration   75
Team Approach   75
  Early Contractor Involvement   75   *Proactive Versus Reactive Response*   76
The Design-Build Entity   78
  Contractor-Led Design-Build   78   *Designer-Led Design-Build*   84
  Integrated Design-Build Firm   85   *Other Models*   85
  The Concept of the Controlling Design-Build Party   85
Case Study: To the Rescue   79
  The Lead-In   79   *The Story*   79   *The Close*   83
Managing Risks   87
Performance Requirements   89
  Communicating Performance   90   *Design Specification Evolution*   90
  Verifying Performance   91   *Reducing Waste*   92
Who Controls the Details of the Design   92
The Relationship Between Design and Cost   96
  Developing Costs in Design-Build   96   *Estimating Formats in Design-Build*   97
The Integrated Design-Build Schedule   97
Contracting Approach   98
  Governing Documents   98   *Design-Build Contract Issues*   100
  Relational Contracting   102   *High Performance Contracts*   102
A New Paradigm   103

CHAPTER 4

BUYING DESIGN-BUILD SERVICES   107

Acquisition Strategies   107
  Go Slow to Go Fast   108   *Mining for the Gold*   109
Source Selection   112
  Low-Bid Procurement   113   *Qualifications-Based Selection*   114
  Best Value   116
How the RFQ – RFP Process Works   117
  Qualifications Phase   118   *Best Value Phase*   118
Project Design as a Competitive Factor   119
How the Proposal is Scored   124

Other Evaluation Methods   126
  Project Enhancement Approach   129   ▪   *Negotiated Design-Build*   132
Case Study: Icing on the Cake   134
  The Lead-In   134   ▪   *The Story*   134   ▪   *The Close*   137
Amount of Design in the RFP   138
Selecting a Contract Type   141
  Incentivizing for Performance   143   ▪   *Applying Award Fees*   144
You Get What You Plan For   147

CHAPTER 5

GETTING READY TO COMPETE   151

Know the Owner   152
  Reading between the Lines   152   ▪   *Preemptive Marketing*   153   ▪   *Procurement Workshops*   155   ▪   *Show Off Your Best Qualities*   156
Selecting the Design-Build Team   157
  Blending Cultures   160   ▪   *Stepping Outside Your Comfort Zone*   162
Go or No Go   163
  Step 1 – Analyze the RFQ   170   ▪   *Step 2 – Respond to the RFQ*   171   ▪
  Step 3 – Make the Short List   171   ▪   *Step 4 – Analyze the RFP*   172   ▪
  Step 5 – Conduct a Design Charrette   172   ▪   *Step 6 – Produce the Proposal Documents*   174   ▪   *Step 7 – Deliver the Proposal*   175   ▪   *Step 8 – Prepare the Oral Presentation*   175   ▪   *Step 9 – Deliver the Presentation*   176   ▪
  Step 10 – Win the Award   178
Case Study: Dancing on the Edge   164
  The Lead-In   164   ▪   *The Story*   164   ▪   *The Close*   169
Delivering on the Promise   179

CHAPTER 6

DEVELOPING DESIGN-BUILD ESTIMATES   183

Pricing the Unknown   184
Choosing Value and Quality Over Low Price   185
Conceptual Estimating in Design-Build   186
  The Designers' Role   189   ▪   *Filling in All the Holes*   190
Cost as an Element of Design   192
Designing to Budget: Five Stages   192
  Programming Stage   193   ▪   *Conceptual Design Stage*   193   ▪   *Schematic Design Stage*   193   ▪   *Design Development Stage*   194   ▪   *Construction Document Stage*   194
Developing the Budget   195
  Purpose of the Cost Model   203   ▪   *How the Cost Model Evolves*   204
Case Study: Balancing Act   196
  The Lead-In   196   ▪   *The Story*   196   ▪   *The Close*   201
Progressive Estimating Phases   204
  Feasibility Estimate   205   ▪   *Concept Estimate*   206   ▪   *Schematic Estimate*   206   ▪   *Design Development Estimate*   207   ▪   *Detailed Estimate*   207   ▪   *The Estimating Churn*   208   ▪   *Beyond Milestones and Phases*   208

Qualifying the Numbers 210
Accounting for the Unknown 211
   Contingency Development 212 ▪ Assessing Design Risks 213 ▪
   Avoiding Redundant Contingencies 215 ▪ Owner's Contingency 216 ▪
   Displaying Contingencies in the Estimate 216
Adding Value as You Go 218
Role of the Estimator in Design-Build 221
Embracing Uncertainty 222

CHAPTER 7   MANAGING THE DESIGN-BUILD PROCESS 229

Design-Build as a Process 230
   Shifting Paradigms 230
Managing Design in Design-Build 231
   The Design Process 237 ▪ The Design-Cost Interface 240 ▪
   The Design-Construct Interface 241 ▪ The Design-Performance Interface 243
Case Study: All Aboard 232
   The Lead-In 232 ▪ The Story 232 ▪ The Close 236
Establishing Standard Practice 244
Team, Techniques, and Tools 245
   The Design-Build Team – Who 245 ▪ The Design-Build Techniques –
   What 245 ▪ The Design-Build Tools – How 245
Design-Build Management Stages 245
Proposal Stage 246
   Management Task: Establishing Team Focus and Alignment 246
Postaward Stage 252
   Management Task: Proposal and Design Validation 252 ▪ Management Task:
   Design Process Planning 255
Design-Development Stage 259
   Management Task: Design-Cost Reconciliation 259 ▪ Management Task:
   Design-Performance Reconciliation 262
Design-Construct Stage 267
   Management Task: Design Delivery Schedule Alignment 268
Postconstruction Stage 270
   Management Task: Postconstruction Evaluation 271
The Design-Build Management Team 273
   Roles and Responsibilities 276 ▪ The Design-Build Project Manager 278 ▪
   The Design-Build Superintendent 279
But Are You Ready? 280

CHAPTER 8   THE DESIGN-BUILD TEAM 285

The No Gaps Game 286
Making the Mental Shift 286
Teaming for Results 290
   Not Just Any Team 291

Who should be on the Team? 292
   Strategic Partners 293
Establishing the Ground Rules 294
   Proposal Development Phase 294 ▪ *Post-Award Phase* 296
Building Trust 300
   Distinguishing Trust 301
Getting the Right People on the Bus 306
   Design-Builder Characteristics 306
Advocating for the Project 311
Case Study: Practicing What You Preach 312
   The Lead-In 312 ▪ *The Story* 313 ▪ *The Close* 316
Communication, Collaboration, and Integration 317
   Communication 317 ▪ *Some Useful Techniques and Tools for Communication* 319 ▪ *Collaboration and Integration* 321
Leading the Design-Build Team 323
Learning How to Dance 325
   Not for Everyone 326

CHAPTER 9

IMPLICATIONS FOR THE FUTURE 331

The Perfect Storm 331
Integrated Project Delivery 332
Building Information Modeling 334
Case Study: BIM and Beyond 338
   The Lead-In 338 ▪ *The Story* 338 ▪ *The Close* 342
Sustainable Design 344
Lean Design and Construction 346
The Collaboration Economy 348
The Next Generation 352
   Implications for Education 354 ▪ *Expanding the Design-Build Curriculum* 354
Getting Owners on Board 355
The Tipping Point 356

ENDNOTES 360

GLOSSARY 364

CONTRIBUTORS 374

INDEX 404

PREFACE

According to the Design-Build Institute of America, design-build project delivery currently accounts for more than 40% of all non-residential design and construction in the United States. Forty-nine states have adopted procurement legislation allowing contractors to use design-build project delivery for public projects. All major construction firms identify the pursuit and management of design-build projects as part of their overall business plan.

Design-build is and will remain a primary project delivery method in the United States. As its popularity and implementation continue to increase, so does the likelihood that contractors, architects, and engineers will be required to deliver projects using design-build methods. We are seeing more and more owners listing design-build education as part of their RFP evaluation criteria.

THE DESIGN-BUILD LIBRARY

I wrote this series to convey my passion for design-build practice, to encourage industry practitioners who want to develop successful design-build teams, and to encourage instructors who want to support successful design-build programs. The Design-Build Library introduces and thoroughly develops the management concepts these professionals will need to understand in order to mitigate risks, optimize results, and succeed in the design-build arena. The four-volume library provides a complete tutorial on design-build practice and project execution:

>   Volume I: Design-Build Essentials
>   Volume II: Design-Build: The Integrated Project Team
>   Volume III: Design-Build Project Management
>   Volume IV: Design-Build: An Owner's Reference

Whether you are using this series as a student in a construction management program, as an instructor in a college or professional development program, as a practitioner, or as an owner, it will provide the tools and techniques you need and the confidence to use them.

Features of the Design-Build Library

Virtually everything about the design-build process is different from the design-bid-build process. The solicitation process, evaluation and selection process, team make-up, business relationships, design process, estimating procedures, design-cost interface, and project management process all change when you adopt the highly collaborative and integrative design-build model. However, because design and construction professionals are so accustomed to doing business the traditional way, they often find it hard to appreciate the differences between the two approaches. Thus, they fail to capitalize on the potential for the design-build process to transform their business. This four-book series will focus on doing design-build right, explaining how to establish standard practices for ensuring consistent design-build project delivery and management.

In keeping with the dynamic nature of design-build, I wanted everything about this series to be different from standard textbooks. These are not stuffy, academic books written by a non-practitioner. Each book in the Design-Build Library talks about design-build in a conversational way. Each chapter includes industry input that shows the faces and tells the stories of real contractors, architects, engineers, specialty contractors, superintendents, vendors, and owners. Readers will encounter real examples of projects described from multiple perspectives.

I want people to read the personal stories, testimonials, and case studies in these books and talk about them. I want these books to serve both the AEC industry and the university markets. Most of all, I want people who have read these books and tried this delivery method to recommend Design-Build to other industry practitioners.

Design-build is the future for young constructors, architects, and engineers. University construction management, architecture, and engineering programs are gearing up to incorporate design-build material into their curriculum. Construction management programs in particular have recognized the need to teach design-build and other integrated project delivery models as a distinct segment of their curriculum. Several four-year universities have already established Design-Build Masters or Integrated Project Delivery programs. I designed this series to teach more than just what design-build is and how it works. This series will help students learn how design-builders think, how they communicate, and how they interact to solve problems together.

DESIGN-BUILD ESSENTIALS: A CAREFULLY DESIGNED RESOURCE

As the first volume in the Design-Build Library, *Design-Build Essentials* is truly a design-builder's resource. The book begins by describing the industry transformation that is taking place today, identifying the fundamental distinctions of design-build, and distinguishing the collaborative nature of the process. The book then teaches how to manage the design-build process and build an integrated design-build team. It concludes with a look at the future of the industry.

Readers will benefit from the following features in every chapter of *Design-Build Essentials*:

:: Case Studies identify specific design-build challenges in real-world projects ranging from the rebuilding of the United States Pentagon to the design and construction of a new production facility for Krispy Kreme donuts. In each case, readers see the context in which the challenge arose and discover how the design-build team resolved it.

- Pioneer Perspectives highlight the stories behind the individual contractors, architects, engineers and owners who have been at the forefront of the design-build movement in the United States. These ground-breaking practitioners offer valuable historical perspectives linked to specific chapter content. As the design and construction industry continues to move toward a more integrated services model, the insights presented by these pioneers show us the way forward.
- Industry Quotes from contractors, architects, engineers, and owners highlight the opinions and ideas of the people working on the front lines of the design-build business today.
- Mentoring Moments bridge the student-practitioner gap by presenting advice from recent graduates who are working on design-build teams. These newest practitioners offer fresh perspectives on what to study, what to look forward to, and how to prepare.
- Takeaway Tips provide practical advice that can be incorporated into design-build practice immediately.
- Author Editorials explore important issues facing our industry. As part of my research for this text, I interviewed a broad sampling of industry professionals, gathering perspectives from every sector of the market. As I considered the input of stakeholders from across the United States, I was struck by their insights regarding the transformation and future of the construction industry. Thus, the Author Editorials are my attempt to synthesize the current perspectives of some of the top design-builders in the industry today.
- D-B Dictionary Definitions define essential design-build terms. Because the design-build process differs so completely from traditional project delivery, it requires a whole new lexicon. I have spent several years developing this lexicon from the practices and approaches that design-builders use to meet the unique challenges of the design-build process. The D-B Dictionary Definitions clearly highlight and define terms that take on a unique meaning within the design-build context, as well as terms that were coined to facilitate design-build communication.
- Chapter Highlights summarize the key points and best practices developed in the chapters.
- Recommended Readings, at the end of each chapter, encourage readers to expand their understanding of the design-build approach through supplemental reading of various business and professional development books.
- A special Appendix profiles the many firms and individuals who offered their experience, examples, case studies, best practices, and other design-build wisdom from the marketplace.

INSTRUCTOR RESOURCES

Delmar Cengage Learning offers robust Instructor Resources to accompany *Design-Build Essentials*. With this unique resource on CD-ROM, Instructors can spend less time planning and more time teaching. The Instructor Resources CD includes:

- An Instructor's Manual containing instructional outlines and various resources for each chapter of the book, available in Adobe Acrobat PDF® format.
- A Computerized Testbank in ExamView®. With hundreds of questions and different styles to choose from, instructors can create customized assessments for students and add unique questions and print rationales for easy class preparation.
- Customizable instructor support slide presentations in Power Point® format that focus on key points for each chapter.
- An Image Library to enhance instructor support slide presentations, insert art into test questions, or add visuals wherever you need them. These valuable images, which are pulled from the accompanying textbook, are organized by chapter.

ABOUT THE AUTHOR

Barbara J. Jackson, Ph.D., DBIA, is Director of the California Center for Construction Education, an educational outreach organization at California Polytechnic State University in San Luis Obispo, California, which serves the AEC industry and owner groups. She is also a Professor of Construction Management in the College of Architecture and Environmental Design. Dr. Jackson's focus is on integrated project delivery, design-build project management, and interdisciplinary collaboration. Dr. Jackson spent twenty years in the construction industry working as a project estimator and senior project manager and then as President and Owner of Design/Build Services Inc. She holds a B.S. in Housing and Design, a Masters in Construction Management, and a Ph.D. in Education and Human Resources from Colorado State University. She served on the National Board of Directors for DBIA from 2007–2009 and currently serves as the National Education Chairman, received a Design-Build Distinguished Leadership Award in 2001 and 2002, and received the DBIA Distinguished Service Award in 2005. Dr. Jackson has given presentations to or trained a multitude of private companies, professional organizations, and public agencies on integrated project delivery, design-build, and high performance project teams. The list of organizations she has worked with includes: the American Society of Civil Engineers (ASCE), Design-Build Institute of America (DBIA), Construction Specification Institute (CSI), PENREN Renovation Joint Venture (DMJM/3DI), Mortenson Construction, National Guard Bureau, CDM Incorporated, Sundt Construction, The Bannett Group, PCL Construction, 3-D International, Plan One Architects, South Pacific Division-US Army Corp of Engineers, Armada Hoffler, Rosendin Electric, Santa Fe Indian School, Korte Construction, GKK Works, Hawaii Department of Education, Panama Canal Authority, Albuquerque School District, and others. She also provides private consulting services, customized training, and project team facilitation.

ACKNOWLEDGMENTS

I acknowledge with enormous gratitude the many design-build professionals and associates who contributed to this book by supplying interviews, quotations, photographs, illustrations, and samples. Most of you are highlighted in the appendix of this book and I can't thank you enough for your valuable time, expertise, and wisdom.

I extend a heartfelt thanks to my most valuable and capable assistant, Jessica Frazier, without whose help this book would have never been completed. Her persistence in tracking down and organizing the many photographs, illustrations, permissions, and releases necessary in a book like this was Herculean. I also wish to acknowledge my assistant, Tana Anastasia, as well, for helping me keep my schedule and life on track during the writing of this book and for her committed dedication to my success on a daily basis.

It is with great appreciation that I acknowledge several staff members at the Design-Build Institute of America for providing important data and information to support many aspects of the book. Special thanks to Lee Evey, Lisa Washington, Stephanie Zvonkovich, Kate Dodd, Richard Thomas and Patricia Carpio.

I'd also like to thank my editor Mary Clyne and her team Ann Shaffer, Barbara Clemens, and Mary Pat Shaffer for their superb effort and dedication in helping make this an exceptional book.

And last but not least, I thank my husband, Wayne, for his everlasting patience and steadfast support in all of my endeavors. I also want to thank my children, Andrea and Jason. Although grown and out of the house, their phone calls to ask about "the book" were always greatly appreciated.

In addition, the publisher wishes to acknowledge the following reviewers for contributing their invaluable expertise to the development of this book:

Dave Engdahl
Senior Vice President, Chief Architect (retired)
The Haskell Company
Jacksonville, Florida

David Gunderson
Associate Professor
Washington State University
Pullman, Washington

Robert Hartung
Founder/President
Alternative Delivery Solutions LLC
Laguna Niguel, California

Mike Mezo
Associate Professor
Ball State University
Muncie, Indiana

Kelly Strong
Associate Professor
Iowa State University
Ames, Iowa

John Tocco
Professor
Lawrence Technological University
Southfield, Michigan

Peter Tunnicliffe
Corporate Director of Project Development
CDM
Cambridge, Massachusetts

Barbara Wagner
Senior Vice President
Clark Construction Group, LLC
Costa Mesa, California

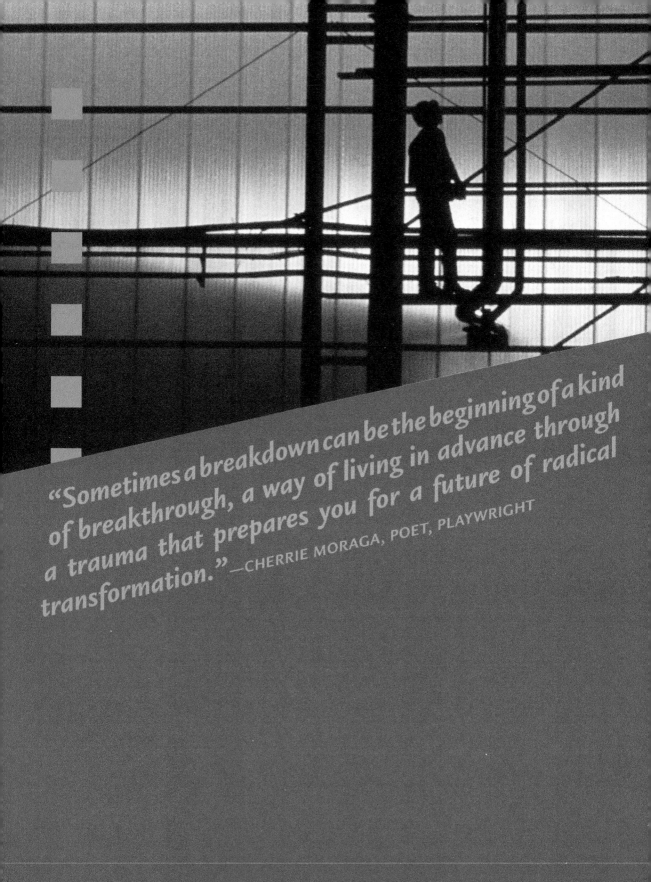

"Sometimes a breakdown can be the beginning of a kind of breakthrough, a way of living in advance through a trauma that prepares you for a future of radical transformation." —CHERRIE MORAGA, POET, PLAYWRIGHT

1 TRANSFORMATION OF AN INDUSTRY

*I*f you look up the word "transformation" in the dictionary you will find a definition that reads something like this: *Transformation – a noticeable change in character, appearance, function or condition.* There is no doubt that the design and construction industry has evolved throughout its history. We have come a long way from building huts with mud and straw, using rocks and timber as our tools. Today advanced materials, methods, and technologies, such as Global Positioning Systems (GPS), computer-aided earth moving systems, robotics, and four- and five-dimensional visualization techniques, as well as high performance steel, super-strength concrete and advanced polymers, can be used on any given project. But these types of industry advances, although remarkable, still represent independent, isolated improvements—in other words, nothing that fundamentally changes the traditional process and approach associated with designing and constructing buildings and structures.

Today, however, we stand on the verge of a revolutionary transformation—and at the heart of this transformation is a progressive merging of minds unlike anything we have seen before. Contractors, architects, engineers, and owners are closer than ever to collaborating across all functions of the design and construction process, with **design-build**, combined with other integration strategies, at the forefront. Combine this merging of minds with new integration technologies, such as Building Information Modeling (BIM), and the "most collaborative" generation of young, tech savvy design and construction professionals in history, and it seems that we may be very close to a tipping point that could alter the way we do business forever—a revolutionary breakthrough. In his book *Leading the Revolution*, Gary Hamel calls this kind of innovation **non-linear innovation**, because it goes beyond incremental improvement into the development of a whole new business concept. According to Hamel, "the gap between what can be imagined and what can be accomplished has never been smaller."[1]

D-B DICTIONARY

Design-Build –
A project delivery method in which design and construction services are provided under a single contract using a team approach to deliver an integrated project solution, which is in contrast to the two contract method of design-bid-build in which designers and contractors deliver their independent services under separate contracts.

Non-linear Innovation –
Innovation that goes beyond predictable improvement to encompass whole new ways of doing business or activities. Examples include using global positioning satellites in surveying instead of transits and total stations, or using digital technology instead of 35mm film in photography.

We have just scratched the surface of the potential capabilities of the design-build approach for delivering highly successful projects. The fact is, traditional roles, thinking, and contracting methods have kept us from taking advantage of the efficiencies, creativity, innovations, and extraordinary results that an integrated design-build process can deliver. In other words, "the way we have always done things" is getting in the way of "getting things done." Design-build, when it is executed and managed correctly, can simultaneously reduce costs, shorten schedules, and improve performance, thereby adding the kinds of value that project owners are clamoring for today.

Many factors have kept the architecture, engineering, and construction (AEC) industry from reaching its collaborative potential, but probably the biggest hindrance is the industry's history itself. So before we take a look at what makes design-build unique and effective, let us take a look back and see how we got off track in the first place.

FROM MASTER BUILDER TO SEGREGATED SERVICES

When people discuss the design-build approach, they often refer to the old Master Builder concept. The term Master Builder has been used to describe the individual responsible for both designing and constructing historic marvels such as the Dome of Santa Maria del Fiore in Florence, Italy, designed and built by the quintessential Master Builder, Filippo Brunelleschi in 1420. (It was the first octagonal dome in history to be built without a wooden supporting frame, and was the largest dome built at the time. It still stands today as the largest masonry dome in the world.)

The Master Builder was responsible for coordinating and integrating all aspects of a project's design and construction, including aesthetics, proportion, quality, and function, along with manpower, materials, costs, and schedules. The Master Builder of ancient times was not an architect, engineer, or general contractor, but was instead a distillation of all three. As he made many decisions required throughout the course of a construction project, his every interaction with artisans, laborers, merchants, and project sponsors filtered through his mind. In other words, all the pieces of the design and construction puzzle were stored in a single location—in the Master Builder's brain, providing a single-source accountability without the complications of conflicting agendas and independent, special interests.

The Master Builder approach held sway from the Middle Ages through the early years of the Renaissance. However, as the Renaissance progressed, the functions of design and construction began to diverge, becoming separate tasks. The Master Builder concept eventually evolved into a segregated services model, with separate individuals performing design and construction as specialized functions without the benefit of a single-minded perspective.

Severing the Ties That Bind

Leon Battista Alberti (born 1404–died 1472) is often referred to as the first modern-day architect. He is credited with initiating the movement to

---

A Master Builder communicated his design intent and construction detailing through a physical 3-D model that was carried to the jobsite. As the industry evolved, two dimensional plans and specifications began to be used. These were far more portable and did not require the designer to be on site to explain all of the particulars. Today, the model is again the preferred communication method—only now we use a computer generated digital 3-D model that can be transmitted via email.

separate the art of design from the craft of building. According to Jeffrey Beard, Michael Loulakis, and Edward Wundram in *Design-Build-Planning through Development*, in the mid-1400s Alberti convinced Pope Eugene IV that he could direct the construction of a new façade on Florence's Gothic church, Santa Maria Novella by way of drawings and models, without directly supervising the work.[2] For the first time in history, these plans and diagrams enabled the "designer" to instruct the "builder," thus overseeing the project from a distance. This division of responsibilities set the stage for the segregated services model that forms the basis for the traditional design-bid-build approach used today.

The Industrial Revolution encouraged further separation between the design and construction functions. Specialized design and construction expertise was needed to address unique production and facility needs. Cities expanded, drawing people from their rural homes. This created the need for new urban housing, as well as infrastructure, buildings of commerce, government facilities, transportation lines, etc. New tools, equipment, and building materials gave rise to even more specialized expertise among the various divisions of the design and construction functions.

As design and construction expertise grew more sophisticated in the mid 1800s, architects and engineers formed separate professional societies in an effort to elevate the status of their professions. The American Society of Civil Engineers (ASCE) was organized in 1852. The American Institute of Architects (AIA) formed in 1857. Contractors eventually followed suit, creating the Associated General Contractors (AGC) in 1918. The establishment of professional identity through independent organizations continued throughout the nineteenth and twentieth centuries to include the Associated Builders and Contractors (ABC), the American Subcontractors Association (ASA), the Mechanical Contractors Association of America (MCAA), and many others.

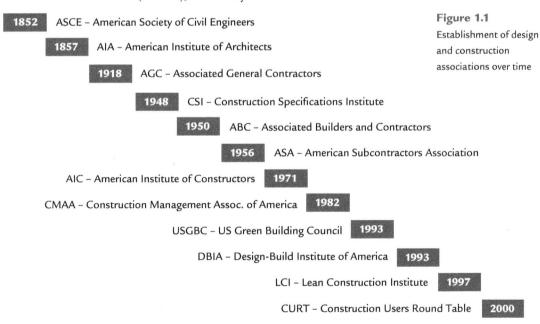

Figure 1.1
Establishment of design and construction associations over time

By the end of the nineteenth century, the notion of the Master Builder was long gone. New facilities and buildings were being designed and constructed by specialists, each of whom focused on an ever-narrowing set of concerns. The first "legal" requirement that effectively separated design from construction occurred in 1935, when the Miller Act went into effect. This important law required contractors to post two surety bonds (one for performance and one for labor and material payment) on federal projects exceeding $100,000. Because designers generally are not in a financial position to reimburse a surety in the event of a default, this law pretty much took designers out of the construction business. Thus, the Miller Act effected a legal separation between design and construction.

But as long as these projects remained fairly simple, and the parties charged with designing and building them remained cooperative with aligned goals and objectives, the new segregated model seemed to work. However, as buildings and structures became more complex, and the associated risks became more contentious, the segregated services model began to break down. Aspects of projects began to fall through the cracks and "finger pointing" became the primary response to accountability issues. Everyone began to look for ways to shield themselves from liability. Architects, in particular, generally risk adverse, began to withdraw from their traditional responsibilities and abdicate authority for the project as a whole.

The Slippery Slope

As the design and construction industry became increasingly disjointed, the collaborative spirit that once characterized the industry began to deteriorate as well. The gap between the Master Builder concept and the segregated services concept further widened. Various economic pressures and outside influences, such as foreign competition and escalating prices, would eventually bring the dysfunctional relationship to a climax, and many industry participants and clients alike began to push back and demand that designers and builders find a way to work together for the mutual benefit of all involved.

As you will learn in the next few pages, the disconnect between design and construction became more pronounced over the years, to the detriment of the overall project goals and objectives. The design and construction industry became one of the most fragmented and litigious industries in the nation. As a result, many attempts have been made in the last twenty years or so to "right our course" as an industry and return it to the collaborative nature embodied by the Master Builder. But before we can examine these corrective attempts, we need to understand one of the primary causes of our inability to align our underlying interests and motivations: the low-bid mentality.

LOW-BID MENTALITY

Most people would agree that cheaper is not always better, and there are many factors that determine the suitability of a product or service. Yet, during the mid-twentieth century, it became law in virtually every state in the United States that construction contracts would be awarded on the basis of lowest price, at least in the public sector. Furthermore, the legislation required that public agencies provide "complete" plans and specifications before soliciting bids from contractors. Private sector owners soon followed suit. Competitive bidding for construction contracts became the norm and a powerful low-bid mentality among owners and clients took hold.

On the other hand, when it came to selecting designers, the Brooks Act, established in 1972, mandated the use of Qualifications Based Selection (QBS) for federal procurement of architecture and engineering services. It became unlawful to select architects or engineers based upon price. Many states adopted the same requirements. This distinction between procurement approaches drove the wedge between design and construction even deeper.

As price became the driving factor for contractor selection, the construction industry responded by working hard to eliminate every extra service or product that might add an extra dollar to the project. If an item or detail was not shown on the presumably complete drawings or in the specifications, then contractors could no longer afford to add cost to their estimates to cover the inevitable minor errors or omissions by the designers—not if they expected to stay in business, that is. Instead, they were forced to point out the errors and submit change orders, hoping to pick up the profits they had left on the table in order to get the job in the first place. This "you only get what you pay for" attitude set the stage for a progressive deterioration of the designer-contractor-owner relationship. Change orders, claims, and litigation became the standard methodology for managing risks and protecting profits.

Back in the 1960s when I started my design career, architects and engineers really relied on the expertise of the contractors to fill in the holes and gaps in their drawings—everyone understood that the design reflected the "intent" of the work to be done. The designers and contractors were very respectful of one another and it was a real supportive environment. When something went wrong, no one pointed fingers at one another. Everyone just put their heads together to fix the problem. Perfection was never the expectation.

Nick Watry, PE, Licensed Architect
Founder (retired)
Watry Design Group
Redwood City, California

On the other side of the table, owners were getting tired of paying for all the "extras" left off the drawings and started pressing the designers to deliver close-to-perfect plans and specifications, while at the same time decreasing the fees they were willing to pay them, in an effort to compensate for the added contract costs caused by change orders. Designers began to look for ways to shed risks and liabilities while still protecting their fees. Now that they had to focus so much attention on ensuring that every little detail on the drawings was correct, instead of relying on the cooperative gestures of contractors to help "pick up" what they may have missed, something else had to give. In response to the lower fees they were receiving from owners, designers soon eliminated jobsite visits and other construction administration services and the chasm between design and construction appeared to be growing even wider.

As a result, contractors, designers, and owners fell into a relationship characterized by skepticism, suspicion, and insecurity. The low-bid mentality initiated through public procurement laws and spilling over into the private sector had created an environment of little or no trust among the parties. Collaborative problem solving was forced to take a back seat to risk mitigation. Both contractors and designers learned quickly how to position and protect themselves by shifting risks to other parties. Contractors shifted from self performing the construction work with their own forces to subcontracting the work to specialty contractors. They shifted from the role of "builder" of the project to "manager" of the contracts. Architects outsourced as much of the design liability as they could to consulting engineers and other design professionals.

ECONOMIC CONDITIONS & OUTSIDE INFLUENCES

Although they could control costs and shift risk, one thing neither the contractors nor the designers could keep at bay was the effects of the overall business culture and economic climate of the 1980s. Industry practitioners identify the 1980s as the most tumultuous and challenging decade for the design and construction business in modern times.

Decade of Decadence

Some of you reading this book will remember the album by the heavy metal band Mötley Crüe entitled *"Decade of Decadence."* (Mötley Crüe is one of the most successful American heavy metal bands, in the United States.) They sang about the decade 1981–1991, which saw rapid advancements in technology that impacted every sector of our economy—including the design and construction industry. Apple introduced the Macintosh® computer, the first desktop computer to use a graphical user interface. This point-and-click technology would forever change the way we do business, personally and professionally. Computing became something that anyone could learn to do without learning a new "language." Cell phones, the size of a large tennis shoe, were slowly appearing on construction jobsites. Fax machines helped speed up the transmission of written communications and handheld video recorders helped document jobsite progress. The industrial age had given way to the information age.

As if the strains of the low-bid mentality of the 1960s and 1970s had not already caused enough havoc for the designer-contractor relationship, the 1980s actually brought a new set of challenges into the mix. The "decade of decadence" was characterized by an extreme consumer appetite for goods and services of all types, and United States companies were experiencing fierce competition—especially from foreign rivals—to produce and deliver these products to the American public. This foreign competition was at the forefront of corporate concerns. Japanese manufacturers, in particular, were not only giving the United States technology companies a run for their money, but they were also having a grave effect on many other industries as well, such as the automobile industry. The ability to produce high-quality products—and to produce them very quickly—became a primary goal of American manufacturers that wanted to stay competitive in the 1980s.

Change on the Horizon

Indeed, drastic changes were taking place in American businesses. Industries and corporations that once managed their own construction projects began to rethink that strategy as "faster, better, cheaper" foreign producers began to eat away at their market share. Many manufacturers, such as Boeing, once had their own construction management departments and construction management professionals on the payroll. Not only did the company design and build airplanes but they also managed the design and construction of the facilities in which they built them. Construction project managers were charged with managing the contracts for multiple corporate projects. In some cases hundreds of these managers swelled corporate payrolls.

As pressure from shareholders and global competition increased, companies had to consider more efficient methodologies for delivering projects. American corporations soon realized that if they were going to be able to compete in the new global marketplace, they had to refocus their attention on their primary business function, increase their own efficiencies and productivity, and get out of the facility-building business. They looked to the expertise available in the design and construction industry for help. However, as they began to fine-tune and "lean up" their own operations to gain a competitive edge, they began to wonder why the construction business was not doing the same. Practically every other industry was finding ways to eliminate waste, streamline processes, improve quality, and add value for their customers. Yet the design and construction business seemed more fragmented than ever.

Owners once again began to express their dissatisfaction with the AEC industry. It became increasingly apparent that the traditional design-bid-build delivery method was insufficient. The biggest complaint was that the process was simply too slow. "Speed to market" was the rallying cry of corporate America in the 1980s. Time, not cost, was the number one priority during these fiercely competitive times, and design-bid-build was not delivering fast enough. Owners needed their design and construction service providers to:

- Deliver projects faster without sacrificing quality.
- Provide earlier guaranteed pricing prior to completion of design.
- Respond quickly to economic fluctuations.
- Provide a single source of responsibility for project performance.
- Deliver both design and construction services under a single contract.

Although it had been a long time coming, the economic conditions of the 1980s and the demand for more efficient processes for delivering facilities set the stage for the impending surge in the popularity and use of design-build project delivery in the United States in the 1990s.

More Bad News

Another event occurred in the early 1980s that altered the very nature and personality of the design and construction industry. On July 17, 1981, two of the four interior walkways inside the Hyatt Regency hotel in Kansas City, Missouri collapsed onto the lobby floor during a dance competition, killing 114 people and injuring more than 200 others. At the time, it was the deadliest structural collapse in U.S. history. At least $140 million was awarded to victims and their families in both judgments and settlements in subsequent civil lawsuits. In the opinion of many contractors, architects, and engineers, this single event had a devastating impact on the industry. At the time, insurance companies were already reeling from losses suffered in the real estate market in the 1970s. Overnight, industry insurance premiums skyrocketed. Errors and omissions insurance premiums and contractor general liability costs went through the roof. Many contractors, architects, and engineers went out of business because they simply could not afford their insurance premiums. In response, industry attorneys were on the lookout for any discrepancies between plans and specifications and constructed details in hopes of mitigating exposures and hedging liabilities. Risk advisors swarmed the industry, advising designers and builders alike to protect themselves by withdrawing from the front lines. The industry's innovation, creativity and problem-solving capabilities were being smothered by the fear of litigation and lawsuits. Contractors started shifting risks to the subcontractor community and designers stopped visiting jobsites. Traditional architectural accountabilities, such as shop drawing approvals, were deemed "too risky" by insurance companies. As a result, fabricators, vendors, and subcontractors were left holding the bag. The architects' and engineers' "APPROVAL" stamp was replaced with a "COMMENTS" or "NO-COMMENTS" stamp, or similar verbiage intended to shed the risk and liability onto someone else. Instead of moving toward collaboration and integration, the AEC industry was continuing to go in the opposite direction and just about everyone involved was unhappy with the situation.

> When I worked for an engineering firm back in the 1970s, we had a construction supervision department, and I went out on the job site and worked directly with the contractors to help run the job. But because of the growing litigious nature of the business, my job went from construction supervision, to construction administration, to construction observation, to construction related services. It got to the point where I really couldn't do anything or say anything because it became too risky. I remember telling my boss back then: I feel like I'm now watching the job from a knot hole in the fence. What kind of leadership and support can I provide from there?

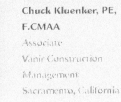

Chuck Kluenker, PE, F.CMAA
Associate
Vanir Construction Management
Sacramento, California

The industry was forced once again to react to these outside pressures with a survival mentality. Conducting business at arm's length, with limited interaction and engagement, seemed like the safest way to respond. Pointing fingers and blaming others seemed like a logical way to shift risk to the other side. This protection mechanism became so ingrained, that when more collaborative methodologies began to emerge, many contractors, designers, and owners failed to behave any differently. Over a three-decade period, an industry once characterized by collaboration and cooperation was gaining a new reputation as one of the most fragmented, inefficient, and litigious business environments in the United States.

> **TAKEAWAY TIPS**
>
> *Making History*
>
> Consider documenting your own company history and tracking your own transformation as you move from a hard bid, design-bid-build organization to a team-oriented, best value, design-build organization. This will help you identify a clear and deliberate change in philosophy and intention, and help define a new path forward. Check to make sure that your company vision, mission, values, goals, and objectives still align with the new intention.

IN SEARCH OF A BETTER WAY

Design-bid-build project delivery was not delivering—at least not at the "faster, better, cheaper" levels the marketplace demanded. Projects were getting larger and more complex and owners were looking for ways to better control project costs, schedules, and quality. They had begun to look for **alternative project delivery** methods as far back as the mid-1960s. One of the first new approaches coming out of these efforts was professional construction management (CM). Construction management as a professional service takes the project management functions from both the contractor and the designer and places it in the hands of an independent entity—the construction manager. Under a related approach, the agency CM model, the construction manager acts as an agent or advisor to the owner. When first initiated, it was often coupled with the multiple prime project delivery method, in which separate contracts are issued to the various trade or specialty contractors. The construction manager, working for the owner, is responsible for coordinating, scheduling, and

> **D-B DICTIONARY**
>
> **Alternative Project Delivery** – Any project delivery method that is not design-bid-build. Design-build and CM-at-risk are examples of alternative project delivery methods.

D-B DICTIONARY

Preconstruction Services – Services that affect project design; usually entail services such as design and drawing reviews, scope definition assistance, preliminary and schematic budgeting, constructability reviews, and value engineering. Early contractor selection allows for these services to be incorporated into the design-build process.

managing the various contracts on the owner's behalf, in essence taking the place of the general contractor under the traditional design-bid-build method. Furthermore, the construction manager could provide **preconstruction services** to the owner as well. If employed, the construction manager can offer up-front planning services such as constructability reviews, conceptual estimating, and value engineering by working directly with the designer without jeopardizing the competitive low-bid process. (The various project delivery methods will be discussed in detail in Chapter 2.)

Throughout the 1970s the construction management approach was used successfully on a number of projects. The continued pressure applied by owners to the AEC industry was bearing some fruit. An article appearing in the *Harvard Business Review* in 1973 stated it this way:

> *The construction management concept has been touted by some of its more evangelical proponents as a major breakthrough in the history of building, but it is really not a single new discovery. Instead, it is a mixed bag of techniques and procedures, dating back to antiquity, which have been fused together under the pressure of the present building crisis. The catalyst for this fusion has been the owner's determination to force the construction industry to regard a highly fragmented series of discrete decisions and events as a single process. In other words, the owners have argued that the building process can be effectively managed.*[3]

In 1972 the AGC adopted guidelines for the use of construction management and developed some standard construction management contract forms. In 1976 the American Institute of Architects (AIA) followed suit and embraced construction management as an alternative to traditional design-bid-build project delivery. They too developed a set of standard construction management contract documents. In 1982, the Construction Management Association of America (CMAA) was formed.

Although construction management provided some needed improvements, it still did not address three important owner concerns:

1. Agency CM is not actually a project delivery method; it does not deliver either the design or the construction. It is simply a service tacked onto a project delivery method, such as multiple prime or design-bid-build, and purchased for a fee. The construction manager does not hold any financial liability for either the design or the construction.

2. Unless the owner takes advantage of the preconstruction services component and uses the multiple prime project delivery method, the construction manager is doing the same job that the general contractor would be doing under design-bid-build. Many owners see this as adding an unnecessary extra layer of overhead expense to the project.

3. Instead of having two contracts to manage, one for the design and one for the construction, the owner now has three contracts to manage. Furthermore, the owner is still in the middle, between the designers and the contractors, and is still left to deal with the consequences when something goes wrong.

Today general contractors provide preconstruction services for a fee, even when they are not involved in the construction contract. Ultimately, a construction management hybrid emerged, combining the preconstruction services and the construction management services under the general contractor's contract. Known as CM-at-risk, this is considered an alternative project delivery method and is often seen as a stepping stone to design-build. Although CM-at-risk adds the preconstruction component, it still does not provide the single source responsibility that many owners seek.

CONSTRUCTOR AS A PROFESSIONAL

In 1971, about the same time that Construction Management emerged on the scene, a group of about thirty contractors from across the country gathered to create an organization that would promote, recognize, and define the "constructor as a professional." The group formed the American Institute of Constructors (AIC) and attracted a network of contractor members. In the early 1990s the AIC established an independent Constructor Certification Commission that developed a certification program to distinguish the "Certified Professional Constructor" (CPC) as a recognizable entity nationwide. According to the AIC website:

> *The Professional Constructor is an individual who commits to serve the construction industry in a professional and ethical manner and engages in the continued development of his/her skills and education to meet increasing industry challenges and changes.*[4]

At the center of the Institute's mission and goals is the AIC Code of Ethics:

1. A member shall have full regard to the public interest in fulfilling his or her responsibilities to the employer or client.
2. A member shall not engage in any deceptive practice, or in any practice that creates an unfair advantage for the member or another.
3. A member shall not maliciously or recklessly injure or attempt to injure, whether directly or indirectly, the professional reputation of others.
4. A member shall ensure that when providing a service which includes advice, such advice shall be fair and unbiased.

> Today there are approximately 800 Certified Professional Constructors (CPCs) in the United States and 4500 Associate Constructors (ACs). Currently the average number of undergraduate and graduate students taking the Associate Constructor exam each year before leaving college is approximately 1900. Consequently, the number of CPCs will grow exponentially as these graduates move into the workforce and obtain the required experience to pursue the higher credential. You can learn more about the AIC and the Professional Constructor certification process at *http://www.aicnet.org*.

5   A member shall not divulge to any person, firm, or company, information of a confidential nature acquired during the course of professional activities.

6   A member shall carry out responsibilities in accordance with current professional practice, so far as it lies within his or her power.

7   A member shall keep informed of new thought and development in the construction process appropriate to the type and level of his or her responsibilities and shall support research and the educational processes associated with the construction profession. [5]

The concept of regarding contractors as professionals (with or without the certification) makes a huge difference in how they are perceived on a design-build team. The contractor of old and the contractor of today are very different roles. In the past, contractors were the primary builders of projects. They wore tool bags and performed physical work in the field. Today the contractor's primary role is that of a project manager, overseeing multiple trade contracts and coordinating every aspect of the construction process. They must have a keen sense of both business and the building process, and understand all types of project risks and how best to mitigate them—not only for themselves, but also for their clients. That is why their input is so valuable to the design-build team. This evolution represents one of the most significant changes in the AEC industry over the past thirty to forty years and often goes unnoticed, which keeps old perceptions and prejudices in place that only hinder the effectiveness of the design-build process.

There was a time when being a contractor meant being a builder—an actual skilled craft worker who performed the physical work in the field with the aid of a few laborers. Today's general contractor, although aware of how all the parts and pieces of a building or structure go together, is typically not a tradesperson, but rather a business person and project manager who performs numerous preconstruction and planning services such as constructability reviews, feasibility studies, conceptual estimating, site analysis, and due diligence. He also manages the budgets, purchases, schedules, labor, materials, equipment, safety, and contracts of multiple specialty contractors and all construction-related risks on a project. Today's commercial contractor typically has at least a four-year degree in construction management, construction

---

Kim Lum
Executive Vice President
Charles Pankow
Builders, Ltd.
Pasadena, California

The ability to meaningfully participate in the preconstruction phase of the project is what distinguishes a true design-builder from a general contractor who hires an architect. This means bringing our scheduling, estimating, and coordination skills to the table every day of the process, not just during those periods when the design team throws their current design "over the fence" for us to evaluate. Being able to provide real-time, immediate feedback to the team about their daily decisions keeps projects from veering too far off track.

engineering, construction science, building science, civil engineering, or a similar discipline and is a true professional. Many also hold advanced degrees in business, engineering, or law. The contractor's project management functions are typically two-fold—management of field operations (actual construction by subcontracted forces) and the management of the overall project performance. Although some general contractors may still have their own production crews to "self-perform" portions of the physical work in the field, many have opted to shift the risks to specialty contractors who actually perform the work.

Generally, design only constitutes 5–10% of the project's cost, and the contractor is typically responsible for managing and overseeing approximately 90–95% of the cost of every project they are involved in. The contractor is an extremely valuable team member in design-build, so it is critically important that he be brought on board at the very onset of a project. His expertise and knowledge is invaluable to the design team and highly beneficial to the owner in their decision-making process.

TAKEAWAY TIPS

Certification Savvy

Whether you are a seasoned veteran, mid career, or new to the construction industry, consider pursuing a certified designation such as *"Certified Professional Constructor"* through the American Institute of Constructors or the *"Certified Construction Manager"* through the Construction Management Association of America to better differentiate you as a "construction professional" and help elevate and correct the image of the construction industry.

Construction Management is an interdisciplinary profession by nature. The four-year construction management degree combines business management principles and practices with the art and science of building, engineering and architecture, to create a unique educational experience. Today there are approximately 125 four-year colleges and universities listed with the Associated Schools of Construction (ASC) that offer accredited Construction Management degree programs.

RE-EMERGENCE OF DESIGN-BUILD

Throughout the 1960s and 1970s, a few owners and several contractors and engineers were experimenting with another alternative project delivery approach—design-build. Although the concept was nothing new, it was not commonly used on public projects because public procurement laws would not allow it. Furthermore, back then, the American Institute of Architects (AIA) Code of Ethics prohibited their members from having any financial affiliation with contractors, so to participate with contractors under a design-build contract was seen as unethical by the AIA.

However, according to Beard, Loulakis, and Wundram (*Design-Build: Planning through Development*), it is believed that the first use of public funds involving the design-build process occurred in Indiana in 1968.[6] A small community in the southern part of the state had a school project designed and constructed by the design-build method. Several other school districts in the state followed suit and used competitive design-build procurement for their new educational facilities. Beard, Loulakis, and Wundram go on to report that in 1980, the Portland, Oregon city council made a bold decision to accept a design-build proposal from a team of designers and builders to design and construct "the first post-modern public building of any consequence."[7] The controversial project received an Honor Award from the AIA at its annual design awards program in 1983.

The AIA's first Code of Ethics, adopted in 1909, forbade its members from participating in design-build projects due to a perceived conflict of interest. In 1986 the AIA changed its Code of Ethics to allow its members to participate in design-build contracts.

MENTORING MOMENT

Ready, Set, Go

Matt Dahlberg, AIA, DBIA, LEED AP
Architect
The Haskell Company
Jacksonville, Florida
Education: Cal Poly State University, San Luis Obispo
Major: Architecture

I was one of the lucky ones who were exposed to a design-build education while still in college, and having that education has opened a wealth of opportunities for me. I think the collaborative environment of design-build is a perfect fit for many young designers and constructors. Traditional design education has focused on two different methods, the "theory approach" and the "practical approach." But design education in a design-build format offers students a hyper-practical approach that includes all the realistic problems like cost, schedule and logistics into their everyday design theory and history lessons to create a truly holistic understanding. Construction Management education in a design-build setting not only offers students core-competency training in estimating, scheduling, methods and materials but also provides a context in which the constructor can be involved in the entire process—from the start of design through project completion. In this scenario, students learn to take "ownership" and responsibility for the complete project and not just their narrow part.

Fortunately, we are now seeing more forward-thinking educators who are creating courses and developing curricula that represent the "new world" of integrated project delivery and design-build. Preparing students for such a changed work environment gives young professionals many intangible benefits that will serve them for their entire careers. It just makes sense for higher education to address the future that design and construction students will enter into. Such an education provides an opportunity to learn by absorption, through side-by-side problem solving. Coupling design development with immediate constructability reviews allows everyone to see and hear the impacts of even small decisions. The process of organizing teams around different backgrounds and strengths allows each student to become familiar with the rich contribution that such diversity brings. By understanding the specific needs and challenges of each team member, the design-build student (and practitioner) can better anticipate the "gaps" that cause the inefficient use of time and money and resolve issues before they become real problems.

Young design-builders are quickly seizing the opportunities that a design-build education opens up for them—taking positions and responsibilities they would typically have to wait many years for. The industry paradigm shift has opened up new avenues of success for those entering the workforce, and they are ready to make the most of it. In turn, companies will have to make changes in their recruiting strategies so they can retain these unique young professionals.

Design-Build Gains Prominence

One of the earliest practitioners of design-build in the United States, the Austin Company, first offered the one-stop shop, integrated approach way back in 1907 in Cleveland, Ohio. However, the individual most often credited with advocating for the widespread acceptance of design-build as the preferred project delivery method is Preston Haskell. After witnessing firsthand how ineffective the design and construction process was in his own career as a young engineer, he set out to create a company that would place contractors and engineers, and eventually architects, under the same roof, thus allowing the company to deliver both design and construction as integrated services. He had experienced the problem-solving capabilities of contractors and engineers when they worked as a team, with the same goals and objectives, focused on delivering a project in a timely fashion, within budget, and with superior quality. Doing that consistently, he figured, would give his company a competitive edge, and the best way to do it was by putting the entire team under the same roof. In 1965 he founded the Haskell Company, a full service, integrated design-build firm dedicated to delivering comprehensive facility solutions.

The Haskell Company was not the only firm trying to market design-build services in the 1960s. There were several companies offering design-build services, including Gray Construction in Lexington, KY, Pankow Builders in San Francisco, CA, the Opus Group, in Minneapolis, MN, Suitt Construction in Greenville, SC, Korte Construction of Highland, IL, and Ryan Companies, in Minneapolis, MN, to name just a few.

But Preston Haskell launched a movement that would open opportunities in the marketplace, both public and private, that would move the AEC industry closer to achieving the level of collaboration and integration of which it is capable.

The Austin Company refers to itself as the "father of design-build project delivery." They eventually registered their single-source project delivery method as *The Austin Method*®. The Austin Company was purchased by Kajima Corporation in January, 2006.

PIONEER PERSPECTIVES

Finding a Better Way

When I established the Haskell Company in the mid-1960s, we were initially construction, what would be known today, as a construction manager. That term did not exist in 1965, but that's what we were doing. We were selected on a qualifications basis simultaneously with the architect. We worked with the architect preconstruction to develop budget and cost and value engineering (again a term we didn't know in those days, but that's what it was). We were charged by the owner to work collaboratively with the architect and bring the project in on budget and on schedule and still meet all of the elements of program. So it was really design-build with an outside architect. That continued for a couple of years, and then we began actually doing design-build projects where we had the prime contract for both design and construction with the owner. The architect was still an outside architect, but contracted directly with Haskell. So, effectively, we were doing genuine unified design-build. The concept was there, the culture was there, the objective was there, in terms of having a small, but effective and integrated core of design-build professionals.

In 1970 I became personally registered as an engineer and that permitted me to sign and seal drawings; so I was then able to bring at least the engineering and shortly thereafter the architecture in-house. Florida law permitted the corporate practice of engineering if the principal of the company was a registered engineer. I readily took advantage of that statute and licensed the company for the corporate practice of engineering, and of course did the same thing with the corporate practice of contracting. In 1978, the law was changed to permit the corporate practice of architecture. So as long as we

Preston Haskell, PE, DBIA
Chairman,
The Haskell Company
Founder,
Design-Build Institute of America
Jacksonville, Florida

PIONEER PERSPECTIVES *(Continued)*

had a responsible officer of the company who was a registered architect, we were then able to present ourselves as architects, engineers, and contractors and that became our mantra, our advertising tagline—integrated in-house delivery of all three disciplines.

In those early years we were selling on reputation, and on price, and on schedule, and the ability to get it done—to deliver a project on time and below what it would cost using other delivery systems or other competitors. However, the climate was kind of hostile toward design-build at that time. While I was able to sell it to owners, the market out there did not yet embrace design-build to anywhere the degree that it does today. We were sort of in the wilderness. The architectural profession was particularly hostile to design-build. It wasn't until around 1980 that the American Institute of Architecture really accepted design-build and began to develop practice policies regarding design-build. Prior to that time they simply opposed it as it intruded upon their historical position. Back in those years, the contractor had little to do with the owner. The architect was indeed the supreme figure in the delivery process. That position, of course, is totally different today. The architects are more members of the team, but no longer control the process to the extent that they did at mid-century. Legislatively, there were many battles to fight as well. Today, architects with the Haskell Company are officers and leaders in the AIA at every level and things have changed dramatically. In Florida, owners have enjoyed the option to procure their design and construction services through a qualifications-based design-build selection process. We have made some real progress.

I point out some of these struggles just to underscore the difficulty and hostility that we underwent in the 70s, and even into the 80s, in attempting to sell our services and to have design-build recognized and accepted as a legitimate professional project delivery process. It's not always been an easy task—but worth every ounce of effort. ∎

The Design-Build Institute of America

In 1992, Preston Haskell read an article written by Jeffrey Beard entitled *Design-Build in the Public Sector*. The article was published by the American Society of Civil Engineers (ASCE). Beard worked for the ASCE organization at the time; Haskell called Beard to request reprints of the article. Haskell asked for, and funded, 1,000 copies of that article to be reprinted. He kept 500 and asked Beard to distribute the rest as he saw fit. He began wondering what it would take to create an organization that was solely dedicated to promoting the design-build industry and profession. He recognized that there were strong, well-established organizations such as AGC for contractors and the AIA for architects. There were also a variety of engineering societies for each of the branches of engineering. Around this same time the CMAA had been established to advance the interests of construction management. But none of these organizations had a singular focus on design-build, even though some of them were beginning to show interest in it. The AIA reluctantly began creating guidelines for architects who wanted to participate in development, finance, and construction; and the AGC was beginning to recognize the fact that a lot of its members were interested in delivering design-build as well. So Haskell met with the AGC and the AIA leadership to discuss the idea of creating a special "practice community" within their organizations to support practitioners who wanted to pursue design-build. However, these attempts were unsuccessful. Instead, he started to investigate starting an organization from scratch. He met with Jeff Beard in Washington, D.C. to discuss the idea further. His next step was to contact his friend and professional associate, Jim Gray with Gray Construction in Lexington, Kentucky, who had also been an early proponent of the design-build approach.

Gray Construction was also an integrated design-build firm. He invited Gray and some of his colleagues to the Haskell office in Jacksonville for an all-day meeting. The two exchanged their design-build concepts and best practices. They opened their books to one another to share some of the statistics and financial metrics used to measure design-build success and determine if there were any foundational concepts and procedures about how they did business that was worth promoting. They found that they shared similar philosophies, processes, and techniques and discussed the possibility of creating an association of like-minded practitioners. The two men made a list of about a dozen others who they thought might share their vision and began calling them one by one with an invitation to join them in Washington, D.C. for a two-day exploratory meeting.

In February 1993, approximately 10 of the invited list arrived at the Ritz Carlton in Pentagon City near Ronald Reagan Washington National Airport. Haskell had invited Jeffrey Beard to sit in on the meeting as well. The first day was an open exploratory session to see if there was any merit to the idea. The second day was an all-day working session to establish an initial steering committee and assign various organizational tasks. At that meeting the founding members pooled their money and established a budget. After considering several options, they decided upon the organization's name—the Design-Build Institute of America (DBIA). In the spring of 1993, Jeff Beard, along with an administrative assistant, was hired as the first President of DBIA. The organization chose Washington, D.C. as the location for its headquarters.

The steering committee met roughly monthly for the next four or five months and recruited additional members to expand their membership base. They were successful in selling DBIA and these initial corporate members, paying $5,000 per year each, were the founding members of the new organization. Throughout 1993 and 1994, they focused on broadening the membership to include practitioners from around the

> The initial steering committee, forming the founding membership of the Design-Build Institute of America, consisted of representatives from five companies—Preston Haskell of the Haskell Company, Jim Gray of Gray Construction, Rik Kunnath of Pankow Builders, Don Warren of Suitt Construction, and Kraig Kreikemeier of Sverdrup Engineering. Each of these corporate leaders volunteered their time to help establish the organization and develop many of the resources currently available to DBIA members today.

> The DBIA is a unique membership organization in that it welcomes all discipline and stakeholder interests. The DBIA represents an interdisciplinary membership of architects, engineers, general contractors, public owners, private owners, specialty contractors, vendors and suppliers, insurance and surety professionals, university students and faculty, legal and finance professionals, and many other AEC affiliates. DBIA offers its members opportunities for education, networking, information, and advocacy through its web site, bookstore, certification program, courses, conferences, chapter events, and its monthly publication. To learn more go to *http://www.dbia.org*.

country. The invitation to join was no longer limited to strictly integrated design-build firms, but was extended to large single-discipline contracting and design companies who had made design-build a part of their overall business plans. Haskell credits Al McNeil of Turner Construction and Charlie Davidson of J.A. Jones Construction as early supporters of the DBIA.

In late 1993, the first official meeting of the DBIA was held in Chicago in a hotel near the O'Hare Airport. Jeff Beard presided as president and the twenty-five inaugural members adopted bylaws, created a charter, and established a board of directors. Haskell was elected the first chairman of the board. He served as chairman throughout the remainder of 1993 and all of 1994.

In the fall of 1994, the first DBIA Annual Conference was held in San Francisco and Rik Kunnath of Pankow Builders was elected the second chairman of the organization. During 1995 and 1996, two significant documents were published; the *DBIA Manual of Practice* and the *DBIA Code of Ethics*. Several members contributed best practices and procedures to the *Manual of Practice* and Ed Wundram of the Design-Build Consulting Group in Beaverton, Oregon was retained as a consultant to assist with that effort. Around the same time, the DBIA published its own family of Design-Build Contracts with the help of Michael Loulakis, an attorney with Wickwire Gavin, PC at the time, and Bennett Greenberg with Seyfarth Shaw LLP. Within the same time frame, DBIA entered into an alliance with McGraw-Hill to produce the first design-build magazine which McGraw-Hill published for several years. That magazine then became an in-house publication that is now the official publication of the DBIA—the *Design-Build Dateline* magazine.

At the urging of many public and private owner groups, the DBIA created a Certified Design-Build Professional™ program in 2002. The Certification Program is an opportunity for design-build professionals to distinguish themselves as having both educational mastery and documented experience in the practice and execution of design-build project delivery and design-build project management. You will learn more about the DBIA Designation Program in Chapter 9.

Legislative Action

Haskell's involvement in promoting and championing design-build project delivery did not stop with the formation of his own company or the founding of the DBIA. He knew that as long as there were legislative restrictions and a perceived fear of impropriety among architects, owners would not have easy access to a process that he believed was capable of delivering exactly what they said they were looking for. He continued to advocate for design-build at every opportunity and became progressively more involved in the legislative process. One of the biggest stumbling blocks to design-build becoming a nationally recognized, primary project delivery method was prohibiting legislation at both the state and federal levels. So Haskell became directly involved in three procurement legislation initiatives:

1. Haskell led the lobbying effort that led to the Florida legislature enacting a statute giving state agencies and local governments the authority to procure public works through best-value design-build competition in 1989. Haskell noted, "That was a real breakthrough. I believe it was one of the first, if not the first, in the nation. It was widely and successfully used from the outset. The Florida Department of Transportation and school districts have been among the biggest users. Other states have since followed course."

2. In 1994–1995, Haskell led an eight-member consortium (DBIA, ASCE, National Society of Professional Engineers (NSPE), AGC, AIA, American Council of Engineering Companies (ACEC), ABC and the Construction Industry Presidents Forum (CIPF)) to obtain enactment of a Florida-like procurement statute at the federal level. According to Haskell, "We had many ups and downs but ultimately prevailed. I give Chet Widham, who was the very enlightened president of AIA in 1994, high marks for his courage and cooperation on that effort."

3. In 1997, the Florida legislature enacted another groundbreaking statute, one that gave state agencies and local governments the authority to procure design-build services by the **qualifications-based selection** (QBS) method. Haskell was at the forefront of this legislative effort as well. Arizona has since developed similar legislation and other states are in various stages of adopting comparable statures.

Haskell is currently in the early stages of pursuing legislation that would allow QBS of design-build services for Federal Projects. You will learn more about qualifications-based selection and procurement in Chapter 4.

> **D-B DICTIONARY**
>
> **Qualifications Based Selection (QBS)** – A selection methodology traditionally used by owners to select architects and engineers based on the design professional's qualifications in relation to the project. In some states, procurement laws now allow the QBS method to be used to select contractors as well. Price is not a consideration in a qualifications-based selection method.

Federal Acquisition Regulations Modification

In 1997, another major breakthrough for design-build occurred at the federal level. The Federal Acquisition Regulations (FAR) were modified to include new regulations for design-build procurement. The Army Corp of Engineers, the Navy, and the National Guard have all embraced design-build as their project delivery method of choice. These modifications are part of what made it possible for Lee Evey, at the time the program manager for the Pentagon Renovation project, to choose design-build as the project delivery method to rebuild the Pentagon after the September 11, 2001 attacks. That rebuilding effort, originally slated to take approximately four years, was completed in 364 days, on September 10, 2002, one day shy of a year.

CASE STUDY
THE POWER OF A PROMISE

THE FACTS
Project Name:
The Pentagon - Phoenix Project
Project Type:
Government High-Security Facility
Location:
Washington, D.C.
Description:
Demolition and rebuilding of 1.6 million square feet
Cost:
$526 million
Duration:
1 year
Completion Date:
September 10, 2002

THE TEAM
Owner:
United States Government, Department of Defense
General Contractors:
Hensel Phelps Construction, AMEC Construction
Architects:
HDR Architecture, Shalom Baranes Associates, and Studios Architecture
Consultants:
KCE Structural Engineers, and DMJM, 3D/I Joint Venture.
Specialty Contractors:
There were a total of 87 different contractors involved

THE LEAD-IN

Every one of us remembers where we were on September 11, 2001. It is a day that just stood still for most of us. We sat glued to our television sets for days—first trying to figure out what the heck was going on and then sitting dumbfounded in disbelief of what had happened. On that same day, while you and I were rendered numb, a group of equally stunned contractors and engineers were trying to figure out how to deal with the very personal human loss and suffering; while at the same time begin cleaning up the twisted steel and concrete and planning for the rebuilding and repair of almost two million square feet of the Pentagon in Washington, D.C.

THE STORY

Just a few years earlier in November 1997, Lee Evey was named new program manager for the twenty-year, $4 billion Pentagon Renovation project. The total project would eventually include the complete renovation of approximately 6.5 million square feet while maintaining security and keeping all systems operational. On top of that, a total of 25,000 people would have to be temporarily relocated during the

process. The Pentagon had not had a major remodel in fifty-eight years. The project would entail the replacement of all building systems, the removal of all hazardous materials, the replacement of all windows, improving the energy efficiency, bringing the entire building up to code, improving the vertical mobility, enhancing security, improving pedestrian and vehicle traffic flow, and preserving or restoring all historical features.

The Pentagon Renovation Project, known as PENREN, had already been going on, using the traditional design-bid-build approach, for about ten years when Evey came on board—mostly consisting of planning activities and construction work in the basements. The project had a dismal performance record. Evey was brought in to clean up the program and get it on track. If he failed, it was going to be canceled. The project was notorious by this time for its schedule delays, cost overruns, and litigation. According to Evey, "Everything that could go wrong had gone wrong." Within a short time after his arrival, the design-bid-build solicitation for renovating Wedge 1 was scheduled to be released. (A wedge is one-fifth of the above ground Pentagon. The five wedges, each configured in a chevron shape, when combined total 6.5 million square feet.) Evey had just enough time to initiate some modifications to the RFP before it hit the streets. Although he maintained the lump-sum fixed-price concept of the original solicitation, he was able to tweak a few of the other details in an effort to remedy some of the issues that had consistently plagued the job. For example, he added an award fee feature and conducted a best value source selection instead of asking contractors to compete on price alone. AMEC Inc. of New York was awarded the design-bid-build contract on Wedge 1 and started construction in early 1998.

He then went to work trying to sort out what had gone so terribly wrong over the ten-year period before his arrival. He looked closely at the organizational structure of the internal team and started assessing the fit—literally, person by person. He ended up cutting the workforce (in some areas by as much as two-thirds), making sure he had the most capable people placed precisely where they needed to be on the team. Next he went about changing the rules of the game altogether.

The first step was to take a good, hard look in the mirror. As he went about interviewing the various project participants, he found the construction contractors to be the most candid in sharing their opinions. They suggested that bidding the project on a low-bid basis was at the root of many of the problems. The contractors revealed that the only way they could win the job was to bid below what they knew the project would cost and hope to make up the difference later, through change orders, when problems arose because of all of the gaps, holes, uncertainty, red tape, and bureaucracy. Then they reminded Evey that the owner, the government in this case, had set up all the rules of the game in the first place and therefore they were the ones who could make the biggest difference in changing them. Evey had to face the truth. They could not change the way contractors behaved and responded unless the government was willing to change first.

Shortly thereafter, at breakfast, Evey was reading an article in the *Washington Post* about how to go about building a new home. The article said that one of the biggest mistakes that people make is in selecting their architect and letting them create the design before bringing on the contractor. To Evey, this moment was an epiphany. Knowing nothing about design-build at the time, he set out to find alternative approaches to procuring design and construction services that placed both the designers and the constructors on the same team at the onset of the project. What he discovered was design-build. He and his staff attended a DBIA conference to learn more about the process and tried to figure out the best way to structure the remaining 80% of the work to be done under the program. This included the renovation of Wedges 2 through 5 and some outlying ancillary projects. Evey knew that he needed an integrated process that was highly collaborative and able to demonstrate flexibility, accountability, and creatively in order to respond to the huge challenges of the program; design-build was it.

The first project to come online under the new integrated design-build model was a 250,000-square-foot Remote Delivery Facility to correct the security risk associated with the receipt of building supplies and mail, and the internal supplies and material delivery system inside the Pentagon itself. It turned out to be a good first test of the process. The design-build team of Hensel Phelps and HDR design was awarded the contract in May 1999. The team delivered the project right on schedule and under budget in August of 2000. As a matter of fact, the Hensel Phelps team gave the government back a series of building improvements as well as a substantial financial refund when it was all said and done.

In the meantime, Evey went to work to streamline the actual RFP documentation that would govern the solicitation for the next phase of the program—the renovation of Wedges 2 through 5. The original design-bid-build RFP requirements documentation for Wedge 1 was 3,500 pages long. When the PENREN team was done, the document for Wedges 2 through 5 was reduced to only sixteen pages of performance requirements and standards that Evey expected to be met when the renovation of the Pentagon was complete. Taking stock of what he had learned over the course of his twenty-year career with the Air Force, the Department of Energy, and NASA, he knew that it was virtually impossible for laypersons (owners), to delineate, capture, and instruct experienced, professional providers of goods and services regarding all of the parts, pieces, steps, activities, and minute details necessary to achieve the ultimate project objectives without leaving something out or making a mistake. In other words, a patient should not try to micromanage their surgeon; it won't turn out well. Evey believed the same applied when it came to design and construction services. "Why spend 3500 pages trying to tell the professionals how to do their jobs. Just tell them what results you expect in terms of performance, and charge them with figuring out the most efficient way to do that." Completing the triage on this massive document took some time, but the team determined that the time savings inherent in the design-build approach would more than make up for the time spent improving the RFP.

By early fall 2001, the renovation of Wedge 1 was just weeks away from completion and the exacting design-build competition for the renovation of Wedges 2 through 5 had concluded. The design-build team of Hensel Phelps Construction, HDR Architecture, and Shalom Baranes Associates & Studios Architecture had won the design-build contract for the balance of the renovation. The winning team was scheduled to be officially announced to the public on September 11, 2001.

At approximately 9:37 a.m. on September 11, 2001, a terrorist-commandeered Boeing 737 jetliner plunged nose first into the southeast elevation of the Pentagon, impacting both the nearly complete Wedge 1, and the yet-to-be-renovated Wedge 2. The plane penetrated the E, D, and C rings of the building, affecting 1,600,000 square feet of office space and damaging 5 floors. In a matter of minutes, Wedge 1, a project that had just taken over three years and cost $258 million, was now a burning pile of rubble. Evey and his PENREN team had worked very hard to turn the PENREN Project around and they had succeeded. Now, they had a whole new set of challenges that would test the mettle of every participant on the project.

The rebuilding effort became identified as the Phoenix Project, named after the mythical bird that represents rebirth and immortality. The team's slogan, proudly displayed on shirt sleeve patches and a jobsite sign, "Let's Roll," replicated the heroic command given by Todd Beamer on Flight 93 just before it crashed on September 11 in a Pennsylvania field.

And "roll" they did. The initial estimates for demolishing, reconstructing and repairing the damaged Wedge 1 and Wedge 2 facilities came in at four years and $800 million. Demolition alone was estimated to take eight months. However, the PENREN/Phoenix team would have none of it. The cleanup and removal of 50,000 tons of debris and rubble was completed in just thirty-two days. The team publicly declared that they would complete the rebuilding in one year. On that promise, Evey proceeded to announce to the world that on Sept 11, 2002 there would be a Worker Appreciation Ceremony, exactly at the site of the attack on the building, at the Pentagon to honor the 3,000 people and 87 contractors and subcontractors and their families who had worked so very hard to achieve what many believed was the impossible. He went on to promise that, as a result of the reconstruction efforts, people would be sitting at their desks, inside the building, at the point where the aircraft hit, watching the ceremony on September 11, 2002 through their office windows.

THE CLOSE

As promised, on September 11, 2002, the entire E-ring of Wedge 1 was reoccupied and operational. A total of 3,000 Pentagon tenants were returned to their once-damaged offices. The Phoenix project was completed in 364 days, one day shy of a year, and three years earlier than projected. The total cost for reconstruction was $526

million, some $274 million below projections. Lee Evey resigned from the federal government on September 15, 2002. Today he is the president of the DBIA. He is a man who used his common sense, respect, and trust of others to figure out how to do projects right. After an already stellar twenty-six-year career, he spent another extraordinary six years of his life fine-tuning the design-build process to accommodate one of the most bureaucracy-laden projects ever, under arduous circumstances, and delivered extraordinary results.

From the Owner:

"Design-build works. If you put the right team together, give them the resources they need, trust them, and let them use their expertise, knowledge, and creativity to solve your problems, you will literally get everything you want and often more. Owners have all the power. They need to start being part of the solution and stop blaming the industry for a flawed process that does not work. It takes hard work, commitment, and downright courage. The bottom line is no guts, no glory."

Walker Lee Evey, Former Program Manager (1997–2002),

PENREN Project / Phoenix Project

There are many people involved in trying to make procurement legislation at the state level more design-build friendly and you'll learn more about those efforts in Chapter 2. And although design-build has been practiced informally by various multidisciplined firms as far back as the early 1900s, it really has only been defined in modern project delivery terms since the mid-1960s. However, it has come a long way since the first days of the Haskell Company and the first days of the DBIA. And now it is about to be redefined for the next generation.

INCREASED COLLABORATION AND INTEGRATION

Part of the evolution of design-build has entailed an increased emphasis on collaboration and integration, particularly among contractors and designers. Many have recognized that the real power of the process is the ability to leverage the various perspectives and collective intelligence of the primary design-build players. By working together in a collaborative fashion, the owner has the greatest opportunity to achieve superior results, the project team will optimize efforts, and risks are reduced for all concerned. Two pivotal publications have drawn increased industry attention to these important concepts.

The Collaborative Process Institute White Paper

In 1996 a group of concerned owners, contractors, engineers, and architects met in San Francisco to discuss the deteriorated state of the industry. The outcome was the publication of a white paper entitled *"Collaboration in the Building Process"* (October 8, 1996), and the creation of the Collaborative Process Institute (CPI). This organization, founded by some of the key AEC innovation leaders in the nation, laid out in the open what many people had been feeling for years. No one was satisfied with the way things were going in the industry—not owners, not designers, and not contractors. The CPI declared that the industry's woes could not be solved by tweaking project delivery methods alone. It would take work on the people side of the equation as well.

At the core of the Collaborative Process is the assertion that optimized projects result when the design and construction process seeks an integration of people *and* systems. The group identified seven maxims which underpin the Collaborative Process:

1. Integrity and trust are essential for true collaboration.
2. The long run is more important than the short run.
3. Teams make better choices than individuals.
4. In building a team, prequalify firms and select people.
5. True creativity focuses on option generation.
6. Change is inevitable: be prepared for it.
7. The basis for decision making should be facts and reason, not opinions and emotion.[8]

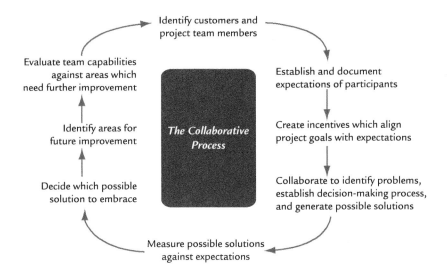

Figure 1.2
Overview of the collaboration process. Used by permission from The Collaborative Process Institute

Although the organization has since disbanded, the ideas and concepts outlined in their 1996 white paper are a perfect fit with the philosophy of design-build. Many of the people who were involved with creating the Collaborative Process Institute are still working hard, through different venues, to continue the efforts that the group initiated. They recognized early on that the potential of the AEC industry was being hindered by the continued fragmentation and adversarial environment that surrounded the design and construction process. The group realized that the full benefit of collaboration could only be achieved when both the designer and the contractor were selected early in the process when the scope definition, budget, and schedule were being established. They also suggested that contracts be negotiated on an open-book basis whereby all project participants could see the true costs associated with the project. Furthermore, the CPI insisted that the senior management of all parties extend their full commitment and support to the process. Most former CPI members attribute the suspension of the organization to a lack of owner involvement and support.

The white paper published by the Collaborative Process Group in 1996 may be found on the DPR Construction website at *http://dprconstruction.com/news-and-events/technical-papers/tech-collabor.cfm*.

CURT White Papers

In 2000 a group of owners organized to create the Construction Users Round Table (CURT)—an offshoot of the Construction Committee of the Business Round Table which had been around for over thirty years. Identifying itself as *"The Owners Voice to the Construction Industry,"* the members of CURT represent construction and engineering executives of some of the largest corporations in America. The mission of the Construction User's Roundtable, according to their website, is to help create competitive advantage for construction users in the marketplace. As an organization, CURT is dedicated to promoting excellence in the creation of capital assets by providing aggressive leadership.[9]

A special CURT task force was assigned to address the perception of inadequate, poorly coordinated AE (architecture and engineering) drawings, errors and omissions, and similar challenges and difficulties commonly experienced in the design and construction process. The task force determined that these issues were actually indicators of a much larger problem within the industry.[10] In August of 2004, CURT published a white paper entitled "Collaboration, Integrated Information, and the Project Lifecycle in Building Design, Construction and Operation." In this paper, CURT concluded that:

> *The difficulties experienced in typical construction projects, including those identified by CURT members, are artifacts of a construction process fraught by lack of cooperation and poor information integration. The goal of everyone in the industry should be better, faster, more capable project delivery created by fully integrated, collaborative teams. Owners must be the ones to drive this change, by leading the creation of collaborative, cross-functional teams comprised of design, construction, and facility management professionals.*[11]

In July 2006, CURT published a second white paper entitled "Optimizing the Construction Process: An Implementation Strategy." In this document CURT concluded that wholesale industry change is necessary to achieve the levels of collaboration and team integration necessary to optimize project results.[12]

The CURT white papers represent a breakthrough in the transformation of the AEC industry. There are many individual practitioners, firms, and organizations like the CPI that have worked tirelessly to integrate the design and construction industries over the years—long before there was a CURT organization. So efforts to improve collaboration, integration,

---

Gregory L. Sizemore, Esq.
Executive Vice President
The Construction Users Roundtable
Cincinnati, Ohio

The AEC industry continues to experience rapid changes. Whether in the design professions, the construction arena, insurance, or law, the multitude of disciplines that converge to make up the complex world of the built environment are experiencing new challenges every day. The Construction Users Round Table is about driving change—about giving voice to the owner community and about spurring innovation. As a result of our efforts, the old way of doing things is starting to evolve. Not only are owner organizations undergoing a tremendous metamorphosis, but the entire industry is on the move as we look for new and better ways to get projects completed while maintaining cost effectiveness. Architects, engineers, and builders are no exception. With the use of technology and an emphasis on integrated project delivery models like design-build and others, they are now seeking ways to collaborate at levels never before seen on the industry landscape. In today's economy more than ever, it is critical that we all work together to find approaches that will add efficiency, reduce waste, save time, and enhance the value of every dollar spent on real estate development and capital improvements.

and project optimization have been made in previous approaches. What is new is that for the first time in the industry's history, it appears that all stakeholders are finally looking in the same direction. The primary reason for the failure of the 1996 CPI initiatives, lack of owner participation, may now be changing: Owners are now recognizing that they play a huge part in achieving the full collaboration necessary for an integrated solution. Without their participation, all other efforts will most likely fall short.

You can read more about CURT and the collaboration and integration concepts presented in their whitepapers by visiting their website at *http://www.curt.org*.

THE ROLE OF TECHNOLOGY

As much as collaboration and integration are now gaining acceptance in the industry, another primary driver for this trend has actually been technology. The Computer Aided Design (CAD) systems of the past have evolved into sophisticated modeling systems that are already contributing to increased industry integration. And the technology is gaining acceptance with key players, particularly architects.

Building Information Modeling

In its most basic form, Building Information Modeling (BIM) is 3-D modeling software that can be with a design and construction database, allowing integration of the estimating and scheduling tasks into the model as it is built, thereby creating a 5-D information model. The model acts as a single repository for all the bits and pieces of information across all the disciplines and functions necessary to construct the project. According to the National Building Information Model Standard committee:

A Building Information Model (BIM) is a digital representation of physical and functional characteristics of a facility. As such it serves as a shared knowledge resource for information about a facility forming a reliable basis for decisions during its life cycle from inception onward. A basic premise of BIM is collaboration by different stakeholders at different phases of the life cycle of a facility to insert, extract, update or modify information in the BIM process to support and reflect the roles of that

Technology isn't an end in itself, but it's a facilitator and certainly for design-build there's nothing that's going to integrate design and construction more than Building Information Modeling. BIM is not just another version of CAD; you have the ability in the technology to build as detailed a virtual model of the building that you want, down to the sheet rock screws and square feet of glue if you want. You can back up and include in that model any level of detail for any area that you want. BIM is made to order for design-build but it still doesn't guarantee the collaboration.

Dave Engdahl, AIA, DBIA
Senior Vice President,
Chief Architect (retired)
The Haskell Company
Jacksonville, Florida

Figure 1.3
Building information modeling examples. Courtesy of NTD Architecture, CA

stakeholder. The BIM is a shared digital representation founded on open standards for interoperability.[13]

There is a great deal of excitement around BIM and its application potential in the design and building business. The AEC industry is buzzing with anticipation. Many firms, which only a few years ago were "waiting to see" where this technology was headed, already have BIM systems up and running or have implementation plans in place. Although the industry has used various 3-D modeling programs for years, many people believe that the integration capabilities of the BIM technologies will revolutionize the way we design and construct buildings and structures.

Architects Getting on Board

Owners have not been the only ones dragging their feet when it comes to the notion of integrated project delivery and collaboration. As a group, architects have been the most hesitant to embrace the notion of design-build and an integrated design and construction process, particularly as it is implemented in BIM. They cite a high initial cost of BIM systems, and also legal concerns. Some architects fear that sharing BIM information with a contractor may expose them to liability. In addition, they are concerned about possible file compatibility concerns when working with contractors and others involved in the process. Yet BIM has also been the catalyst that has prompted significant activity within the architecture community. In recent years some real progress has been made. In March 2007 the Design-Build Knowledge Community of the national organization of the American Institute of Architects (AIA) convened its first AIA Design-Build Summit in Kansas City. This two-day event was a breakthrough, as architects, contractors, and engineers discussed best practices in design-build, including the use of BIM. The event acknowledged that design-build was becoming a preferred project delivery method for owners, and that architects, by resisting making the transition, were missing out on opportunities to provide leadership. The two-day event focused on how architects could more fully participate in the design-build process and why they should.

> Many architects thought that design-build was only appropriate for warehouses or parking garages. But its future is in its potential for all project types. Just look at the Chicago Public Library, or Denver's Downtown Aquarium, or Caltrans District 7 Headquarters in Los Angeles. They are design-build exemplars. More and more architects are finally participating. Currently, more than 8,000 AIA members or approximately 10% of the membership belong to the Design-Build Knowledge Community. Trends for design-build projects compared to design-bid-build projects indicate that by 2010 an equal number of projects will be completed by each. Beyond that point, design-build will dominate the construction delivery process.
>
> **Bill Quatman, Esq., FAIA, DBIA**
> Attorney at Law
> Shughart, Thomson, & Kilroy, PC
> Kansas City, Missouri

NEW INITIATIVES, NEW STRATEGIES

Over the last twenty-five years, a number of other initiatives have emerged that clearly signal the move toward more collaborative approaches to the design and construction business. Different factors have spurred these developments, but they all point in the same direction—toward more efficient and effective ways to deliver design and construction services to the marketplace. Four initiatives, in particular, complement the design-build approach and philosophy:

1. Integrated Project Delivery

There is a great deal of buzz around the concept of Integrated Project Delivery (IPD) today, and for good reason. According to the AIA California Council's (AIACC) working definition document:

> *Integrated Project Delivery (IPD) is a project delivery approach that integrates people, systems, business structures, and practices into a process that collaboratively harnesses the talents and insights of all participants to reduce waste and optimize efficiency through all phases of design, fabrication and construction. Integrated Project Delivery principles can be applied to a variety of contractual arrangements and Integrated Project Delivery teams will usually include members well beyond the basic triad of owner, architect, and contractor. At a minimum, though, an Integrated Project includes tight collaboration between the owner, the architect, and the general contractor ultimately responsible for construction of the project, from early design through project handover.[14]*

The AIACC's Integrated Project Delivery model, although not specifically identified as design-build in the definition document, is similar in character and principle to the philosophy represented by the design-build model. It recommends early contractor involvement, expanded project teams to include specialty contractors and vendors, and high levels of

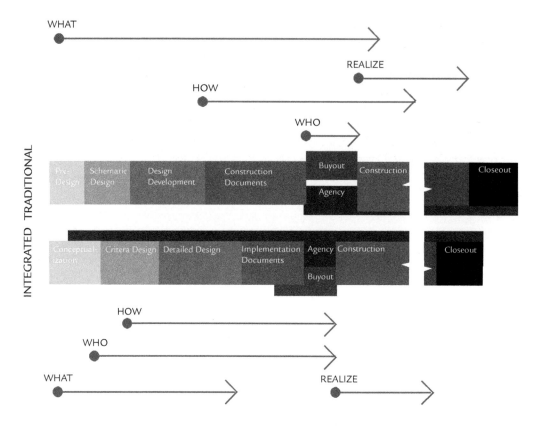

Figure 1.4
Integrated project delivery model. Published with permission from The American Institute of Architects, California Council

You can access both the AIA-California Council's document entitled "A Working Definition: Integrated Project Delivery" and the National AIA's "Integrated Project Delivery: A Guide" at http://www.ipd-ca.net/IPD Definition.htm.

collaboration, communication, and integration. The clear difference is that the design-build model requires a single source responsibility—one contract with the owner. The AIACC definition document identifies design-build, design-bid-build, and CM-at-Risk as contracting methods only, and suggests that you can apply the Integrated Project Delivery principles to any one of these contracting methods. However, it doesn't appear that design-bid-build can meet the requirements identified in the definition; the contractor ultimately responsible for building the project is not involved with the architect at the early stages of design.

2. Lean Construction

One of the most hopeful initiatives within the design and construction industry is the development of Lean Construction techniques and strategies for eliminating waste in the design and construction process and increasing efficiencies. It is a systems approach to project delivery and considers the work of the team to be more process-management-oriented then project-management-oriented, which is also the case with design-build project management. According to the Lean Construction Institute (LCI) website:

> *Lean Construction is a production management based delivery system emphasizing the reliable and speedy delivery of value. It challenges the generally accepted belief that there is always a trade between time, cost and quality.*[15]

Much of the work associated with the Lean Construction movement has been the result of efforts by the LCI's Greg Howell (cofounder and managing director) and Glenn Ballard (cofounder and research director). As consultants, they began to study productivity in the construction industry and began to recognize just how inefficient and unreliable traditional methods and management processes were. They started looking at construction management from a manufacturing perspective, specifically at the principles applied by the automobile manufacturer, Toyota. Certain they were onto something, they organized the LCI in August of 1997. According to Howell, "although the effort started as more of a consulting enterprise, it then morphed into a not-for profit research entity, and today is probably best described as a think tank."

The fundamental idea behind Lean Construction is that a project should be managed as a whole process instead of a collection of independent elements and activities. Instead of viewing project management as a procedure for managing contracts, Lean Construction considers project management to be a procedure for managing production. One of the techniques developed by the LCI is a scheduling process, or more appropriately referred to as a production control process, called the Last Planner™. The Last Planner concept addresses work flow reliability and the concept of a pull schedule versus a push schedule.[16] A pull schedule releases work assignments and materials onto a project based upon the readiness of the project to receive it. Onsite personnel determine when the project is ready for the scheduled work. Whereas the traditional push schedule releases work assignments and materials onto a project based upon pre-assigned due dates from a master schedule, typically prepared by project managers at the macro level without the benefit of frontline knowledge regarding readiness.

3. Relational Contracting

LCI, and several other groups around the world, such as Alliance Contracting IQ and the Serco Institute in the UK, also promote the concept of **relational contracting**, a contracting methodology that focuses on the relationship between the parties to the contract, as well as the transactional aspects of the contract, such as products and services. It is similar to **alliance contracting**, a cooperative contracting model that is characterized by a collaborative process that aligns values, goals, and objectives among the project owners, designers, and contractors. Recently Will Lichtig, an attorney with Sutter Health and an associate of the LCI, developed a new relational contract instrument called the Integrated Form of Agreement (IFOA). Sutter Health has implemented this new agreement on a few of its hospital projects in California. This agreement is based on the concept of a "core team" organizing itself to function as a single entity with unified goals and objectives. The core team members, usually the owner, the contractor, and the architect, sign a single contract document together and

---

You can learn more about Lean Construction techniques by visiting the LCI website at *http://www.leanconstruction.org/*.

D-B DICTIONARY

Relational contracting – A contracting methodology that focuses on the relationship between the parties to the contract, as well as the transactional aspects of the contract, such as products and services. Responsibilities, risks, and benefits are intended to be apportioned fairly and transparently, establishing a basis for trust and cooperativeness.

Alliance contracting – A cooperative contracting model that is characterized by a collaborative process establishing an alignment of values, goals, and objectives among the project owners, designers, and contractors. This contracting approach is meant to promote openness, trust, and risk sharing among the project participants.

jointly agree to common terms and aligned interests. The IFOA not only addresses the traditional business transactions associated with the project, but also addresses the relationship imperatives as well. In essence, everyone on board agrees to look after one another and support the interests of the "whole." It will be interesting to see how this contracting approach works out. Although the contract has a long way to go before it is considered mainstream, it is another indicator that the industry transformation is an ongoing process.

4. Leadership in Energy and Environmental Design—LEED

In 1993, the United States Green Building Council (USGBC) was formed as a nonprofit, nongovernmental organization with a diverse membership of architects, contractors, owners, product manufacturers, environmentalists, and others who are interested in the promotion of green building in the U.S.

In 1998, the organization launched a nationally-recognized rating system for benchmarking the design, construction, and operation of "green buildings." The system, referred to as Leadership in Energy and Environmental Design (LEED), is a voluntary rating system that identifies a definitive standard for what constitutes "green building." The goal of LEED is to evaluate the environmental performance of a building from a holistic perspective, and measure that performance over the building's life cycle.[17]

Although the rating system is voluntary, the LEED evaluation system has become popular across the country. For example, in the state of California, all new state buildings and major renovations over 10,000 square feet and subject to Title 24 Energy Code will be designed, constructed, and certified at a rating of LEED-Silver or higher.[18]

Achieving LEED certification is not easy. LEED promotes a whole building approach to sustainability. According to several LEED-accredited professionals, it is almost impossible to achieve LEED certification using a project delivery system other than design-build. The entire team—owner reps, architects, contractors, engineers, product vendors, subcontractors, everybody needs to be in sync right from the very beginning of the design to properly monitor and document the various "green" attributes of the project. It takes a tremendous amount of collaboration and integration to get it right. Otherwise, the effort is at great risk for failing at some level.

The design-build methodology and mentality is well-suited to the intent and spirit of each of these initiatives. They are not competing strategies, but rather highly complementary. The best way to serve the industry is to encourage dialogue among the various groups to find ways to integrate approaches and share techniques in order to better serve the industry and the built environment as a whole. The goal is to find the attributes within each of them that will benefit the design and construction process and deliver the best value for the client. While companies are still exploring the practical applications of these approaches and techniques, the future will tell how the relationships among them will eventually play out.

BACK TO THE FUTURE

A common thread runs through all of the advances of the past forty years—a search for better, more efficient ways to deliver design and construction services to the public. It has been a bumpy road at times. But we have come full circle and almost everyone involved in the design and construction process, contractors, architects, engineers, and owners, are recognizing that the industry's dysfunction can no longer be tolerated. Many of our traditional policies, procedures, and prejudices are hindering the integration process. The major forces that can influence how we do business in the future are finally beginning to align.

It has been a long time since the days of the Master Builder and the idea that a single person could anticipate all of the needs and challenges associated with a complex project and plan, design, manage, and construct a building or facility single-handedly. Today's Master Builder is the Integrated Design-Build Team. Although that team needs the skills, knowledge, and experience to deal with the technical aspects of a project, more importantly, they need to accompany them with the more difficult soft skills of communication, facilitation, and collaboration in order to fully integrate the project solution. This part of the transformation process has not been easy to achieve. For too long, the complex role that people play in project delivery has been ignored or misunderstood. But in the end, it's the "people portion" of the equation that will make the biggest difference.

The New Master Builder

These are exciting times in the AEC industry. For many, they are also unsettling times. The old foundations underlying our traditional ways of doing business, our traditional roles, our traditional ways of interacting with one another, are crumbling under the weight of necessity, change and opportunity. The demands and challenges of the marketplace coupled with technological advances are causing barriers to fall and lines to be blurred.

Those who have not grasped the magnitude of this industry transformation will be unprepared to keep up with the pace of change that is sure to occur over the next decade. In the book *The Next Architect: A New Twist on the Future of Design* (2007) by James P. Cramer, founding editor of DesignIntelligence, and Scott Simpson, president and CEO of Stubbins Associates, the authors conclude:

> As owners seek more streamlined delivery of their projects, they will increasingly turn to design-build, contracting with a single entity to provide integrated design and construction services. This will encourage the traditionally factional and fractional design and construction industry to find new ways of collaborating. The shift to design-build will also substantially affect issues of risk management, since so many formally competitive

entities will be linked by a common contractual bond. Essentially, design-build is a return to the traditional role of the Master Builder. [19]

Whether you are a seasoned practitioner or a young student about to graduate from the university, everyone will be affected by this transformation; everyone will have a role to play and a contribution to make as we redefine how we do business together as design and construction professionals. Design-build is a model that provides an opportunity to bring out the best in each of us, but only when all key players understand and execute it correctly. The purpose of this book is to provide an understanding of the fundamental concepts that are unique to design-build and that allow this approach to work at its highest potential.

Even if you feel you already know how to do design-build, this book can still provide you with valuable information about the practices and techniques to help you maximize the process. If you think design-build is not much different than design-bid-build, this book will demonstrate design-build's unique requirements and character. If you think design-build simply means that now the contractor gets to "control" the architect, you will find the approach is much more complex than that. As you will learn in Chapter 3, there's very little about the design-build process that is like design-bid-build. But before we discuss that, we will first examine the basic project delivery and procurement options available, and why design-build has become so popular.

CHAPTER HIGHLIGHTS

1. Design-build is often compared to the ancient Master Builder concept used to build many great wonders of the world before and during the early Renaissance. The Master Builder was not a contractor, an architect, or an engineer. The Master Builder was the embodiment of all three disciplines. The Master Builder concept depicts the earliest model for delivering design and construction as an integrated service.

2. After the Renaissance, design and construction began to separate. Individuals began to identify themselves with either the design tasks or the construction tasks associated with a project. Professional associations and societies were established as early as the late 1800s and early 1900s, further distinguishing designers from contractors. Specialization continued to support a more segregated services model for design and construction throughout the Industrial Revolution.

3. Continued specialization among designers and contractors throughout the Industrial Revolution sustained the ongoing transformation of the design and construction industry. The Master Builder concept was quickly being replaced by a more segregated services project delivery model known as design-bid-build.

4. In the mid-twentieth century, it became law that public construction contracts would be awarded on the basis of lowest price. The

private sector also embraced this practice. Competitive bidding for construction contracts became the norm and a low-bid mentality characterized traditional design-bid-build project delivery.

5   Over the years, the design and construction industry has degenerated from a culture of cooperation and teamwork to one of fragmentation and inefficiency. There are many historical reasons for this deterioration but many organizations are making efforts to restore it to its collaborative roots.

6   The design and construction industry is currently in the midst of a major transformation, driven by owner demand, huge work backlogs, and advanced technologies such as building information modeling. The traditional low-bid mentality associated with design-bid-build is giving way to alternative project delivery methods.

7   Owners are the major drivers for change in the industry and they are becoming more and more vocal about the need for team collaboration and integration to improve efficiencies and reduce waste in the construction industry.

8   Design-build project delivery is intended to be a highly collaborative, fully integrated process and is at the forefront of this transformation. Although there is still much to be learned about how best to implement design-build, it is being used more and more to deliver projects faster, better, and at a lower cost.

9   There are many current initiatives associated with design and construction and the built environment that complement design-build. Examples of these initiatives include Integrated Project Delivery, Lean Construction, Relational Contracting, and Leadership in Energy and Environmental Design (LEED).

10 The emergence of Building Information Modeling (BIM) is one of the major drivers behind the current interest and pursuit by owners, designers, and contractors for project delivery strategies that facilitate and demonstrate more cooperation, collaboration, communication, and project integration.

RECOMMENDED READING

:: *Construction – Craft to Industry*, by Gyula Sebestyén, E & FN Spon, 1998.
:: *Leading the Revolution: How to Thrive in Turbulent Times by Making Innovation a Way of Life*, by Gary Hamel, Harvard Business School Press, 2000.
:: *The Next Architect: A New Twist on the Future of Design*, by James P. Cramer and Scott Simpson, Greenway Communications, Ostberg, 2007.

"If your project doesn't work, look for the part that you didn't think was important." —ARTHUR BLOCH, AUTHOR, MURPHY'S LAW AND OTHER REASONS WHY THINGS GO WRONG

2

PLANNING, DESIGNING, AND CONSTRUCTING PROJECTS

In any new project, every owner must make several fundamental decisions. One of the first is choosing a project delivery method. **Project delivery** is the comprehensive, start-to-finish plan for organizing, implementing, executing, and completing the design, construction, and startup of a building, facility or structure.

Unlike commodity products such as computers, or tennis shoes, construction is a professional service that, like architecture and engineering, delivers a unique "project" as its end product. In doing so, it presents its own challenges in the production process.

When manufacturers build computers, tennis shoes, or cars, they typically construct a prototype first, followed by a trial model. Then they thoroughly test and retest the model to work out any "bugs" before sending the product into production. They develop dies and molds so they can mass-produce identical products. Furthermore, this production work is typically performed under controlled, indoor climate conditions, using state-of-the-art technology that usually includes mechanical systems and robotics. The men and women producing these products comprise a relatively consistent workforce. They systematically manipulate the various parts needed to assemble the products, without having to consider varying criteria or changing conditions. There are no Requests for Information (RFIs), submittals, pages of drawings or design details to interpret. The process is consistent, well-scripted, and routine, and produces a standardized and consistent product. The computers, tennis shoes, and cars are then sold in the marketplace, where they are considered a commodity, with price being the primary differentiator.

Unfortunately, construction has often been placed in the same category as these products, with price being the major differentiator. But there are significant differences between commodity products and construction projects. Construction is not a commodity. Every construction project is unique—built as a one-of-a-kind facility or structure, each on a different building site, under unpredictable weather conditions, and

> **D-B DICTIONARY**
>
> **Project Delivery—**
> The comprehensive, start-to-finish plan for organizing, implementing, executing, and completing the design, construction, and startup of a building, facility or structure.

> Owners must make three fundamental decisions before embarking on a new construction project. They need to select a project delivery method, a source selection method, and a type of contract. The first decision is covered in this chapter. The next two decisions will be covered in Chapter 4.

on varying topography with inconsistent soil. Every community and municipality imposes its own regulatory restrictions, environmental requirements, and building codes. Each project represents an untested prototype in and of itself. The "bugs" get worked out as a project's design and construction progress. A construction job's skilled and semiskilled workforce is primarily transient, moving from job to job, with an assortment of tradespeople entering and leaving the process at various stages. In construction, most labor is still performed by the hand of the craftsman or tradesman—forming and placing concrete, setting steel, or pulling wire. The uncertain nature and unique circumstances that surround every construction project demand careful consideration when planning, designing, and constructing all of the elements that make up the building project.

How we deliver design and construction has evolved over time. The integrated Master Builder approach used for centuries gradually evolved into a process in which design and construction tasks were separated into distinct, isolated activities. These independent tasks emerged as the design-bid-build project delivery method. The name itself reflects the method's sequential, linear nature. Design-bid-build is often referred to as the "traditional" project delivery method, and is still the most frequently used process today for nonresidential construction projects—but by a very small margin, and that margin is shrinking. According to the *2007–2008 Design/Build Survey of Design and Construction Firms* (ZweigWhite Information Services, 2007), nearly three-quarters (72%) of firms responding to the survey predicted an overall increase in the use of design-build project delivery over the next five years.[1] The reasons for this increase become evident if we examine the other principal project delivery methods in use today.

PROJECT DELIVERY METHODS

The three most common project delivery methods are design-bid-build, CM-at-Risk, and design-build. Regardless of which method is used, three primary players have major roles: an owner, a contractor, and a lead designer. For buildings, the lead designer is an architect; for highways or bridges, the lead designer is an engineer. These three project delivery methods differ significantly in the type of contractual relationship that exists among these three players. Furthermore, each method has advantages and disadvantages, and one of the keys to successful project delivery is to select the method that is best suited to a project's special challenges, needs, and goals.

Design-Bid-Build

Because design-bid-build is the "traditional" project delivery method, all owners, contractors, and designers are familiar with it, as are subcontractors, suppliers, financers, insurers, bonding companies, regulatory agencies, and other project participants. In a design-bid-build project, the owner holds and manages two separate contracts; one for design and one for construction, as shown in Figure 2.1.

The designer and the contractor operate independently and have no contractual relationship with or obligation to each other. The owner acts as the "filter" and handles most discrepancies and differences of opinions.

Design-bid-build is a linear process, in which each function typically ends before the next function begins, as shown in Figure 2.2.

Because it is a sequential process, design-bid-build is the slowest project delivery method of all. However, design-bid-build still has its advantages.

Figure 2.1
Design-bid-build contract configuration

Design-Bid-Build Advantages

Perhaps the greatest advantage to design-bid-build project delivery is that everybody knows how to do it—all players are familiar with how it works and what their roles are under this method.

Owners provide programming, scope definition, and project financing. Designers provide traditional design services and document development (plans and specifications), and contractors provide construction services. Because the method is familiar, players find comfort in it. However, comfort may no longer be the driving force behind project delivery selection. In today's competitive marketplace, companies must find higher efficiencies, less waste, and optimized project results. As familiar and comfortable as the design-bid-build approach is, it may no longer meet the needs of the marketplace.

Another advantage to the design-bid-build method is that it is well suited to the competitive bidding process, in which contracts are awarded to the contractor with the lowest price. No legal barriers or procurement restrictions prohibit its use on either private or public projects. All of the rules and regulations, policies and procedures, insurance and bonding instruments, and contract documentation associated with design-bid-build are well established. Every state in the United States has procurement legislation that supports competitive bidding. These laws were enacted long ago to prevent the abuse and misuse of public funds, and until recently design-bid-build

Traditional education curricula associated with construction management, architecture, and engineering programs at universities have all been based on the design-bid-build project delivery model.

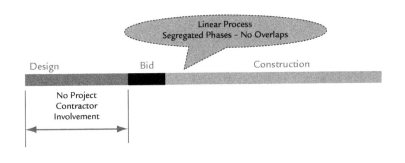

Figure 2.2
Design-bid-build sequence

and competitive bidding were synonymous. Today, a "best value" procurement approach may be used with design-bid-build in several states and at the Federal level. (This topic is covered in more detail in Chapter 3.)

A final advantage of design-bid-build is that the traditional relationship between the designer and the owner remains intact. The owner still holds a direct contract with the designer. Owners are accustomed to being engaged directly with their architects and find comfort in this relationship. They can react with suspicion toward any project delivery process that disrupts this traditional relationship.

Design-Bid-Build Disadvantages

One of the biggest disadvantages of the traditional design-bid-build delivery process is that there is no engagement or communication between the builder and the designers during the design stages; the contractor is not brought onboard until the design is complete. Therefore, the benefit of contractor input regarding issues such as constructability, fabrication, and production is not available early in the process. Problems associated with these issues are typically not discovered until after construction has begun, when making corrections is costly. Additionally, any added value, creativity, or innovation that the construction team could bring to the project is lost in the design-bid-build project delivery model because the process is not set up to accommodate this interaction.

Furthermore, this segregated services model often results in project budgets being missed at the first bid opening because there was no contractor input or adequate cost estimating done during the design process. When this happens, the designers are asked to go back to the drawing board and adjust their designs to bring the project in line with the intended construction budget. By this time, design budgets have been expended and the time and energy that this "re-design" process takes is costly to all parties involved. The project is often delayed and the revised design outcome can be a disappointment.

Ed Palmer, AIA, LEED AP
Principal, Senior Associate
Niles Bolton Associates
San Jose, California

One of the reasons I grew to appreciate design-build years ago was a result of the constant frustration I felt when our nearly complete designs were "value engineered" to the point of being unrecognizable. We would spend countless hours developing project designs to meet our client's functional and aesthetic requirements, only to learn that our solutions fell outside of their budget expectations and required last-minute, often ill-conceived, changes to meet budget.

One obvious way to keep this from happening was to work very closely with the contractors up front, in the beginning of the project, and throughout the design process so our design decisions could be closely correlated with cost factors that met the owner's budget. The design-build process allows this to happen. It works.

In many cases, contractors are asked to "value engineer" the project after the bid opening and bid award. This usually entails the contractors looking at the high-dollar items associated with the design and finding cheaper ways to achieve the same function. However, architects argue that this frequently means that aesthetics and sometimes quality are sacrificed because the "value engineering" process is being conducted as an afterthought, in an isolated fashion, and is not integrated with the design process. What started out as a Cadillac is now a Ford—it is still a car, but one that fits within the budget. This experience at the onset of the project can be disheartening and sets the project's tone. It is a hit-and-miss approach that often leaves all parties—the owner, the designer, and the contractor—feeling dissatisfied.

One of the most common complaints about design-bid-build is the finger-pointing that can go on between designers and contractors. Because the owner is expected to mediate all disputes and disagreements between the design side and the construction side of the project, both parties look to the owner to reconcile their opposing viewpoints. This burden has prompted much of the outcry described in the CURT white papers discussed in Chapter 1. However, the segregated services nature of design-bid-build creates this situation and naturally encourages self-serving agendas. Each party focuses on the independent piece of the project for which they are accountable, with little regard or concern for the other parts of the project, especially if things start going wrong. And given the industry's highly litigious environment in recent years, this attitude is understandable. It is the way the design-bid-build system has been set up and designed.

> True value engineering is an integrated practice that is intended to take place as part of the design process. In true value engineering, the design and construction professionals, along with subcontractors, suppliers, and fabricators, have the opportunity to discuss multiple project goals such as cost, quality, aesthetics, and constructability simultaneously, and in a collaborative fashion.

Often in the design-bid-build world, there is a dysfunctional structure that is caused by the owner's desire to keep the initial planning and design costs down and to secure the lowest construction bids possible, thereby limiting fees to designers and forcing contractors to look for faults in the documents. The owner thinks the contractor is out to rob him, the architect thinks the contractor is out to make him look bad, and the contractor thinks the architect is incompetent (drawings are not perfect) and the owner is cheap (she doesn't want to pay for what's wrong or not shown). Often an owner will hire a third party to manage the process who doesn't have any skin in the game and assumes a position to make others look bad in order to justify their fees. This sets up another adversarial relationship which can cost a project dearly. I see this occurring day in and day out.

Greg Gidez, AIA DBIA LEED AP
Corporate Design Manager
Hensel Phelps Construction Co.
Greeley, Colorado

And finally, one of the disadvantages associated with design-bid-build is that even though the owner has hired a designer to provide the design plans and specifications for their project, the owner ultimately carries the risk for that design when it goes forward to the contractors for bidding and construction. Because the designer works for the owner, the plans and specs are provided to the contractors by the owner. If there are flaws, errors, or items left out of the design that need to be corrected, then the owner bears the costs of those corrections, which are then processed through change orders. So the tradition and comfort of a direct contract between the owner and the designer may actually be a disadvantage, depending on what the owner's ultimate project goals are.

Primary Reasons for Choosing Design-Bid-Build

Even with the significant disadvantages of the design-bid-build approach, there are still some practical reasons why an owner might choose to use it as their project delivery method. Because design-bid-build is assumed in a competitive bidding process and contracts are awarded to the builder with the lowest initial price, then design-bid-build might be a logical choice if low cost is the owner's primary goal. However, initial low bids and final contract costs are rarely, if ever, the same. In competitive bidding the contractor is required to price the project in strict accordance with the completed plans and specs—nothing more and nothing less. But because the plans and specs for the project have been developed without the benefit of contractor input, there are usually many change orders that will ultimately add cost to the project. And even though change orders and cost overruns are a huge complaint by owners, some would rather stick with a method they know and understand. On the other side, the cost overrun issue is one of the biggest reasons why owners have sought out alternative project delivery methods.

John McGarva
President and CEO
Western Water
Constructors, Inc.
Santa Rosa, California

When projects are awarded on the basis of price alone, the life cycle value [benefits and savings associated with the building's entire operation and maintenance and not just initial design and construction cost] of completed projects is diminished. Low bids are often the result of incompetence in bid procedures, lack of experience, application of inaccurate or nonexistent job cost data, and taking unreasonable risks in pricing and production estimating. The solution to the industry's image problem is the solution to the procurement procedure problem. We must move towards "best value" as the priority criteria—not lowest price. The image of contractors will change to the extent the prevailing procurement methods change.

Another reason given for using design-bid-build is that some states and municipalities have legislative restrictions that forbid the use of design-build for public projects.

Design-bid-build may also be well suited for projects in which the owner wants to have a heavy hand in the design process. Because she has a separate contract with the designer, the owner can be as involved as she chooses. However, on a design-build project, such extensive owner involvement can sabotage the attributes of the integrated methodology, especially when the owner insists upon the use of products or methods that the design-build team finds inferior. The owner should always have some involvement in the design process, no matter which project delivery method, but gauging the correct level of owner involvement is a key question when deciding which method to choose.

And finally, design-bid-build is a logical project delivery choice if the project is simple. A highly collaborative, integrated process is not necessary if the project parameters are straightforward and the design and construction challenges and risks are minimal. However, even with simple projects, if time is of the essence, then the linear nature of design-bid-build may still deliver a project too slowly.

CM-at-Risk

As you learned in Chapter 1, the creation of construction management as a distinct discipline came about as an attempt to remedy some of the ills of the design-bid-build project delivery method, such as constructability conflicts and cost overruns. Construction management-at-risk (CM-at-Risk) is performed by contractors who can also offer preconstruction services such as feasibility and due diligence studies, constructability reviews, conceptual estimating, design-development estimates, and value engineering. CM-at-Risk takes advantage of some of the benefits of both design-bid-build and design-build.

With CM-at-Risk, the owner still holds two separate contracts, as in the design-bid-build model. The design contract includes the services typically provided by architects and engineers and maintains the traditional designer-owner relationship. However, the CM-at-Risk contract includes two distinct services. These services are typically identified in the contract as Part A and Part B as shown in Figure 2.3.

Part A of the contract provides preconstruction services. When contracted by the owner early in the process, these services provide the design team with the contractor input and cost information necessary to better manage the project scope and cost throughout the

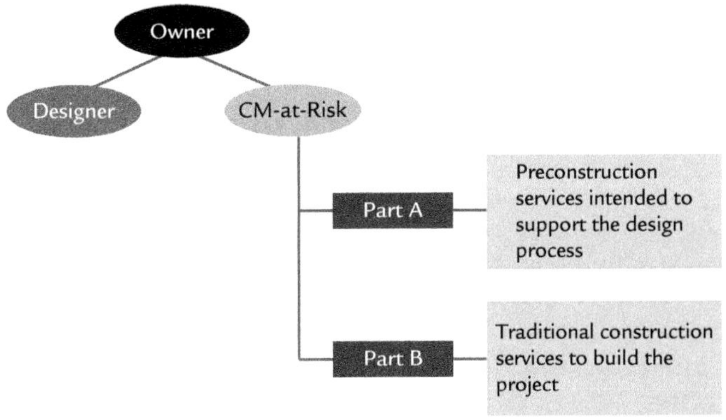

Ironically, in the early years of design-build, it was most often used on simple projects such as warehouses and manufacturing facilities. In fact, initial design-build efforts garnered a negative reputation for the nondescript, big-box architecture the methodology usually delivered. To this day, there are architects who resist design-build based upon these early design deficiencies. However, as design-build practices evolved and became more sophisticated, so did its capability to deliver high-end design, along with the other attributes associated with the integrated process.

Many construction companies now have dedicated preconstruction departments and personnel. For some of these firms, the preconstruction, or what is commonly referred to as "precon services" department, operates as its own profit center. In some cases companies now provide these types of services to owners without ever pursuing the actual construction work.

Figure 2.3
CM-at-Risk contract configuration

design stages, helping to ensure that the project is not over budget by the time the design is completed. When this information is provided to the designers up front, it is invaluable. The contractor is typically paid a flat fee for these services. Part A of the CM-at-Risk contract concludes once the design is complete enough for the contractor to establish a firm fixed or guaranteed maximum price for construction.

In the CM-at-Risk project delivery model, the owner must be willing to play a significant leadership role if the process is to achieve optimal results. The owner must first decide at what point to bring the contractor into the mix—obviously, given the value of their contribution, the sooner the better. In some cases the owner may select the designer first, as in traditional design-bid-build, and must decide how early in the design process the contractor would provide the greatest benefit. However, in other instances, the owner may choose to select the contractor first and the designer second.

Either way, it is important that the owner establish the ground rules for how the parties are to interact and who will have the final authority when differences of opinion arise regarding the design details, particularly as they relate to project costs. In this noncontractual engagement between the contractor and the designer, who is in charge? If the owner fails to clearly set expectations for this working relationship, then the collaboration may all be for naught. Just because the contractor is providing information and making suggestions does not mean the designer will pay any attention to them or use them in making decisions. The owner will always have the final say under this two-contract system because the owner still carries the ultimate risk for the design in CM-at-Risk, just as in design-bid-build. The contractor provides preconstruction advice only and has no liability associated with the design. The "at-risk" component of the obligation relates to the contractor being responsible for the construction cost, quality, and schedule in addition to providing preconstruction advice. Likewise, the designer has no responsibility for the construction under a CM-at-Risk model either.

The Part B component of the CM-at-Risk contract includes a final cost estimate and the traditional construction services to build the project. One of the unique characteristics of the CM-at-Risk contract is that the owner is not obligated to execute Part B with the same contractor and may opt out of executing the Part B component. Once the plans are complete, the owner can theoretically put them out to competitive bid. However, the more common approach is to work with the same contractor through both Parts A and B. Usually it is in the best interest of both parties to negotiate any differences that they may have. By the time the design is complete, the contractor understands the project extremely well, which will be of great benefit in expediting the project. However, if the contractor fails to perform on the Part A deliverables to the satisfaction of the owner, then it would be in everyone's best interest to abandon the execution of the Part B section with this contractor.

Sidebar notes:

If the contractor is brought on board first, she often plays a role in helping the owner select an A/E professional—and vice versa.

CM-at-Risk is also referred to as CMAR or CM-GC. For the most part these terms are interchangeable, although some areas of the country, laws make fine distinctions among the different terms.

CM-at-Risk project delivery is often chosen by owners before moving to full-blown design-build project delivery. CM-at-Risk brings the benefit of early contractor involvement while still maintaining the traditional owner-designer relationship. For this reason, many design-builders also pursue CM-at-Risk contracts as a way to show owners how much value they can bring to the table when engaged early in the process.

CM-at-Risk should not be confused with CM-as-Agent or CM-as-Advisor services. Both of these approaches are "add on" services, rather than project delivery methods. In these approaches, an owner will hire a construction manager or construction management firm to act on their behalf within the parameters of a given project delivery method. The Agent CM or Advisor CM typically works for the owner for a fee and holds no liability for either the design or the construction. These CM Advisor or CM Agent services may be employed with any of the project delivery methods, including design-build. See Figure 2.4.

On a fast-tracked project, construction begins before the design is complete. Building permits are issued in stages to match the design packages released. But not all projects can be fast-tracked. For example, permitting or funding constraints may not allow the construction to begin until the design is complete.

CM-at-Risk Advantages

There are several significant advantages to CM-at-Risk project delivery. It is a more integrated process then design-bid-build and allows for early contractor involvement, which is beneficial. Because the contractor is involved early, the owner will have cost information earlier in the process, before the design is complete. When managed correctly, the CM-at-Risk process avoids budget overruns at the end of design that often occur with design-bid-build. The early contractor involvement also mitigates potential schedule and delivery problems by providing the designers with pertinent subcontractor, supplier, or manufacturer information that helps them make smarter material and or product selections. Faster product delivery leads to faster project delivery as well. CM-at-Risk also allows for the design and construction to overlap, leading to even faster project delivery. Better delivery dates

TAKEAWAY TIPS

CM-at-Risk as a Transition to Design-Build
If you are a contractor trying to make the transition from design-bid-build to design-build, CM-at-Risk is a good place to start. CM-at-Risk will provide an opening to begin working with designers for the benefit of the project while still maintaining the traditional contractor role for the construction.

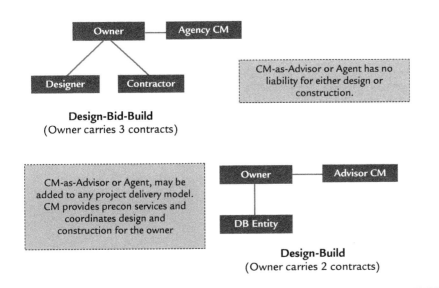

Figure 2.4
Agency CM configuration

are often one of the biggest drivers for selecting CM-at-Risk for projects that face challenging schedules. In recent years CM-at-Risk has become increasingly popular.

CM-at-Risk Disadvantages

Although CM-at-Risk seems to offer many advantages, the owner still has to deal with managing two separate contracts. All disputes and disagreements must still filter through her. And because of the early engagement of designers and contractors, managing the relationships may actually end up being a bigger challenge than expected. Even though the owner's goal may be to have the two parties work as a team, the parties may still have conflicting agendas because they are only obligated to deliver their independent parts of the contract; the segregated services mentality may still be firmly in place. And there is nothing that prevents the designers from ignoring the input that the contractors provide, no matter how valuable it might be. Change orders may still be common on CM-at-Risk projects and the owner is obliged to pay for them, even though she attempted to avoid them through early contractor involvement. The owner still bears the risks associated with

---

With CM-at-Risk many of our clients feel at ease knowing they "control" the design but have a designer and contractor working collaboratively early in the design phase of their project. Budget and schedule goals are constantly monitored and in an iterative design-estimate cycle that never allows the design to exceed an owner's financial commitment to the project. However, this outcome is only achieved if the owner stays engaged and ensures that the valuable cost and constructability input furnished by the contractor are embraced and integrated into the design by the Architect/Engineer [the designer]. Some designers will not compromise or adjust the design in response to contractor input. Such reluctance will occur at the owner's peril. In CM-at-Risk the owner must stay engaged and do some refereeing to gain the full benefit of contractor and designer expertise and collaboration. Optimally, the contractor can ensure the design development keeps pace with the procurement and construction schedule requirements of the project. Importantly, the contractor is able to protect the schedule, which benefits the owner and prevents erosion of profitability for the general contractor and subcontractors.

In a number of ways CM-at-Risk is symbiotic with design-build. DB [design-build] also gives an owner confidence that collaboration between its contractor and the designer are allowing for on-time, on-budget project delivery. In circumstances where an owner can comfortably surrender some control over the design in order to meet cost and schedule imperatives, an owner may take comfort in the design-builder's contractual commitment to the project success. The contractual shift of risk in DB will be attractive to some owners who don't have a real or perceived need to retain strict control over project design.

Bart Dickson, LEED AP
Project Manager
Hoffman Construction Company
Portland, Oregon

the adequacy of the design. And although in many cases, CM-at-Risk is faster than design-bid-build, it has generally been found to be slower then design-build.

Primary Reasons for Choosing CM-at-Risk

The primary reasons for choosing CM-at-Risk have to do with the benefits realized by a more collaborative process. Projects that possess a high level of technical complexity, have significant schedule constraints, or require complex phasing are all good candidates for CM-at-Risk project delivery. These are all project characteristics that would benefit from designers and contractors working together from a multidiscipline perspective in order to meet unique challenges. CM-at-Risk might be the right choice on projects where price guarantees are needed or strict budgeting is required. And any project that could benefit from value engineering could take advantage of the early contractor engagement. Essentially any project can benefit from using CM-at-Risk and all of the advantages that it brings. However, it is important to remember that the method does require a knowledgeable and experienced owner who understands the leadership role that she must play in managing independent service providers.

Although CM-at-Risk offers owners added assurance that the designs developed by their architects and engineers are, for the most part buildable and cost effective, the CM-at-Risk process still lacks the single point of responsibility that many clients desire.

Design-Build

Design-build project delivery is unique when compared to either design-bid-build or CM-at-Risk. From a contractual standpoint, the primary distinguishing characteristic of design-build is that the owner holds only one contract with a **design-build entity** (see Figure 2.5) and that entity is responsible for delivering both the design and the construction for a project.

This concept, often referred to as **single source responsibility**, means that the owner is no longer the go-between in the designer-builder relationship. The design and construction functions are now linked, packaged, and delivered together under a single contract. The intention is that the designer and the contractor work as a team in an integrated fashion to prepare the design from the concept stage to final construction documents while simultaneously constructing the project as the design concludes. This design-build team approach allows for optimum contractor involvement and supports the constructability review process and budget containment effort missing when using design-bid-build. Because of the improved communication that is possible when working as a team, this overlapping design and construction process results in faster project delivery. (See Figure 2.6.)

Any discrepancies or disputes between the designer and the contractor no longer filter through the owner. Instead, they must be handled within the confines of the design-build entity; any liability associated

D-B DICTIONARY

Design-Build (DB) Entity –
An organizational unit comprised of design and construction professionals paired together to provide design and construction services. The DB entity may be organizationally configured in several different ways.

Single-source Responsibility –
A procurement concept that places the responsibility for both the design and the construction of a project with one party under a single contract.

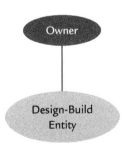

Figure 2.5
Design-build contract configuration

Figure 2.6
Design-build sequence

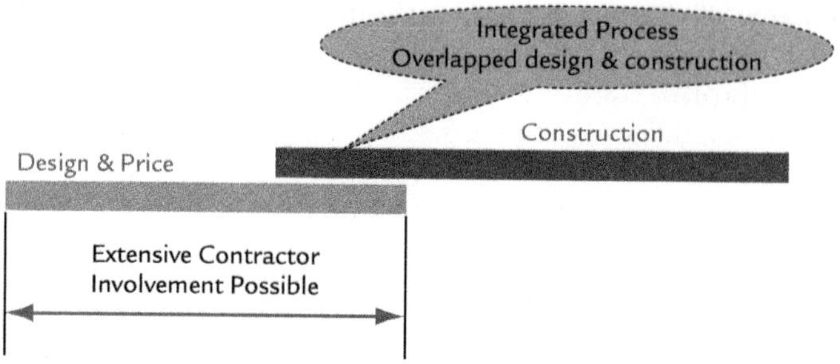

Design & Price
Extensive Contractor Involvement Possible
Construction
Integrated Process Overlapped design & construction

with design errors or omissions now lies with the design-build entity. There are no change orders crisscrossing in front of the owner. There are no remedies to pursue with the owner. Any insufficiencies, defects, or flaws in the design are borne by the design-builder. This is a major differentiator between design-build and the other two project delivery methods and one of the reasons that design-build has grown in popularity. Most of the risk associated with the design now lies with the parties who can best manage that risk—the designer and the contractor working together as a team.

Design-Build Advantages

Although there are many advantages to design-build project delivery, the most important one is the single-source responsibility available when both the design and construction is provided under a single contract, as noted in the *Design-Build Survey of Design and Construction Firms, 2007–2008* conducted by ZweigWhite Information Services.[2] Under the traditional method, owners often find themselves caught in the middle of disputes between their architects, engineers, and contractors as they point fingers and assign blame for flawed details, missed items, budget overruns, or poor facility performance. Owners are expected to act as referees as these disputes often deteriorate into change orders, claims, and even lawsuits, resulting in cost overruns, delayed schedules, and thwarted expectations.

The single-source responsibility inherent in design-build removes the owner from the middle and forces the designers and contractors to work together more closely to mitigate risks that they must now bear themselves if there are discrepancies, errors, omissions, or constructability issues with the design and construction. It has eliminated much of the frustration and risk associated with design-bid-build for the owner.

According to the ZweigWhite Survey, fast delivery and speed to market is the second most noted advantage to design-build project delivery. As you learned in Chapter 1, during the 1980s in particular, getting a project designed and built as quickly as possible was the number one goal for many owners. Even cost was not as big a factor as speed. The 1980s saw huge increases in competition across global markets and design-bid-build appeared to be the slowest and most inefficient method for delivering the capital projects that owners needed to produce goods and services in an increasingly competitive marketplace. And although it is not always possible to fast-track a design-build project due to procurement or regulatory restrictions, particularly on public projects, the inherent "conflicts resolution" that can exist within the DB team environment still reduces overall project time. RFIs are handled more expediently in design-build because the contractors are working directly with the designers as they design and there are fewer RFIs when the project gets to the field. A recent study led by David R. Riley, Ph.D., an associate professor at Pennsylvania State University, which compared design-build projects to other delivery methods, found that:

> *Design/Build projects ended up with 90% fewer field-generated change orders than design-bid-build projects. This resulted in projected "cost-growth" savings of 98%, or more than $1.7 million. This study provides objective evidence that using the Design/Build approach... can decrease both the occurrence and size of change orders. In many cases, it can practically eliminate field-generated change orders.*[3]

Furthermore, the design-build project delivery method appears to be producing far fewer claims and lawsuits as well.

> *Data from some of the largest professional liability insurers in the country say that the number of claims in design-build is far fewer than with traditional project delivery. Claims for errors and omissions or for time delays tend to disappear, because the design-build team would have no one to blame for these shortcomings but itself.*[4]

Another major advantage associated with design-build is the capability of having early knowledge of project cost. (See Chapter 3 for a more complete information about costs under design-build.) One of the biggest frustrations for owners, architects, and contractors occurs when they have spent thousands or even millions of dollars to complete a design, only to learn that the project was over budget at the time of bidding. Everyone pays dearly for this late knowledge of cost, in more ways than one. The owner's project is delayed, contractors who had planned for the project may now have to drop out of the competition because their resources are committed elsewhere, and material and subcontractor costs may escalate. The project then has to be redesigned; aesthetics and quality are often sacrificed in order to bring the project budget back in line, and everyone usually walks away disappointed. Unfortunately this is a frequent problem with the design-bid-build delivery method and even with CM-at-Risk. Figure 2.7 shows how the design-bid-build project scenario often ends up, a process often referred to as design-bid-build-bust.

In contrast, the design-build process allows for costs to be established early in the design phase, before a lot of time and dollars have been wasted on a design that cannot meet budget. A knowledgeable design-build team can skillfully manage that design to completion while maintaining the established budget.

In addition to these advantages of design-build project delivery, the ZweigWhite Survey also noted cost savings, improved risk management, increased quality, and more accurate cost estimates as other benefits to using design-build.[5]

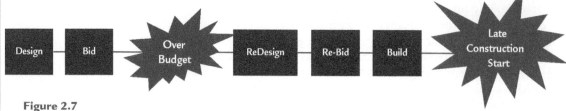

Figure 2.7
Design-bid-build bust

Design-Build Disadvantages

The primary disadvantage of using design-build is that it is not the established way of doing business. The design-build process differs significantly from the more traditional low bid approaches. Because design-build shares neither procedures or outcomes with design-bid-build, some participants have to go up a steep learning curve, which may create resistance among those who are new to the process, do not understand it, or more importantly do not trust it.

Ironically, one of the biggest advantages to design-build is potentially also one of its disadvantages. Faster project delivery requires faster owner responses. With design-build, owners are pushed to make earlier and more timely decisions. This can actually create problems on projects where the owner is not prepared for the fast pace and high level of engagement typical of design-build projects.

One of the disadvantages associated with design-build is the restrictive procurement regulations and restrictions that still exist in a couple of states. In 1993, only nine states had enacted legislation that allowed design-build to be used for the design and construction of their public projects. As of 2009 there were a total of forty-nine states that can use design-build procurement in some form or fashion (see Figure 2.8).

Although design-build has made great progress on the legislative front, some state statues still make it difficult to achieve the full potential of design-build project delivery. In sixteen of those forty-eight states, design-build is only a limited option for public projects, but in the remaining thirty-two it is more widely permitted, even to the point of being allowed for all agencies for any type of public project. Rhode Island is the only remaining state that had no statutory design-build authority for public procurement as of October 2009.

The states that have the most progressive design-build legislation currently are Arizona and Florida. Both states allow the selection of the design-builder to be based upon qualifications only, with no price competition.

Primary Reasons for Choosing Design-Build

Similar to CM-at-Risk, design-build offers all of the benefits of a more collaborative process. As an added bonus, the team is contractually connected to serve the project from a comprehensive, holistic perspective, featuring joint authority and responsibility to advocate for the project, rather than for each party's separate interests. Consequently, any projects that can benefit from the interactive engagement of an integrated team to reduce risk and optimize project results are all good candidates

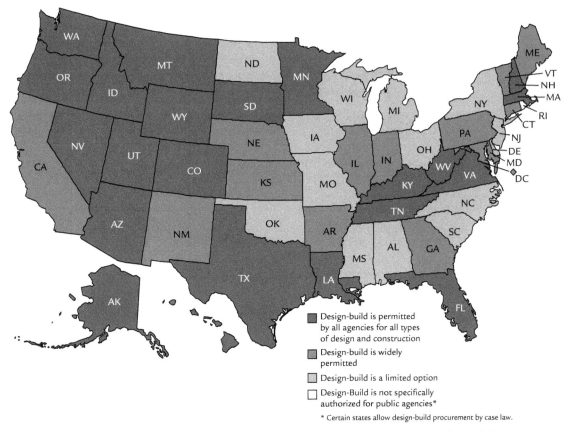

Figure 2.8
2009 DBIA legislation map. Courtesy of DBIA

for the design-build project delivery method, as are projects where design and construction are under tight schedule constraints. Speed to market continues to be a primary driver for the increased use of design-build on both private and public projects.

Another reason for selecting design-build is when owners need an early cost commitment or assistance in achieving budget targets through value engineering, requiring innovative solutions early in the design process. There are many project examples in the marketplace that have benefitted from this approach. The case study, "The Sweet Smell of Success," later in this chapter outlines a creative design-build solution for a seemingly intractable design problem.

And finally, design-build project delivery best suits owners whose priority is best value over the lowest bid. The concept of **best value** means that factors such as qualifications, past performance, design solutions, and overall management approaches are considered in addition to cost. (The concept of best value will be discussed in greater detail in Chapter 3.) According to Dr. Dean Kashiwagi of the Performance Based Studies Research Group at Arizona State University:

> *A major problem with the low bid process is that it considers contractors as a commodity. It assumes that if there is a set amount of material, resources,*

D-B DICTIONARY

Best Value –
A procurement approach that considers both quantitative (price) and qualitative or technical (design, schedule, past performance, management plan, etc.) factors when selecting the design-build team.

and time, no matter what contractor performs the job, the owner will always receive the same product. If an owner bases an award on the idea that all contractors are capable of the same performance...[t]hey are viewed as equal options.... In the low bid environment, expertise and skill commonly become insignificant and immaterial. The quality of the work is then directed by the specifications and other contract documents.[6]

Although the concept of best value can be incorporated with design-bid-build as well as design-build, design-build's up-front collaboration of owners, contractors, and designers extends through the design and construction phases, optimizing the potential for creating best value results.

Early design-build projects featured structures that were simple in design and function. However, today, design-build can be used for any type of project—simple or complex, private or public. In fact, the more complex the project, the more appropriate its use. For example, few projects are more complex than hospitals but design-build has become a favorite approach among healthcare clients, as noted in the February 2009 edition of the California Constructor magazine:

> *The design-build project delivery system in particular has gained favor with some owners seeking ways to mitigate the many challenges posed by healthcare projects. Design-build's advantages include helping to expedite the design and construction process, foster collaboration and teamwork and bring the contractor and key stakeholders in from the outset to ensure the design is complete, fully coordinated and buildable.[7]*

Design-build is being used in all sectors of the market, including commercial, industrial, heavy civil, water and wastewater, transportation, health care, and education. Even on public projects, although much more of a challenge to accommodate the high degree of flexibility that allows design-build to work at its best, design-build has become a prominent project delivery method. It is predicted to overtake design-bid-build as the preferred method for nonresidential projects by 2015 according to the Design-Build Institute of America (DBIA). (See Figure 2.9.)

Jeff Hooghouse, AIA, DBIA, AVS
Deputy Chief Architect
of the Corps
United States Army Corps
of Engineers Headquarters
Washington, D.C.

Historically the Army Corps of Engineers' typical approach to project delivery was execution driven and our construction contracts were awarded based upon low bid. In the last decade or so, circumstances have caused us to become more outcomes driven and we have been utilizing design-build more and more. Right now we have a huge amount of work coming our way, over $40 billion over the next five years. These projects are on a strict timeline and have to be delivered fast and with an emphasis on quality. The only way we can deliver this amount of work is by setting fixed budgets, giving the design-build teams the leeway they need to be innovative and creative, and then get out of their way. This is a total departure from how we used to do business and making the change isn't easy, but the results have been well worth it.

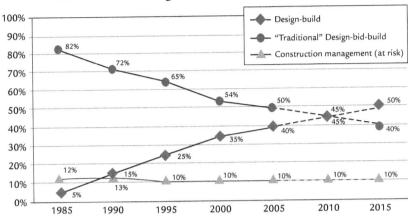

Figure 2.9
Project delivery trend line

In addition to the three major project delivery methods, there are variations on each one. For example, one design-build approach combines single-source accountability with the Part A/Part B concept of CM-at-Risk. Multiple prime project delivery is the same as design-bid-build in that the owner has a separate and distinct contract with the designer for traditional services, but then foregoes the services of a general contractor in favor of contracting directing with each specialty trade contractor—treating each one as if it were a prime contract, for example, for site work and excavation, concrete, and structural steel. This methodology, although still in use, and actually required in some states for public projects—in New York and Pennsylvania for example—it is not as common as the other three.

COMPARING THE PROJECT DELIVERY METHODS

One of the most comprehensive studies to date comparing the three major project delivery methods (design-build, design-bid-build, and CM-at-Risk) was conducted in 1999 as a Construction Industry Institute (CII) research project by Victor Sanvido and Mark Konchar. The authors published the results of that study in a book entitled *Selecting Project Delivery Systems* (The Project Delivery Institute, Fairfax, Virginia, 1999). The study revealed that design-build outperformed both design-bid-build and CM-at-Risk in five primary categories; unit cost, construction speed, delivery speed, cost growth, and schedule grown.[8] (See Figures 2.10 to 2.13.) The results indicated that, on average, the unit cost of design-build projects was 6.1% less than projects built under the design-bid-build method and 4.5% less than projects built using CM-at-Risk.[9]

Construction speed using design-build was found to be 12% faster than design-bid-build and 7% faster than CM-at-Risk.[10]

However, when delivery speed (the time for both design and construction) was considered the results were remarkable. The average delivery

Figure 2.10
Unit cost

Unit Cost Ranking by Project Delivery System

Figure 2.11
Construction speed

Construction Speed Ranking by Project Delivery System

Figure 2.12
Delivery speed

Delivery Speed Ranking by Project Delivery System

Figure 2.13
Comparisons summary

Metric	DB vs. DBB	CM@R vs. DBB	DB vs. CM@R	Level of Certainty
Unit Cost	6.1% lower	1.6% lower	4.5% lower	99%
Construction Speed	12% faster	5.8% faster	7% faster	89%
Delivery Speed	33.5% faster	13.3% faster	23.5% faster	88%

Reprinted with permission from *Selecting Project Delivery Systems*, by Victor Sanvido and Mark Konchar (Fairfax, VA: The Project Delivery Institute, 1999).

speed when using the design-build method is 33% faster than design-bid-build and 23.5% faster than CM-at-Risk.[11]

And finally, the study compared two critical project control items: cost growth and schedule growth. The results revealed that cost growth on a design-build project was, on average, 5.2% less than design-bid-build and 12.6% less than with CM-at-Risk. Likewise, schedule growth on projects utilizing design-build was 11.4% less than projects using design-bid-build and 2.2% less than projects built under CM-at-Risk.[12]

However, cost and schedule are not the only two critical indicators of project success; quality is an equally significant measure. The same study measured performance of each project delivery system for individual quality elements. It found that design-build projects achieved equal, if not slightly better quality results on average than design-bid-build and CM-at-Risk.[13]

At UCI we have found that with design-build the overall project duration can be reduced by 25 to 35%. This means that if a $20 million, four-year project is reduced by just 25%; it could slash a year off the schedule and result in a $2-million savings, assuming an annual escalation of 10%. Using design-build also translates to a more efficient coordination of drawings and a subsequent decrease in the likelihood of legal claims. Therefore, a 5% construction contingency on a normal project that does not result in litigation drops as low as 1.5 to 2%, producing additional savings. Furthermore, the integrated team delivery supported by design-build creates not only savings, but also new ideas, the development of an integrated team, and enhanced problem solving. While these benefits may be a bit difficult to quantify, they are an important part of the equation for successful project completion. The University typically saves between five and 10% in incremental innovations.

Rebekah G. Gladson,
FAIA, AUA
Associate Vice Chancellor
and Campus Architect
University of California –
Irvine, Design &
Construction Services
Irvine, California

DELIVERING BEYOND EXPECTATIONS

From a practical standpoint, cost, schedule, and quality are the foundational measures of project success. According to the CII study, design-build delivers those three project elements better than either design-bid-build or CM-at-Risk.

But design-build delivers another valuable project feature not mentioned in the CII study, one that is rarely mentioned because it is difficult to measure. It is the ability of the design-build process to deliver innovative solutions that otherwise could have never been envisioned at the start of the project. Although studies cannot directly measure innovation, case studies (including those in this book) show examples of just this phenomenon. By virtue of its highly collaborative, integrated, multi-disciplined approach, design-bid-build can realize solutions that could only be discovered through the synergistic engagement of the team itself. The diverse nature of this multi-disciplined team (architects, engineers, contractors, facility managers, specialty contractors, vendors, end users, and the like) and the experience, expertise, and knowledge that they bring to the project, fosters innovative, solutions not only in design, but extending to management, methods, materials, equipment, and technology.

CASE STUDY
THE SWEET SMELL OF SUCCESS

THE FACTS
Project Name:
Morris Foreman Sewage Plant
Project Type:
Environmental-Waste Water Facility
Location:
Louisville, Kentucky
Description:
Major Sewage Plant Upgrade
Cost: Original Budget:
Less than $60 million
Original Base Bids: In excess of $100 million
Final Contract Cost: $64.6 million
Duration:
38 months
Completion Date:
August 2003 (one year early)

THE TEAM
Owner:
Jefferson County Metropolitan Sewer District
Design Builder:
Black & Veatch and J.S. Alberici Constructors Joint Venture

THE LEAD-IN

Do you ever think about what happens to the water that runs down your sink drain after you do the dishes? How about when you flush the toilet? Modern day sewage systems and wastewater treatment plants are not something that you and I may think about very often, but the people of Louisville, Kentucky had no choice for a number of years but to think about it every single day. For over twenty years, neighboring residents of the Morris Forman wastewater treatment plant in Louisville, Kentucky had endured a daily irritant that most of us would prefer *not* to imagine.

THE STORY

Since 1976, the aging Morris Forman sewage plant, located on the Ohio River less than 10 miles from downtown, had used a thermal conditioning process to treat waste solids that generated such a foul odor that the residents could not even sit outside in their yards.

The odor was so bad that the city mandated that the sewage plant be closed down each year at least one week before and after the Kentucky Derby. Louisville was not going to put its guests and tourists through the anguish that its own residents had to endure on a daily basis—at least not on Derby day. The situation had festered for years, resulting in escalating tension and controversy. An ongoing conflict between the community and the Jefferson County Metropolitan Sewer District (MSD) often resulted in daily protests that played out in the local news on a regular basis. And it

Courtesy of CDM

wasn't that the MSD didn't try. The MSD had spent more than $200 million on plant improvements over a fifteen-year period trying to remedy the problem. When conventional wet scrubbers, bio-filters, and chemical masking agents didn't help, they built a four-story facility to house two new incinerators in hopes of solving problem. Unfortunately it did not. The incinerators ended up being both a financial and operational failure. In essence, every "fix-it" dollar spent on trying to mitigate the problem was

Courtesy of CDM

money down the drain. The incinerators were eventually shut down and the building that contained them sat vacant for about twenty-five years.

The unbearable stench was not the only problem that plagued the antiquated system. Solid wastes were being shipped in from the MSD's other wastewater treatment plants to be processed at Morris Forman, pushing the facility toward maximum capacity. Furthermore, after the heat treatment and de-watering process, the remaining solids had to be loaded onto trucks and shipped off to the landfill, adding significant dollars in transportation costs, not to mention the impact on the landfill capacity. Even without the quality of life issues associated with the putrid smell, the Morris Forman Wastewater Treatment facility was facing *big* problems which kept getting bigger.

During the mid-1990s, the MSD attempted to initiate two more project upgrade proposals, both using the design-bid-build approach. But they too were abandoned, primarily due to excessive costs overruns and stalled negotiations. Then in 1998, a third attempt was made, only this time the MSD decided to give design-build a try. The county implemented a standard two-step design-build procurement process consisting of an RFQ and an RFP. After over a year of wrangling, three design-builders were finally shortlisted and asked to submit a proposal on a 30% design issued in the RFP consisting of a two-phase digester and de-watering method that could be incorporated into the county's existing process for the management of the bio-solids.

However, the owner's proposed design required some logistical gymnastics that seemed absurd. For one thing, between the existing plant site and the new building site stood an Exxon gas processing facility. The new treatment method meant that sludge solids from the old site would have to be pumped to the new site either through exposed pipes lying on top of the ground or by an underground tunnel. Next, the effluent left over after de-watering would have to be pumped back across the Exxon property to the existing plant for treatment. This crisscrossing procedure seemed inefficient and unnecessarily costly. On top of this convoluted situation, consider the additional cost to maintain two separate facilities, along with the expense still associated with transporting the bio-solids to the landfill. The owner's proposed design solution did not seem like an acceptable solution.

By all accounts the proposed RFP design looked like it was going to cost in excess of $100 million. Unfortunately, the Jefferson County MSD only had a budget of less than $60 million. So it looked like this third attempt to solve the problems was going to meet with failure as well. By this time, one of the three shortlisted contractors dropped out of the competition, leaving only two to compete. Fearing that the effort would once again fail, the MSD sent out an addendum to the RFP stating that although they still wanted a base bid on the initial design, the county would now consider and review "alternate designs." This opened the door for creativity and innovation. The B&V/Alberici design-build team decided that this was the opportunity that they needed in order to find a solution that would work and still be competitive.

One week before the issuance of the addendum, the joint venture team leader, Blake Childress, attended a presentation on a new waste treatment method by which bio-solids are dried in rotary drums after de-watering—replacing the old sludge by-product with an odor free "class A bio-solid" pellet type product that can be used as fertilizer and/or a soil additive. The MSD had actually looked into a similar technology years earlier but deemed it to be too expensive. If the B&V/Alberici team could come up with a way to cost effectively use this system, they could solve the odor problem and also reduce landfill volume, lower overall energy consumption, and transform a smelly sludge by-product into a potentially marketable, environmentally friendly commodity.

With this new knowledge still swirling around in his mind, Childress decided to visit the project site again. While walking around inside the dingy, dirty abandoned incinerator facility, an idea began to emerge. "What if we could eliminate a major portion of the expense to incorporate this new sludge drying technology by re-using the old facility?" At four stories high, it was big enough, and it was built like a bunker, yet it was going to waste. If the team could figure out how to make the existing building work, it would eliminate the capital expenditure needed to construct a new building, as well as the need to maintain two sites and crisscross the Exxon property. Furthermore, it would significantly reduce or eliminate all together the transportation costs back and forth to the landfill, and mitigate the landfill impacts. The design-build team started planning how they could gut the old building, get the old equipment out, clean up the facility, and install the new sludge drying system. After days of brainstorming and measuring every piece of equipment that would come out and go in, the team decided that they could cut a relatively small hole in the exterior wall of the old building, about sixteen by sixteen feet, to remove the old incinerators and de-watering equipment and then squeeze the new equipment through the exact same opening—in the end closing it off with an overhead door. After the B&V/Alberici team worked out the strategy and calculated the cost, they submitted a price on their alternate design of $64.6 million. Their base bid for the original RFP design was just over $100 million. Once presented, the MSD jumped on the solution. The joint venture design-build team was eventually awarded the contract based upon their alternate design.

The reuse of the existing building saved the client at least $25 million. Eliminating the piping and tunneling across the Exxon property resulted in additional savings. In the end, the design-build team had saved Jefferson County over $32 million dollars in capital expenditures as well as adding significantly to the operating efficiencies. Not only had the team solved the capacity and transportation issues but also resolved the single most important issue of the residents of Louisville, KY. Once the new state-of-the-art sludge drying system was up and operational, the odor was gone! The final solution not only met the owner's budget but it also eliminated an entire year from the schedule.

Design-build opened up an opportunity to produce a solution that otherwise could have never been envisioned. When the owner finally allowed the design-build team to develop "alternate designs," an extraordinary solution was discovered and the design-build team was able to add value far beyond the owner's expectations. The resulting effort not only resolved the ongoing conflict and tension between the MSD and its neighbors but it also vastly reduced the county's transportation and disposal costs. Furthermore, the MSD realized an unexpected revenue stream in the pelleted bio-solid product dubbed "Louisville Green." The pellets are sold to local farmers who pick them up in their own vehicles, adding nothing to the county's transportation costs. Design-build, along with an extraordinary team, delivered a win-win solution for all parties concerned.

THE CLOSE

Today, if you visit Louisville, guests and residents alike now enjoy the sweet smell of bluegrass and goldenrod on Derby day, or any other day for that matter. The foul odor that once plagued the city has been completely remedied. The Morris Forman Wastewater Treatment Plant is a showcase for innovation and a testament to what is possible when a design-build team is empowered to solving a problem using a solution that the owner could not have envisioned alone.

From the Owner:

By thinking outside the box, Black & Veatch looked at possibilities beyond our base proposal. B&V's alternative approach subsequently proved to be a winning vision for MSD and the constituency we serve. By combining environmental principals with bold engineering and design, Black & Veatch delivered a new state-of-the-art facility within the existing, fully operational plant. This approach proved a judicious use of public funds, and resulted in a $32 million capital savings as well as significant operating cost efficiencies.
[The owner quotation is from the January 5, 2005 Letter of Recommendation from MSD to ACEC (American Council of Engineering Companies) for the 2005 Engineering Excellence Award.]

D-B DICTIONARY

Design-Build Plus – A term used when services are added to the design-build contract beyond design and construction such as financing, operations, or maintenance.

When a design-build team is properly managed, with the owner on board as part of that team, the design-build method can produce solutions that neither of the other two project delivery methods can provide. And some clients have found ways to integrate far more than just design and construction on their projects.

Total Facility Solutions

The bundling of design and construction services under a single contract opens the door for owners who want to shift their project risks beyond the traditional design and construction functions. Some owners have linked other services to the design-build contract. For example, project financing, project operations, project maintenance, and even project ownership have been packaged together as a single, sole source solicitation. Today this approach is used by public and private clients alike, and is often referred to as **design-build plus**. The most common add-on services today are:

:: Design-Build – *Finance*
:: Design-Build – *Operate*
:: Design-Build – *Maintain*

It is also not uncommon to see multiple add-on services combined such as:

:: Design-Build – Finance-Own-Lease
:: Design-Build – Operate-Maintain

Under these comprehensive, total-facility-solution types of contracts, individuals from numerous professional and technical organizations join forces to provide a wide range of services under the design-build approach. Many integrated design-build firms have planners, landscape architects, interior designers, facility operators, and commissioning experts on staff along with their engineers, architects, builders, and construction managers. It is not unusual to find school and/or hospital administrators, teachers, physicians, and nurses on the payroll as well, depending on a design-build company's marketplace niche. It may seem odd that nondesign and construction personnel would be hired by design-build firms, but who better to advise the designers, engineers, and contractors that design and build hospitals then a nurse, doctor, or administrator who understands the ins and outs of patient care and hospital administration. The same is true of schools, which might have teachers or administrators on the design-build team, or production facilities, which might include facility managers.

The truth is, just about any number of services can be bundled together under a single design-build contract. For example, in 1997 the Haskell Company of Jacksonville, Florida, an integrated design-build firm, not only provided the City of Pembroke Pines, Florida with the design and construction services needed to construct the Pembroke Pines Charter Schools, but they also advised the owner on financing options along with providing the administration services to manage the entire campus—including curriculum development, text book selection, furnishings, and faculty hiring and management. At that time Haskell had an in-house educational services division staffed with professional educators including the former superintendent of Dade County Public Schools. Haskell contracted for the administration services and they provided these as part of their design-build total-facility solution. Three years later, the education management group spun off their own company, but today Haskell still provides facilities maintenance services to the City through a permanent workforce of 125 staff. These facilities management personnel maintain all city buildings including a charter school system with four campuses with a total of 5,600 students. It is an extraordinary example of just how much value can be delivered with design-build-plus.

When nonintegrated design-build entities team up to provide such a broad menu of services, they often do so through a series of custom subcontracts designed to fit the unique needs of the project. Sometimes the controlling entity or prime contractor provides overall management of all services, but usually the service contract is transferred to an independent, professional manager or facilitator with expertise in the area of service. However, the controlling entity may still maintain the direct liability associated with the contract as the sole source provider to the owner. There is no single model that fits this comprehensive services design-build approach. The same creativity and innovation that are hallmarks of the design-build method also hold true for contracting arrangements. The high performing design-build team is expected to find solutions to all kinds of project challenges and problems. The old days of thinking like a traditional contractor, architect, or engineer are gone. To manage a design-build team, you must be a highly resourceful facilitator and project integrator in addition to bringing technical know-how to the problem-solving team.

MANAGING DESIGN-BUILD PROJECTS

Neither design-bid-build nor CM-at-Risk offers the broad range of services that is available through design-build. Experience as a contractor, engineer, or architect working under the design-bid-build model, or even CM-at-Risk, does not necessarily prepare you for working under the design-build project delivery model. In particular, such experience fails to prepare you to *manage* an integrated design-build team under this sole source method. Design-build project management is vastly different from traditional construction management, and anyone who does not understand the distinctions might find themselves ill-equipped to manage the design-build project and the risks associated with it. This is especially true when those challenges and risks spill beyond design and construction. For example, in some instances you may also be responsible for the performance of a facility such as a manufacturing plant where you must guarantee that a certain number of widgets will be produced per a designated recording cycle. While traditional construction management skills and talent are important, the manager's mental attitude is critical to direct multiple project functions that extend beyond one's expertise. The design-build project manager is really a process manager and a project integrator whose job is not to solve all the problems themselves, but rather to facilitate a solution, using whatever means or resources are necessary. You will learn more about this unique process in Chapter 7.

MENTORING MOMENT

Getting Out of Our Comfort Zones

Erin Brozovich
Project Engineer
CDM Constructors Inc.
Location: Rancho Cucamonga, California
Education: Ohio State University, graduated 2003
Major: Civil Engineering

The design-build process is very fast and quite often designers do not have time to put together a complete design and specification package. This can make them feel quite uncomfortable—especially if they aren't used to working in design-build. They don't understand that the details that they usually draw just aren't needed by the contractors. Experienced contractors know how to build the details. But for design-build to work, the designers have to trust the contractors to "do the right thing." Designers are so accustomed to serving as the "watchdog" in design-bid-build that it takes them awhile to understand that everyone is on the same team in design-build and that the contractor will ultimately pay the price if the design is not built correctly and doesn't perform as intended. Unfortunately, building that trust takes time, even in an integrated firm like ours. But I think it involves both parties "getting out of their comfort zones." The engineer or architect needs to get out in the field and observe the construction, and the contractor needs to be involved from day one of the design. The key is constant communication—a picture might be worth a thousand words, but sometimes there just isn't time to draw that picture. Instead, you have to rely on words to get your point across. That's why designers really should spend time in the field working with the contractors, answering questions when they arise, in real time. This would also allow them to see that their contractors do know what they are doing and are committed to building things right.

I definitely think that design-build is easier for my generation. For one thing we have less baggage. Besides I think it's hard to train people to "unlearn" what they know. I see it every day, working with people that are so used to the traditional hard bid construction. The engineers or architects and contractors are separate entities and in some regards, they view one another as the enemy. But people like me see how well the design-build process works; we don't know any other way. We're taught the design-build process at the beginning of our careers, and we honestly don't have anything to compare it to. It just makes sense to us because it is so efficient. It's quite difficult to understand the distrust between people who are supposed to be on the same team. It can be quite frustrating at times but you do begin to see changes in attitudes after awhile, especially when you get to work with the same people on multiple projects. Eventually the barriers come down—you just have to be patient.

CHAPTER 2   *Planning, Designing, and Constructing Projects*

MAKING SMART PROJECT DELIVERY DECISIONS

Selecting the best project delivery method is no easy task. Unfortunately, not all clients consider the different factors and criteria that impact this decision. Whether the project is simple or complex, it is important that owners take the time to apply the proper due diligence to evaluate the project from multiple perspectives before making their project delivery choice.

There are three fundamental factors to consider before selecting a project delivery method:

- **Project Goals and Objectives**: These are usually the most obvious factors, like budgets, schedule, quality, performance, image, appearance, aesthetics, and function. For example, if schedule is the most critical of all the project requirements, then design-build project delivery might well become the first choice, presuming that all other factors are compatible.

- **Constraints and Limitations**: These factors often appear as hindrances such as zoning regulations or licensing requirements. Others might pose outright obstacles such as public contracting laws or environmental restrictions. This category can also include limitations or restrictions that encourage the selection of a given project delivery method—such as finance assistance for under funded projects or critically sensitive schedules tied to business performance. In these instances, the project delivery method may actually mitigate or eliminate the constraint altogether. Other constraint type factors include the owners' in-house capabilities, design review committees, tedious protocol or chain of command issues, and the like.

- **Intangible Impacts**: These factors are more difficult to pinpoint and are often overlooked or not given the consideration that they deserve. They include such issues as the corporate culture or an individual owner's or manager's personality relative to expected design involvement, risk comfort, decision-making ability, or overall project control issues. The prevailing political environment is another less-tangible influence that may have a significant impact on the project delivery selection. Market conditions, union influence, and availability of talent will also influence the owners' project delivery choice, and there are many other intangible factors that can come into play.

Some owners and public agencies have sophisticated processes for evaluating their project delivery choices. Others use a more informal process. And still others may start with electronic selection tools such as the *Design-Build Selector Predictive Tool* developed by Dr. Keith Molenaar at the University of Colorado (*http://www.colorado.edu/engineering/civil/db/*). For a thorough review of project delivery systems, see the *"Construction Project Delivery Systems: Evaluating the Owner's*

TAKEAWAY TIPS

Developing Project Criteria

Consider developing a checklist, selection matrix, or other analytical tool to help you formally think through the significant criteria associated with your project. It's important to develop a consistent methodology to help you ask all the right questions, prioritize objectives, and cover all the bases. And don't forget to invite your end users to join in, including facilities people, maintenance staff, and anyone else who can give you honest input relative to the overall project goals and expectations.

Figure 2.14
Project delivery selection matrix

Criteria	Criteria Weight	DBB	Multi-Prime	CM-at-Risk	DB	Other
Schedule Flexibility	25	4 / 100	9 / 225	6 / 150	8 / 200	
Owner design control	15	10 / 150	10 / 150	6 / 90	4 / 60	
Awarding on best value	20	3 / 60	3 / 60	8 / 160	10 / 200	
Low initial cost	10	9 / 90	9 / 90	5 / 50	6 / 60	
Promoting team work	15	4 / 60	3 / 45	7 / 105	9 / 135	
Establishing early price	15	5 / 75	2 / 30	8 / 120	9 / 135	
Totals	100	535	600	675	(790)	

Rank each delivery method relative to delivering on criteria on a scale from 1 (low) to 10 (high)

Alternatives" CD-ROM by AEC Training Technologies available at *http://www.aectraining.com/index.html*. Figure 2.14 shows a sample of a Project Delivery Selection Matrix.

Doing Your Homework

When selecting a project delivery process, the owner must conduct an in-depth inquiry into the way the various methods meet the needs of her organization. (You will learn more about this process in Chapter 4.) The deeper the probe, the more accurate the assessment and the more appropriate the project delivery selection. As a potential proposer on a design-build project, be aware of the motivations that have influenced the owner's project delivery choice. It is important for project professionals—contractors, architects, and engineers—to fully understand the magnitude of this due diligence process, because it is directly correlated to the overall success of the project and your ability to influence that success. I regret that on occasion some owners still give me poor reasoning for their project delivery choices. I still hear reasons like, "We're sticking with design-bid-build because it's what we've always used," or conversely, "We're going to go with design-build because we never tried it before," or even worse, "Everybody is using design-build, so we thought we'd give it a shot."

In an ideal world, the owner has conducted a meaningful evaluation to ascertain the best project delivery method for their project. Unfortunately, you can't always count on that happening, so instead, as a design-builder you must hone your own questioning and observational skills so you can get to the true source of the owner's expectations in order to fully meet them and avoid painful surprises and disappointments later. You'll learn more about developing these skills and characteristics in Chapter 5; they are critical to understanding of the initial project delivery decision, because it influences overall project success.

DIFFERENT IN EVERY WAY

Now that you have a good understanding of the major project delivery methods, the rest of this book and the Design-Build Ready Team series will focus exclusively on design-build project delivery. Many people think that design-build is more similar to design-bid-build or traditional project delivery than it is. They do not fully understand just how dissimilar the two processes are. This error can be risky. Chapter 3 will focus on how unique the design-build process is from the others. Even to those who think they already know everything about design-build: Brace yourselves, because, as Dorothy told her little dog in the classic movie *The Wizard of Oz,* "Toto, I've a feeling we're not in Kansas anymore!"[14]

CHAPTER HIGHLIGHTS

1. Construction is not a product, it's a professional service. Every project is unique, with different site, weather, and topography conditions, as well as municipality regulatory restrictions, environmental requirements, and code issues. Project delivery is the comprehensive plan that addresses the uncertain nature and unique circumstances of every construction project

2. The three most common project delivery methods are: design-bid-build, CM-at-Risk, and design-build. The primary players are an owner, a contractor, and a designer. The contractual relationship among these three is what differentiates the delivery methods.

3. Design-bid-build is the "traditional" project delivery method, in which the owner holds separate contracts with the designer and the contractor. Design-bid-build has many advantages: all the players are familiar with the process, it is well suited to the competitive bidding process which is in turn supported by laws in all fifty states, and the traditional relationship between the owner and the designer remains intact.

4. Design-bid-build has several disadvantages, which primarily stem from the contractor not being involved until the design is complete. This often results in "redesign" because the designer didn't have the benefit of the contractor's knowledge during the initial design. In addition, estimates are usually inadequate and because of the contractual setup, the owner bears the costs of flaws and errors in design.

5. Owners may choose design-bid-build for a number of reasons: to use the competitive bidding process, to deal with fewer legislative restrictions, to have extensive involvement in the design process, and for small-scale projects that don't require an integrated process.

6. In CM-at-Risk the owner still holds two separate contracts, but the contractor contract has two parts, one for preconstruction and one

for construction. The owner has a significant leadership role and ultimately holds the risk, just as in design-bid-build.

7. One of the primary advantages of CM-at-Risk is that it provides early contractor input, which helps projects come in on budget and on schedule. The ability of designers and contractors to overlap activities can help a project become fast-tracked, resulting in even faster project delivery. One disadvantage is that because they are under separate contracts, there is no guarantee that the designer and contractor will work as a team. This ultimately leaves the owner bearing the risk if there is inadequacy in the design.

8. Owners may choose CM-at-Risk because it is a more collaborative process that helps particularly in projects that have a high level of complexity, significant schedule constraints, and/or a need for a strict budget.

9. In design-build, the owner holds a single contract with a design-build entity, which is responsible for delivering the design and construction of a project—also called single-source responsibility. This removes the contractor from the middle and forces designers and contractors to work together more closely to mitigate risks that they must now bear themselves if there are discrepancies, errors, omissions, or constructability issues.

10. Design-build allows for faster speed because: construction can begin before the design is complete; RFIs are handled quickly; and there are fewer change orders. These factors also result in lower costs for the project, fewer claims or lawsuits, improved risk management, and increased quality.

11. If design-build is new to you, you face a steep learning curve. Although most states now allow design-build procurement in some form, there are still restrictive procurement regulations in a few states.

12. Design-build offers a more collaborative process with a team that is contractually connected to serve the project from a comprehensive, holistic perspective, characterized by joint authority and responsibility to advocate for the project from an integrated perspective.

13. Results of a study in *Selecting Project Delivery Systems* revealed that design-build outperforms both design-bid-build and CM-at-Risk in five primary categories: unit cost, construction speed, delivery speed, cost growth, and schedule growth.

14. The diverse nature of the multidiscipline design-build team (architects, engineers, contractors, facility managers, specialty contractors, vendors, end users, etc.) and the experience, expertise, and knowledge that they bring to the project creates an environment in which new, innovative solutions can be realized, leading to innovations in design, management, methods, materials, equipment, and technology,.

15. Owners should consider at least three fundamental factors before selecting a project delivery method: 1) project goals and objectives, 2) constraints and limitations, and 3) intangible impacts. An in-depth inquiry into these factors should lead the owner into selecting the right project delivery method for their project.

RECOMMENDED READING

:: ***Selecting Project Delivery Systems***, by Victor Sanvido and Mark Konchar, published by The Project Delivery Institute, Fairfax, Virginia, 1999.

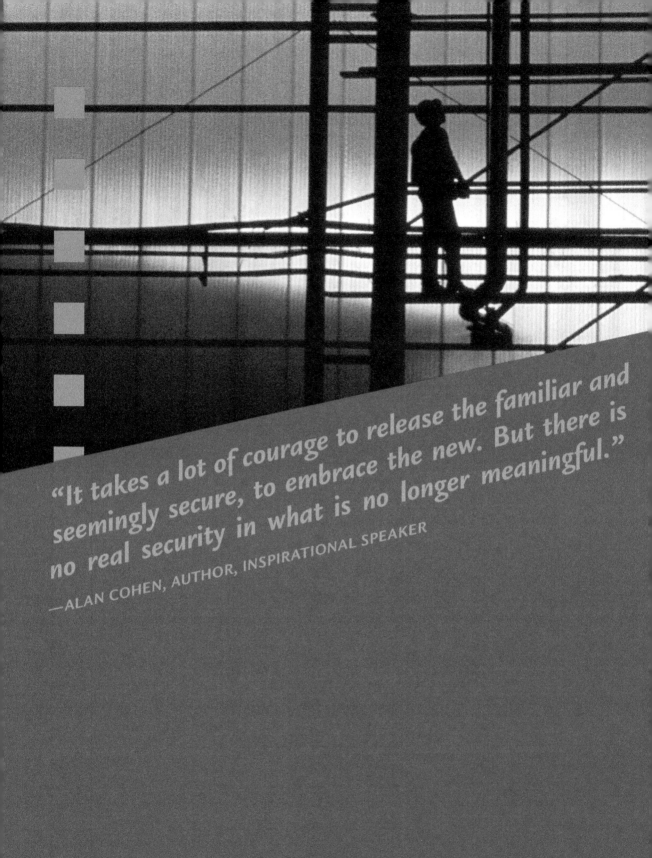

"It takes a lot of courage to release the familiar and seemingly secure, to embrace the new. But there is no real security in what is no longer meaningful."

—ALAN COHEN, AUTHOR, INSPIRATIONAL SPEAKER

3
FUNDAMENTAL CHARACTERISTICS OF DESIGN-BUILD

In the last chapter you learned about the fundamental differences between the three basic project delivery methods. You also learned about some of the advantages and disadvantages of each. This chapter focuses on the specific concepts and characteristics that distinguish design-build as a unique project delivery method. You need to understand these characteristics if you want to make the most of the benefits offered by the design-build project delivery method. In fact, this chapter may actually be the most important chapter in the book, both for long-time practitioners of design-build methods and for newcomers.

Design-build is completely different from design-bid-build. This point is critical because the main mistake contractors, architects, and engineers make when transitioning to design-build is failing to recognize the differences between the design-build process and the design-bid-build process. Design-build is *not* just a slight tweak of the traditional design-bid-build model—it is different in every way.

The fact of the matter is, most current AEC practitioners have not been formally trained in design-build project delivery, nor do they understand how to take advantage of the full potential of the process. It wasn't until 1993-1994 that the Design-Build Institute of America (DBIA) published its first *Design-Build Manual of Practice* and started teaching public courses on Design-Build Project Delivery. Thus it is not surprising that design-build projects often look and perform just like a design-bid-build project. Unless you are new to the field, you were probably trained to some degree in the traditional design-bid-build project delivery approach. Unfortunately, this old knowledge comes with old attitudes and behaviors that can sometimes get in the way of understanding the nature of the design-build approach. The need to leave behind old ideas and adjust our attitudes is one of the most difficult aspects of mastering the design-build approach. The best place to start is to make sure that you fully understand the basic concepts of design-build and how they differ from traditional project delivery.

Nine primary characteristics distinguish design-build from traditional project delivery: 1) solicitation and basis of award; 2) team approach; 3) the design-build entity; 4) risk management; 5) performance requirements; 6) who controls the details of the design; 7) the design-cost relationship; 8) the integrated design-build schedule; and 9) contracting approach. These characteristics are explained in the following sections.

Dave S. Crawford, PE, DBIA
President and COO
Sundt Construction, Inc.
Tempe, Arizona

I decided some time ago that if we are going to do design-build, we were going to do it right or not at all. We're going to put both feet in the water and swim. We are not just going to put one toe in at a time—timidly trying a little bit of this or a little bit of that to see if it works. Design-build doesn't allow it. You're either in or you're out. Not only do you have to trust the people on your team, but you have to trust the process as well.

SOLICITATION AND BASIS OF AWARD

One of the most important characteristics of design-build is the process by which an owner solicits and awards the work. Also, the contractual relationship between an owner and a design-build firm is different from the contractual relationships in the more traditional approach. Let's start with the solicitation and contract award process.

Requests for Proposals (RFP) Versus Invitation for Bids (IFB)

The most common solicitation instrument in design-bid-build is the Invitation for Bids (IFB), with the contract typically awarded to the contractor with the lowest bid. Price is generally the primary determining factor and the design does not even come into play as part of the solicitation for construction.

In design-build, however, the primary solicitation instrument is the **Request for Proposal (RFP)**. The contract is typically awarded to the firm that offers the best value, both quantitatively (price) and qualitatively (design, schedule, past performance, management plan, etc.).

The notion of best value is really at the heart of design-build. In order to provide best value, the design-build team must optimize all facets of the project—design, cost, schedule, quality, and performance—to deliver a project solution that meets or exceeds the owner's expectations. The team's goal is to deliver to the owner the biggest bang for the buck. And how does the owner decide what team does in fact provide the best value? As you will learn in Chapter 4, several evaluation methodologies can be used under the best value approach.

Getting work by responding to an RFP is very different from getting work by responding to an IFB. An inexperienced contractor who is not accustomed to competing in the RFP format will learn very quickly just how different it is. Winning a design-build proposal requires a set of skills and abilities far beyond just coming up with the lowest price. Design-build contractors

> **D-B DICTIONARY**
>
> **Request for Proposal (RFP)**
> The solicitation document that an owner releases to elicit design and construction services in a design-build procurement process. The document typically becomes a component of the initial contract.

and designers need to be well-versed in negotiation skills, presentation skills, public speaking, and a host of soft skills. Not just any contractor, architect, or engineer can be successful in winning and managing design-build projects. (Details regarding team skills are discussed in Chapter 8.)

Keep in mind that different sectors of the industry may have different requirements for responses to an RFP. For example, the way you would respond to a public procurement will vary from how you would respond to a private entity RFP. Design-Build can work for any sector of the industry—public, private, commercial building, transportation, environmental, heavy civil, water/waste-water, infrastructure, and industrial. The fundamental way we respond is the same. However, the evaluation criteria and the amount of flexibility allowed may vary. The key is to be responsive and to demonstrate your team's ability to add value to the project.

Contractual Configuration

As discussed in Chapter 2, there is only one contract between the owner and the design-build entity. This is different from the design-bid-build approach, in which the owner holds separate contracts for design and construction and often finds himself in the undesirable middle position. The single point of responsibility in the design-build approach means that the entire design-build team is in the same boat. If there happens to be a hole in that boat, it is the entire team's responsibility to plug it. Design-build is not about pointing fingers and assigning blame. It is about finding out what works, and then doing it.

TEAM APPROACH

Design-build is a team sport. It is a collaboration of professionals working together with the owner to come up with the best solution for the project. For professionals who are transitioning into the design-build approach, the hardest conceptual leap is often the leap from the segregated services mentality (so common in design-bid-build) to the integrated service mentality. Most team members are just not accustomed to true collaboration. As a matter of fact, just the opposite is often the case. Although most of us try to cooperate in good faith on our projects, in reality we have been trained to be skeptical of one another, to limit our communication, and not to disclose information. Suspicion is the name of the game when you have separate contracts and conflicting agendas. However, to succeed in the design-build arena, team members need to learn to trust one another. Building trust among design-build team members is the toughest nut to crack.

Early Contractor Involvement

One of the more obvious differences between design-bid-build and design-build is that, with design-build, the contractor is involved from the get-go. By contrast, in design-bid-build the contractor is not brought into the picture until after the design is complete. Thus, the contractor has no input or involvement in developing the design. In design-build, the contractor should be involved in the design process from the beginning.

Kim Lum
Executive Vice President
Charles Pankow
Builders, Ltd.
Pasadena, California

Contractors provide a valuable pool of talented people with many of the skills required to lead the design team, especially as more and more trade partners begin to participate. While architects continue to want to be the primary contact with owners, experienced "contractor" project managers have the precise skills needed to provide this leadership. What they need is a better understanding of how the design process works and the patience to deal with professional designers who are not used to working under budget and time pressures. Once they have that understanding, they become true design-build project managers.

In practice, many contractors take on design-build projects and then fail to participate in the design process. This is a huge and sometimes costly mistake. Early contractor involvement in the design development phases is exactly where the benefits of the design-build process can be best realized, as the cost-influence curve in Figure 3.1 illustrates. The early stages of the process provide the greatest opportunity for the contractor and architect to influence the design at the lowest possible cost. However, keep in mind that contractors aren't accustomed to participating in design development and architects aren't accustom to having them participate. This presents yet another team-building challenge for the design-build team, but one that is definitely worth overcoming.

Proactive Versus Reactive Response

Contractors, architects, and owners who are new to design-build often struggle to understand that the entire design-build process is geared toward finding solutions proactively. The idea is to engage the entire team, including the owner, as partners in the problem-solving activities.

Figure 3.1
Cost influence curve

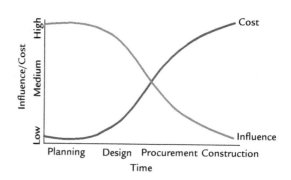

Instead of reactively looking for someone to blame when something goes wrong, the design-build team should proactively engage its collective experience and expertise to attack problems and mitigate their impacts as soon as possible. This is easier said than done; not all teams function at this high level. However, it should be the goal for all design-build teams.

Drinking From a Fire Hose

MENTORING MOMENT

John Parnell, LEED AP
Project Manager
Ryan Companies US, Inc.
Phoenix, Arizona
Education: Cal Poly State University, San Luis Obispo – graduated 2006
Major: Construction Management

Even though I have a degree in construction management and I work in my company's construction department, my job is so much more than being a contractor. All day long I'm looking at the project from different perspectives. I may be trying to step into the designer's shoes at one point during the day; then at other times I'm in the subcontractor's shoes. And you're always trying to see things from the owner's perspective, trying to get it right for him. It's really like the old concept of the Master Builder.

Design-build is not for everyone. All the details are not worked out and you have to work as a team to solve problems on the fly. Some people are really uncomfortable with uncertainty and like somebody else figuring out all of the answers and being told what to do. That's fine, but don't come to work for a design-builder. On the other hand, if it thrills you to be creative and figure things out by working with your team of field experts and designers, in a situation in which everything is not necessarily fully defined, then you should consider the options offered by design-build.

I'm on the phone with my architects, engineers, and subs constantly—just moving along, working things out and fitting things together and just getting it done. It's a whole lot better than wasting time waiting for someone to respond to your RFI. It's a lot easier to document a decision than wasting time shuffling paper through a dozen people's hands before you can get to work solving the problem. Some days it's like drinking from a fire hose, but I love what I'm doing. Design-build is for anyone who doesn't want to be stuck in a box. If you are interested in investing yourself and your skills in the bigger solution, then design-build might be for you.

This proactive approach is useful both when dealing with internal team problems and when working to assist owners in achieving their project goals. The design-build team and the owner need to be flexible, ready to adjust details in order to best serve the project and solve problems. In the early stages, the design should be viewed as a dynamic work in progress. Eventually, the design and all details associated with it become fixed, but the overall nature of design-build is fluid, not rigid. Ideally, neither the team nor the owner should feel shy about suggesting improvements to the project or alerting everyone to a potential problem. When an improvement is suggested or a problem spotted, the automatic response is not a change order. In design-build, the expected response should be, "What can we do to mitigate this problem or implement this improvement without affecting the budget, schedule, or quality in a negative way?"

This is not the traditional thinking in design-bid-build. In design-bid-build, when someone notices a problem, the response is often "Gotcha!" For that reason, many people new to design-build fear that, if they don't get the RFP exactly right and happen to miss something or make a mistake, they are going to "get caught." However, design-build is *not* a gotcha game. Owners, contractors, engineers, architects, and even subcontractors need to understand the proactive nature of the design-build process.

When a design-build team is functioning at its highest level, the owner should be able to count on the team to achieve the project's goals, even if the owner fails to articulate them accurately in the initial solicitation. By the same token, the owner needs to give the design-build team the flexibility it needs to achieve the project goals. You cannot tie the hands of the design-build team or restrict their creativity and options completely and then expect them to be responsive. (You will learn more about what works and what does not work in the solicitation process in Chapter 4.)

The savvy owner will use the design-build team to get a better project. After all, in the design-build approach, the owner has the opportunity to pick the best possible design-build team and, then use that team's collective expertise to achieve the project goals. It is not unusual for an owner to overlook something that the design-build team notices and addresses, thereby ultimately achieving project results beyond the owner's original expectations. An example is illustrated in the case study on page 79.

THE DESIGN-BUILD ENTITY

So far we have considered two of the nine main characteristics of the design-build approach. Now consider the third—the unification of the design firm and the construction firm into a single entity. This design-build entity can be configured in two basic ways—either as a single firm or as two separate design and construction firms joining forces for a specific project. In the latter case, the two firms can be joined either in a legal joint venture or as non-binding partners, in which one firm hires the services of the other.

Once the design and construction teams have joined forces, the resulting single entity is usually organized according to one of the four models described in the following sections.

Contractor-Led Design-Build

Contractor-led design-build is the most prevalent model, with approximately 56% of all design-build projects delivered under this type of contractual arrangement, according

---

TAKEAWAY TIPS

Tracking Enhancements

Consider tracking the project enhancements delivered by your team per project by way of an internal score card. This does two things. First, it gives you a way to show potential clients how you have added value to projects in the past. Secondly, it will influence your employees' thinking, helping them shift from a traditional, reactive approach to the proactive, value-seeking approach preferred in design-build.

CASE STUDY

TO THE RESCUE

THE LEAD-IN

On May 23, 1939, the Unite States Navy submarine Squalus, then America's newest submarine, sank in 250 feet of water during a test dive off the New England coast.[1] If you were a sailor on a submarine in the late 1930s, and it sank, the odds of rescue were nearly zero because the means to carry out a rescue had not yet been developed. But fortunately for the crew of the Squalus, a Navy submariner named Charles B. "Swede" Momsen used a pear-shaped diving bell of his own design to save thirty-three crewmen over a grueling thirty-nine hour period. This is still considered the greatest undersea rescue ever. Momsen's remarkable feat was the impetus for the Navy's development of advanced deep-sea rescue units, leading to vast improvements in submarine design and construction and in the training of submarine crews.[2]

THE STORY

The New London Naval Submarine Base in Groton, CT the oldest such facility in the nation and home to 8,500 sailors and seventeen nuclear submarines, was targeted for closure in May 2005 along with thirty-three other military installations nationwide. But less than four months later, New London won a reprieve from the Department of Defense's Base

Courtesy of M. A. Mortenson Co.

THE FACTS
Project Name:
Submarine Escape Training School
Project Type:
Public Sector Building
Location:
Naval Submarine Base, New London, Groton, CT
Description:
One-of-a-kind, 22,600 square foot training facility, including a deep dive tank (20 feet in diameter, 40 feet deep), classrooms, administration and support spaces, built on a very difficult site
Cost:
$13,565,136
Duration:
26 months
Completion Date:
August 2007

THE TEAM
Owner:
United States Navy
Contractual Design-Build Team Leader:
M. A. Mortenson Company
General Contractor:
M. A. Mortenson Company
Architect-Engineer:
KlingStubbins
Specialty Consultant(s):
McClymont & Rak, Thielsch Engineering
Specialty Contractor(s):
Perini Contracting

THE TEAM (continued)
Base Mechanical, Semac Electric, John Strafach & Son, AC Dellovade, Pittsburg Tank & Tower, Massey's Plate Glass & Aluminum, Fernandes Masonry, United Steel, Acoustics, Inc., Park-Roway Company, KONE, Aero Mechanical, BKM Floorcovering, The Concrete Supplement Co., Schnabel Foundation, Neptune-Benson, The H.R. Hillery Co., Brand Fireproofing

Realignment and Closure Commission (BRAC). The 587-acre base not only survived, but also resumed its ambitious capital construction program that included the one-of-a-kind construction of a $13.5 million submarine escape training facility.

The Submarine Escape Trainer Facility project entailed a two-phase design-build qualification-based selection process. Mortenson was one of five preselected contractors permitted to propose on the project. The RFP was developed by the Navy in a standard format that included performance, prescriptive, and functional requirements, along with all of the standard and relevant Federal Acquisitions Regulations (FARs). The RFP was issued to the pre-selected design-build teams on November 1, 2004. From that point, each design-build team developed its technical solution and pricing and delivered that information to the Navy in mid-December 2004. The Mortenson/KlingStubbins team won the project on a best-value selection criteria that took into account the technical solution, proposed quality, the past performance and reputation of both companies, the proposed schedule, and price.

The submarine escape training facility was named "Momsen Hall" after the mastermind of the Squalus rescue, in recognition of Charles Momsen's pioneering efforts in submarine escape and rescue. A major concern in the design and construction of the Momsen Hall was the training tower, where escape training, which is one of the rites of passage at the sub school, takes place. The tower planned for the new building was intended to replace an earlier version, which was demolished in the

Courtesy of M. A. Mortenson Co.

Courtesy of M. A. Mortenson Co.

1980s. While it lasted, the earlier tower was a landmark; many people felt it was the defining motif of the Groton skyline. It was also a part of the collective memory of the base and left a strong impression on all who trained in it. So rather than designing a semi-anonymous building that blended in with the other campus buildings of the Sub School, the design team sought to express the building's unique and important training function by emphasizing the new tower.

This unique, 22,600-square-foot facility provides a realistic and tightly controllable training environment for practicing escapes and rescues from sunken and disabled submarines. Program requirements included a 20-foot-diameter by 40-foot-deep dive tank, an attached support building for instructional space, infrastructure for related activities such as dive equipment maintenance, and hyperbaric chamber facilities.

The goal of this design-build project was to meet the Navy's goals for the project, while allowing the submarine training portion of the building to be the primary form of the exterior. In order to meet these design goals, significant changes were made to the Navy's schematic design as set forth in the RFP. The government's original design concept called for the building to blend into the rest of the campus. Site conditions and square footage concerns led to an early government concept in which the building consisted of many smaller forms popping out and changing shape to meet the functional requirements. This early design resulted in a complex and unbalanced cluster of forms set awkwardly upon a large hill of fill material.

The Mortenson/KlingStubbins design-build team worked closely with the Navy to analyze the RFP in order to find innovative ways to

Courtesy of M. A. Mortenson Co.

accomplish the Navy's requirements within the project constraints. These constraints included a strict limitation on overall square footage due to the Congressional approval process, and a sloping and rocky site.

A major innovative design solution was to relocate the fifth-floor support area of classrooms and office functions to below the main pool deck level. This revision to the government's initial schematic design allowed the design-build team to significantly reduce the amount of fill needed to level off the upper part of the site. It also minimized the requirements for a large retaining wall on the northern side of the training tower. The relocation of the fifth-floor also placed more emphasis on the training tower portion of the building, allowing the cylindrical form to stand out as the tallest part of the building. The building design was developed into two distinct sections: a two-story support section and a four-story training tower. These sections were visually set off from each other by a section of curtain wall.

These changes, made as part of the design-build process, allowed the submarine training functions to be expressed in building form and to allow the project to stand out among the other building on the base.

Mortenson/KlingStubbins also worked closely with the Navy to completely redesign the pool system, with the goal of shrinking the water storage system and incorporating the tank within the building footprint. This resulted in cost and time savings, and made maintenance of the pool system easier. In addition, the design-build team relocated the second floor training rooms below ground to take advantage of the sloped project site, significantly reducing excavation and fill material requirements. This also provided a lower profile for the building, which in turn reduced heating and cooling costs. At the same time, the room layouts and orientations were revised in response to the end users' preferences, providing a more user-friendly layout. An electric traction elevator, typically not allowed in federal military construction, was used to eliminate a machine room; this, in turn, provided more usable square footage, a critical consideration on a project governed by a total square footage restriction. Finally, the design-build team identified several important ways to enhance the project, including expanding the site drainage and incorporating a radon gas detection system.

THE CLOSE

This project is a good example of how the ingenuity of an innovative design-build team can rescue a mediocre project, turning it into something quite extraordinary—even within the confines of some rigid requirements and program constraints. The design-build team exceeded all project goals by listening to the client and end users, evaluating the site conditions, and involving subcontractors in decisions in which they played a part. The team's willingness to change the design program at its own expense resulted in completely redesigned floor plans, revised equipment systems designs, and more efficient spaces—all within the tight gross square footage and budget allowed by the government.

From the Owner

The Mortenson/KlingStubbins team delivered a high-quality product that fulfilled all contract requirements as well as addressing all end user special needs. The team was always open, available, and willing to go the extra mile to work with us on all aspects of the design and construction while maintaining strict cost, quality, and schedule control. I believe it will be hard to find a better example of how the design-build process, with the right design-build team, can turn a project from good to great.

Robert E. Howland

Project Manager

Department of the Navy

Naval Facilities Engineering Command, Mid-Atlantic

[Case study adapted from Mortenson Construction's submission for the Design Build Institute of America 2008 Design-Build Award Competition, by permission.]

Figure 3.2
Design-build entity configuration

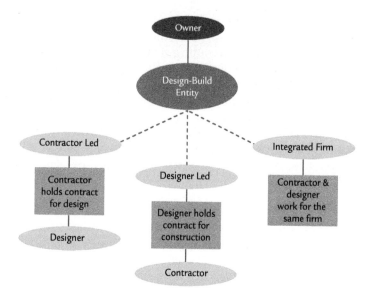

to the 2007–2008 Design/Build Survey published by ZweigWhite.³ As indicated in Figure 3.2, the contractor-led design-build model has the construction firm serving in the prime contract position with the owner. The design-build contractor hires an independent architecture firm to provide design services. The design contract is subordinate to the prime contract. This places the overall responsibility for managing the design in the hands of the contractor. Obviously, the design firm, under contract with the builder, still performs all of the design functions and maintains professional liability for delivering the design. But if there is a problem with the design, the owner will look to the construction firm for remedy. At that point, the contractor and designer must work out a solution and a way to pay for it. This shifting of risks into the hands of the design-build team is one of the main reasons design-build has become so popular among many owners, both public and private.

Designer-Led Design-Build

Designer-led design-build is the opposite of contractor-led design-build in that the design firm holds the prime contract with the owner and hires out the construction services to a general contractor. (See Figure 3.2.) In this arrangement, the design firm is directly responsible to the owner for delivering both the design and the construction. According to the Zweig White Survey, on average only 12% of all design-build projects are led by designers.⁴ Most of the designer-led design-build projects are engineering-intensive projects like power plants, waste water projects, and refineries. Designers tend to be more risk-averse than contractors; only rarely does a design firm, particularly an architecture firm, take on the risk of construction in a design-build contract. The same early contractor participation is required in designer-led design-build as in contractor-led projects.

> Many contractors refer to their design partners as "subcontractors." This is misleading, because designers are not contractors at all. They provide a professional service akin to consulting. In fact, many architects take great offense at the "subcontractor" label. A more respectful term is design partner, specialty consultant, or design consultant. Today even subcontractors prefer being called "specialty" contractors or "trade partners." These terms emphasize their special technical expertise rather than their subordinate position in the contractual hierarchy and better represents true team collaboration.

Integrated Design-Build Firm

Integrated design-build firms, sometimes referred to as full-service design-build firms, make up approximately 27% of all design-build service providers.[5] These entities represent multidiscipled firms that have their own in-house designers and contractors. (See Figure 3.2.) In other words, they all work for the same company. Many times they are even housed under the same roof. Under this organizational structure, the single firm bears the entire risk for the project, with no subordinate contracts to a primary service provider. Subcontracts with specialty contractors and specialty consultants are arranged as needed.

Although this type of arrangement can produce teams that benefit from familiarity and ease of communication, they might also be limited by the group's collective scope of experience. For example, an integrated design-build firm with vast experience in the design and construction of schools, office buildings, and condominium projects may be hard pressed to compete on a hospital or health care facility. However, nothing precludes an integrated design-build firm from partnering with an outside firm to deliver design beyond the firm's range of expertise. Such partner arrangements often arise when an integrated firm decides to go after a project for which it lacks the necessary expertise.

> *It is good for a client to have a single point of responsibility, but if the team is not getting along, the project (and thus the client) is affected, even if they are not contractually involved. As an integrated design-build firm, our focus is always on the client and the project, from both a design perspective and a construction perspective. It provides a true team atmosphere. When an issue arises, the team members focus on their common goals for the project. We don't have to worry about getting bogged down trying to sort through the boundaries imposed when trying to perform under separate service agreements.*

Paul Shea, PE, BCEE, DBIA
President, CDM Constructors Inc
Camp, Dresser, & McKee, Inc. (CDM)
Denver, Colorado

Other Models

Two other, less popular, arrangements for the design-build entity is the joint venture design-build approach and the developer-led design-build approach. These two approaches combined make up only 5% of the design-build organizational variations. However, it is common today to find two or more construction firms forming joint ventures to handle the construction side of a contract, especially on very large projects. When this occurs, if the design-build entity is led by the construction side, the individual construction firms will decide how to divide up the overall management tasks.

The Concept of the Controlling Design-Build Party

For design-build entities made up of multiple firms, one party making up the entity must contract directly with the owner to provide both the

Figure 3.3
Controlling party accountabilities

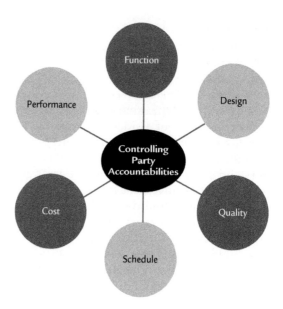

D-B DICTIONARY

Controlling Party
The party holding the prime contract position in a multiple firm design-build contractual relationship.

design and the construction. Thus, this party is the **controlling party**. As illustrated in Figure 3.3, the controlling party needs to think and behave as a true design-builder, taking responsibility for managing the entire process—design, quality, performance, cost, and schedule. Ideally, the controlling party views the design process, cost management, and schedule control as an intertwined, integrated processes. This is not the traditional role of a contractor. Failing to understand the inherent risks associated with design when serving as the controlling party could be disastrous for a contractor new to this role. In Chapter 8 you will learn more about this important role in a design-build project.

AUTHOR EDITORIAL

MIRROR, MIRROR ON THE WALL...

Owners often ask me which of the design-build entity models is best. Is it better to go with an integrated design-build firm or to use separate design and construction firms when executing a design-build project?

Although integrated firms have an advantage in that they all work for the same company and therefore have a mutual interest in the success of their firm, many great multicompany design-build partnerships produce successful projects. The key is to make sure the separate firms share similar core values, with similar corporate cultures. The separate companies should actively work to establish common goals—not just specific project goals but also general business goals. It is common for design-build contractors to create incentives for their architects that reward timely design delivery and quality drawings, for example. Anything that keeps the parties focused on the same prize—that is, a project delivered on time, within budget and with optimized results for everyone. Companies that have teamed together successfully as design-build partners routinely join forces again on other projects.

So when an owner asks me which is best—an integrated firm or multiple firms working as a team—I usually explain that it all depends on the quality of the firms involved. I also explain that the success of a project depends, to a surprising extent, on the owner. One of the most risky variables associated with design-build success is the owner's attitude toward collaboration and open communication. A client who demands a fully integrated team but fails to participate as an equally trusting, communicative partner ends up being part of the problem and not part of the solution. The need for aligned cultures and common goals does not stop on the practitioner side.

I also tell owners that, once they have made sure they are comfortable with the kind of collaboration the design-build approach requires, they should spend some time looking for the right firms. Next, they should hand select individuals from within a particular firm. The key is to make sure you have the right people on the team. When I say the right people, I mean individuals who behave as design-builders, regardless of their training and professional status.

Over the years I have met many individuals in the design and construction business whose main thought is always "What is best for the project?" This does not mean they forfeit self-interest when executing the contract. It means they work a little harder at negotiating a position that serves multiple interests. In essence, that is how a collaborator behaves. They are as concerned with the relationship as they are with the outcome, and are unwilling to forsake one for the other. Ideally, design-build architects and contractors possess what I call interdisciplinary fluency. They can see all facets of a project and listen to their partners' suggestions, or even criticisms, and understand that they are intended to improve the project. But owners cannot assume that every employee of every design-build firm possesses interdisciplinary fluency. Thus, it is important to seek out the best of the best in the design-build firms with which they plan to do business.

MANAGING RISKS

Risk is inherent in any construction project and managing risks is what construction managers do. However, the risks associated with design-build are significantly more complicated than with traditional design-bid-build jobs. In design-bid-build, the contractor is responsible for managing only the construction risks. The design is complete when the contractor puts together a price and schedule, and the construction manager begins his job with a full set of plans and specifications in hand. However, in the more collaborate world of design-build, managing the design is also part of the construction manager's job. Among other things, the construction manager must make sure the design is feasible given the construction schedule and cost constraints, and also ensure that the design meets the performance, quality, and compliance goals as specified in the RFP.

When it comes to assessing and managing risk, design-build is a whole new ballgame. To understand the nature of the risks involved, you need to understand a legal principle known as the Spearin Doctrine. According to this doctrine, a contractor who follows the plans and specifications provided by an owner is not liable to the owner for any loss or damage if the plans and specifications turn out to be defective.

How does this affect risk management? In the traditional world of design-bid-build, the owner contracts separately for design and construction. So once the design is completed by the architect and supplied to the owner, the owner turns around and supplies the contractor with the plans and specifications. Under this scenario, the Spearin Doctrine stipulates that the owner "impliedly warrants" (in other words, guarantees) the plans and specifications to the contractor and maintains that a contractor will not be liable to the owner for loss or damage which results solely from defects in the plans and specifications. In design-bid-build, all risks associated with that design are born by the owner and his designers.

The risk picture is different in the design-build world. Here, the Spearin Doctrine is inapplicable, because the plans and specifications are prepared by the design-builder. Therefore, the design-builder is the one who impliedly warrants the plans and specifications to the owner. Thus any loss associated with insufficiencies or defects in the plans and specifications are born by the design-builder. The contractor cannot seek remedy from the owner for any errors, omissions, or flaws in the design. Consequently, the owner bears no responsibility for the change orders associated with errors, omissions, or flaws in the design. Only changes associated with increased scope or major program deviations that are initiated by the owner require change orders.

For example, if the design-build architect designs a flashing detail that leaks, the design-build team (the contractor and architect) is responsible for making the correction at its own expense, with no charge to the owner. This is the opposite of what happens in design-bid-build. You can see how this changes the entire risk picture. You can also see why many owners are embracing design-build.

Robynne Thaxton Parkinson
Attorney
Robynne Thaxton Parkinson PLLC
Mercer Island, Washington

In design-bid-build, the owner spends too much time telling the contractor how to build something, when the contractor already knows how to do it. This practice costs owners more money than they need to spend. Under the Spearin doctrine, the party providing the design warrants the sufficiency of the design. When owners delegate the responsibility for design, as in design-build, they also delegate the risk associated with the failure of the design. Further, much of the innovation happening in the construction industry comes from the specialty subcontractors who are finding new ways to put together and install the pieces of the project that they manufacture or fabricate. By dictating how an element must be incorporated into a building, the owner limits this innovation. From a legal perspective, many of the specifications for products that are incorporated into a building actually come, not from the architect, but from the manufacturers who will install them; however, because of the liability gap in design-bid-build construction, the owner then becomes responsible for warranting the sufficiency of these same specifications. With design-build, this problem is eliminated.

PERFORMANCE REQUIREMENTS

Another important characteristic of the design-build approach has to do with how the completed project is conceived and, ultimately, evaluated. In design-bid-build, a successful project is one that meets all the specifications prescribed by the owner. In design-build, a successful project is one that performs exactly the way the owner needs it to perform. This distinction is extremely important.

If you look up the word "specification" in a dictionary you will find a definition that reads something like this: "A detailed, exact statement of particulars, especially a statement prescribing materials, dimensions, and quality of work for something to be built, installed, or manufactured."[6] In design-bid-build, the set of completed drawings prepared by the designers are supplemented by a set of **design specifications**. These detailed drawings and specifications explicitly instruct the contractor regarding what to build and how to build it, taking into account project layout and configuration, dimensions, materials, products, manufacturers, model numbers, and the like. In other words, the plans and specifications provide a prescription for the builder to follow. In fact, the specifications that accompany a completed set of drawings in the design-bid-build approach are called **prescriptive specifications.**

Prescriptive specifications are incompatible with the design-build approach, which emphasizes flexibility and innovation. Instead, we start with the owner's description of how he wants the structure to perform when the project is completed. This description takes the form of a set of **performance requirements**. For example, instead of requesting a specific reflective ceiling plan and then prescribing specific light fixtures by manufacturer and model number, the owner writes performance requirements indicating the amount of illuminance expected at desktop height, measured in foot-candles, within a given space. The owner might further clarify the performance requirements by adding other quality or aesthetic descriptors. But specific layouts, fixture types, manufacturers, or model numbers are not prescribed, leaving the design-build team plenty of creative freedom and flexibility. Still, the owner's parameters will ensure that the desired lighting goals are achieved.

Figure 3.4 highlights a few of the distinct differences between performance requirements and prescriptive specifications.

The significant distinction is that, in the traditional design-bid-build model, the owner prescribes the lighting, product, and installation specifications, thereby taking on the risk of the prescribed solution meeting

> **D-B DICTIONARY**
>
> **Design specifications**
> Detailed descriptions of how a product or system must be manufactured and installed or constructed. Also includes identification of specific products, materials, manufacturers, and model numbers. Similar to a prescriptive specification.
>
> **Prescriptive Specifications**
> The traditional method for stipulating materials, products, manufacturers, and techniques in design-bid-build. These stipulations are detailed and explicit.
>
> **Performance Requirements**
> A design requirement expressed in terms of an expected outcome or final result; by contrast, prescriptive specifications dictate specific products, manufacturers and techniques.

Characteristic	Prescriptive Specs	Performance Requirements
Predictability of end product	Maximum	Minimum
Accountability for end product	Owner	Design-Builder
Flexibility and opportunity for innovation	Minimum	Maximum
Responsiveness to unexpected events	Minimum	Maximum

Figure 3.4
Prescriptive specifications versus performance requirements.
Courtesy of DBIA.

It takes practice to write good performance requirements but they are an important ingredient of successful design-build. As an industry, we are accustomed to telling people what to do and what to use; that is a lot easier, but it is also riskier. If we want our design-build teams to have an opportunity to really show us what they can bring to the table in terms of problem solving, value, and creativity, then we have to give them enough leeway to demonstrate their responsiveness and ingenuity while still setting acceptable quality and performance parameters. Performance requirements allow the design-build teams to do that just that.

D-B DICTIONARY

Functional Requirements
Broad descriptions of what a facility or structure needs to be and how it will be used. Also includes an expression of project needs, goals, constraints, or challenges.

the performance goals for the project. If it does not, the owner is left holding the bag, and either has to live with the disappointing results or issue a change order to correct the problem. The contractor simply supplies and installs materials called out in the plans and specs. By contrast, in the design-build approach, the owner can rest assured that the project will ultimately perform as needed.

As you can see, an owner's fears about ceding the details of the design to the design-build team are often misplaced. The reality is that, in many instances, it is far riskier for the owner to proceed with traditional design-bid-build than it is to use design-build.

Communicating Performance

Performance requirements are provided in the RFP document. Occasionally, the RFP also includes some prescribed design. For example, an owner might express a personal preference for very specific items or products (such as a specific brand of lock that matches the locks used at the owner's other facilities). The key is not to carry the design suggestions too far—to the point where the contribution of the design-build team is limited to simply completing the necessary drawings. Otherwise, there is really no point in choosing the design-build approach in the first place and some design-builders will be reluctant to pursue the project. (Chapter 4 has even more to say on the topic of prescribed design in the RFP.)

Design Specification Evolution

When developing performance requirements, it is important to start with the big picture. The first step is to develop the **functional requirements** of the project. In other words, what are the overall goals, constraints, and challenges associated with the project? In the end, what will the facility actually be? How will it be used?

Functional requirements are usually expressed in operational terms. For example, a set of functional requirements might request a six-story multi-use office-retail facility to accommodate 1,100 office workers and ten retail spaces. Before it is possible to articulate a complete set of functional requirements, the owner might have to complete a feasibility study and perform some initial budget projections.

After the project's functional needs have been identified, the performance requirements can be developed. For example, the six-story building mentioned earlier would require mechanical vertical transportation between all six floors and be sized to accommodate 1,100 tenants. This performance requirement would, in turn, evolve into a design specification proposed by the design-builder, supplemented by plans, details, and shop drawings. Ultimately, these would be confirmed as five individual passenger elevators identified by manufacturer and model number. In design-build, all project requirements evolve in this way, from functional requirement, to performance requirement, to design specification. (See Figure 3.5.)

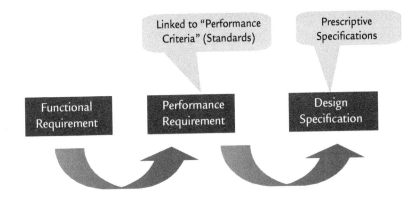

Figure 3.5
Specification evaluation process

Verifying Performance

All this talk about performance requirements is well and good, but how will the owner determine whether the completed project will actually perform as required? From the owner's perspective, that is an important question. From the design-builder's perspective, the question is framed slightly differently: "How will I be able to prove to the owner that I have delivered at least, and possibly more than, the performance that the owner contracted for?"

Ideally, the owner and the design-builder should be able to answer these essential questions by returning to the performance requirements. A good performance requirement comes with a good way to measure it. In other words, a performance requirement must include **performance criteria** that explain how to determine if a project performs as intended. Performance criteria can include a variety of methods to verify compliance, including written standards, field and laboratory tests, third party calculations, on-site mock-ups, manufacturers' warranties, and other measurement or observation techniques. A simple performance criteria for HVAC design and installation is illustrated in Figure 3.6. The referenced standard, formulated by the American Society of Heating, Refrigerating and Air Conditioning Engineers (ASHRAE), identifies how the HVAC system and equipment should perform and how it should be tested to verify that performance.

In most cases the performance criteria are linked to the performance requirements in the RFP. However, if the owner fails to stipulate the standards by which the performance will be measured, the design-builder is certainly not off the hook. On the contrary, if the verification process is not spelled out in the RFP, the design-builder needs to tell the owner how he intends to verify compliance. Verification of performance is ultimately what makes design-build a viable approach.

Research has shown that owners who use design-build for the first time have a tendency to overprescribe design in their RFPs. However, as they gain experience and trust in the design-build process, they decrease the amount of prescribed design and include more performance-oriented requirements in their RFPs. This, in turn, helps attract better design-build teams.

TAKEAWAY TIPS

Hiring Help

If you are considering working with an inexperienced owner who is trying to develop performance requirements or prepare an RFP for the first time, consider suggesting that the owner hire an independent design-build consultant or criteria specialist. Most of these folks come from the professional ranks of architecture, engineering, or construction management, with expertise in the design-build approach. A well written RFP with well written performance requirements will attract the best design-builders.

Figure 3.6
Sample performance criteria

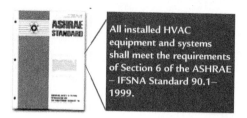

D-B DICTIONARY

Performance Criteria
The means and measures used to verify design and construction performance for compliance to the stipulated performance requirements in an RFP. Performance criteria are comprised of industry and manufacturers' standards, tests and testing, calculations, on-site mock-ups and any other methods used to measure performance.

Figure 3.7
Pentagon renovation project specifications.
Courtesy of DBIA.

Reducing Waste

Ultimately, performance requirements make the construction process more efficient than the traditional design-bid-build approach. For an illuminating example of that fact, let us return to the story of the rebuilding of the Pentagon after September 11, discussed in Chapter 1. The use of performance requirements on that project reduced every kind of waste while delivering extraordinary results. If you recall, the renovation of Wedge 1 of the Pentagon had already started under the design-bid-build model when Lee Evey arrived in 1997. Under the design-bid-build model, the project drawings were accompanied by traditional prescriptive specifications. It took some 3500 pages to communicate all the details of the project. Just think about the energy and time that it took to research, write, print, and read those specs. By contrast, under Lee Evey's supervision, the implementation of the design-build procurement model for the renovation of Wedges 2–5 required an RFP consisting of a mere 16 pages. The illustration in Figure 3.7, adapted from an official Pentagon PowerPoint presentation, underscores the disparity between the two approaches. This graphic makes a very compelling case. When you consider that one of the biggest competitive drivers in the marketplace today is waste reduction and increased efficiency, design-build makes perfect sense.

WHO CONTROLS THE DETAILS OF THE DESIGN

Now we get to one of the most significant differences between design-bid-build and design-build—who controls the design. This difference is often one of the most difficult to grasp.

In design-bid-build, the owner controls the details of the design. The owner hires the designer directly and stipulates or approves every detail.

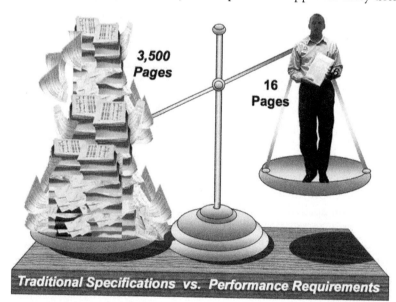

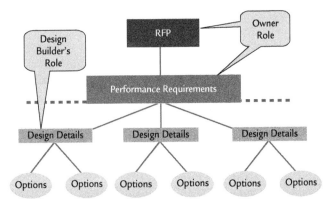

Figure 3.8
Design-builder controls design details

In design-build, it is just the opposite; the design-builder controls the details of the design. (See Figure 3.8.)

The owner prepares the RFP, which includes the performance requirements for the project. It is up to the design-build team to figure out the best way to deliver that performance. In essence, the owner communicates four things to the design-build team in the RFP:

1. I trust that you are the experts in design and construction and I am counting on you to know more than I do about design, form, function, durability, maintainability, sustainability, systems, operation, product performance, construction techniques, material availability, constructability, means, and methods. I also trust that you will work as an integrated team to deliver the best solution, one that meets my budget, my schedule, and my performance requirements.

2. I trust that you will use your skills, knowledge, expertise, talents, creativity, contacts, collaborations, and connections to design and construct a project that meets my performance expectations and also adds value to my project beyond my own knowledge or experience.

3. I trust that you will communicate fully with me and others to get the information you need to understand my performance requirements, even if I have not clearly explained them.

4. I don't necessarily care what means or methods you use, or specifically what materials or products you incorporate—just make sure my project performs in accordance with the quality, industry, manufacturer, and user standards identified in my RFP.

In reality, owners resist allowing design-builders to control the details of their design for two basic reasons. First, they are not accustomed to communicating their needs in terms of performance. They are much more accustomed to specifying four-by-four ceramic floor tiles than describing their flooring needs in terms of durability, ease of cleaning, economy, and aesthetics. Secondly, owners generally do not trust contractors (or designers being directed by contractors). This, of course, can pose a big problem in design-build. Much of this mistrust stems from past design-bid-build experiences, where contractors and designers are

Trust is a critical factor in the design-build equation. Without trust, the design-build approach will not differ substantially from the design-bid-build approach. The topic of trust is discussed in more detail in Chapter 8.

often pitted against one another when it comes to added cost associated with insufficient design. The perception that contractors automatically want to "cheapen" a project just does not hold water in design-build. Because they are involved in the design process, they are in a better position to manage design input that puts the budget at risk; that is exactly what you want them to do if you are the client.

In traditional project delivery, the three main responsibilities (performance and quality, budget requirements, schedule requirements) are segregated, with performance and quality (a result of the details of the design) on the designer side of the fence and budget requirements and schedule requirements on the contractor side of the fence. (See Figure 3.9.) In design-build, the designers and contractors are jointly responsible for all three items and therefore the team will develop and control the details of the design together.

Figure 3.9
Traditional services model versus integrated design-build services model

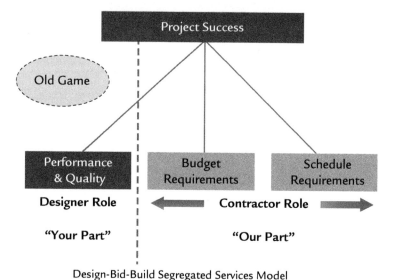

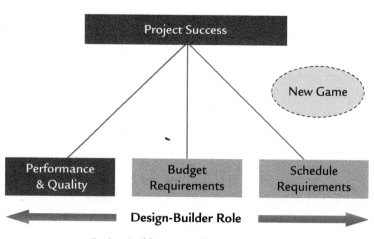

Ideally, as long as the final product performs as required and meets all of the standards spelled out in the RFP, the design-builder should be allowed to modify or change methods, materials, and products as needed to keep the project moving forward. Explicit owner permission is not required. The owner's role in design-build is not to approve the details but rather to confirm that they comply with the performance requirements and standards set forth in the RFP. However, this does not mean that the owner stays out of the discussions as these changes are being considered. The owner must be informed, so his or her concerns can be addressed and incorporated into the overall solution.

For example, a design-builder might propose a concrete structure in the conceptual design. After being awarded the contract, the same design-builder might find that price increases have made that particular design feature too expensive. Therefore, the design-build team, in consultation with the owner, must make an adjustment, perhaps by changing to a steel structure in order to meet the budget requirements. As long as the steel structure delivers the same performance as the concrete structure relative to the owner's RFP requirements, the contractor can contractually make this adjustment in order to keep the project on budget.

Once our design-build team is selected, we conduct a design charrette with the owner and all of the owner's stakeholders. This is a collaborative meeting in which we ask few questions, and instead let the owner and his team open up and give us as much of a download as possible. In these meetings, we want to make sure we are on the right track with our design. We then prepare a control budge+t, which is continually updated as the design evolves through schematics, design-development and construction documents. During the entire design phase, we like to conduct weekly meetings, sometimes bi-weekly, with the owner and the key stakeholders to keep them up-to-date with the design progress, share our documents, budgets, and any other information that is deemed important. This engenders an immediate trust among all the stakeholders, as there are no hidden agendas; everything is transparent with a complete open-book policy. When you have a total team atmosphere, where everyone has one goal in mind, which is to provide the owner with the best possible design that meets his criteria and the budget, you have a win-win situation. That's what makes design-build so much fun.

Stephen Vrabel, DBIA, LEED AP
Pre Construction Manager – Major Projects
PCL Construction Enterprises, Inc.
Denver, Colorado

All too often, an owner will expect the design-build team to hold hard and fast to their original design concept while conditions beyond their control negatively impact costs and other project deliverables. This is like tying someone's hands behind her back and then asking her to catch you when you fall. However, if the team, including the owner, is diligent and committed to meeting or exceeding all project expectations, then there are many ways to adjust, modify, correct, and realign design details while still maintaining performance, quality, cost, and schedule. It all becomes a dynamic dance. It cannot be done without a team!

THE RELATIONSHIP BETWEEN DESIGN AND COST

Every client asks three fundamental questions when considering a project:

1. How long is it going to take?
2. What it will look like?
3. What it is going to cost?

For most clients, that last question is the most important. In the traditional design-bid-build approach, the price is developed in response to the completed design. However, in design-build we do just the opposite—the design is developed in response to the price. (See Figure 3.10.)

In design-build, we start with less than a complete design. The design and all of its details are not worked out completely until after the contract award. However, we are often asked to guarantee a price (and sometimes schedules) based partly upon an incomplete design. After the contract award, we must then manage the design to fit the price and schedule that we have contracted to deliver. This is why flexibility is so critical to the design-build team as it develops the details of the design. With every design iteration, the team works to maintain the balance between design and budget.

Bill Flemming
President and CEO
Skanska USA Building, Inc.
Parsippany, New Jersey

Design and construction firms that have the talent to execute design-build work bring special expertise and understand the teamwork required by all parties to deliver a streamlined, fast-track project. The project team leader must produce financial information early on, determine budgets, and, in many cases, provide a guaranteed maximum or fixed price based on the initial concept design. Such firms can couple their expertise in project execution with the designers' experiences, thus bringing added speed and focus to the schedule and budget. As a result, owners can have the best of both worlds: involvement in critical decisions and early financial guarantees that allow them to get the best possible return on their investment.

D-B DICTIONARY

Conceptual Estimating
An estimating process that develops project costs before design is complete. It is a systems and assemblies approach to estimating, using parametric formulas and historical cost data based upon past completed projects.

Developing Costs in Design-Build

Traditional construction managers and estimators repeatedly ask "How do you estimate the cost of a project when the plans and specifications aren't complete?" Great question. Pricing incomplete design is no easy task. The art and science of **conceptual estimating** requires very experienced estimators and construction managers. In design-bid-build, the detailed estimating process is conducted prior to the contract award; it happens only once, and the goal is to establish the project bid price. In design-build, the estimating process begins in the proposal phase and continues beyond the point at which the contract is awarded, all the way through each of the design phases to the end of the design when the final, detailed estimate is determined. This continual estimating process can be unsettling

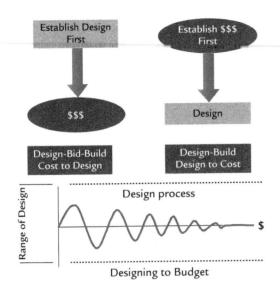

Figure 3.10
Design/cost relationship in design-build

for some traditional estimators. Just as the design is a work in progress, so is the estimate. It is essential that expert estimators perform the unique estimating techniques required in design-build. Some companies go so far as to separate their detailed estimating departments from their conceptual estimating departments.

Estimating Formats in Design-Build

The frameworks for organizing the estimates also differ in design-bid-build and design-build. In design-bid-build, we use the traditional organizational system known as MasterFormat™ to arrange our estimate. The MasterFormat™ is designed around divisions of work such as site work, concrete, masonry, or steel. In design-build, the UniFormat organizational framework is preferred. UniFormat is a systems- or assemblies-based estimating structure ordered around functional elements of a building, such as foundations, substructures, super structures, and exterior enclosures. This organizational framework works well when the details of a design have not yet been determined. Eventually, the UniFormat estimating structure used at the beginning of the design-build process must be converted to the MasterFormat™ structure; this usually happens after the project gets to the buyout stage, when all details of the design have been determined and it is time to place orders with vendors and suppliers and issue contracts to the sub trades. (In Chapter 6, you will learn more about the process of developing costs in a design-build project.)

THE INTEGRATED DESIGN-BUILD SCHEDULE

In addition to the estimate, another critical tool for keeping projects on target is the schedule. In design-bid-build, the contractor is not involved with the design progression, and no coordination between design and construction is required. The only items on the schedule that concern the contractor are the work items related to construction. By contrast, in design-build, the ability to synchronize the delivery of design with the sequencing of construction is an essential task. Instead of having completed design documents in hand from the beginning, the contractor receives the design in pieces, as it is developed. This works as long as the contractor and designer communicate fully to accommodate the most time-sensitive construction activities. Among other things, they must be mindful of lead times required for items that have significant impact on the schedule's critical path. This is tough enough to accomplish even when you have complete drawings and perfect coordination between all parties. It is impossible if the release of a design element is not coordinated with larger needs of the schedule.

Pulling off this type of coordination is tricky because, traditionally, architects design from the top down, whereas contractors build from the bottom up. Also, design is a lot more complex than many contractors realize. Construction is linear in nature; it is easy to see what is needed and what is missing. Design, on the other hand, is iterative in nature and dependent upon many different pieces of information from multiple sources. In Chapter 7 you will be introduced to an analytical design planning technique that allows the design details to be coordinated with construction needs. For now, just keep in mind that, as a first step in design-build scheduling, the team must sit down and integrate the primary design deliverables with the construction sequence in order to see the relationships between the two functions. An example of an integrated design-build schedule is shown in Figure 3.11.

CONTRACTING APPROACH

With any project delivery method, the contracts typically define the relationship and business arrangement between the three primary parties to a construction project—the owner, the designer, and the contractor. In design-build, in which design and construction services are rendered under the same contract, we must be mindful of some unique characteristics regarding the contracts which distinguish them from traditional design-bid-build contracts.

Governing Documents

One thing that makes a design-build contract unique is the group of documents, known as the governing documents, that actually constitute the contract. In design-bid-build, the initial documents that make up and govern the contract are the completed plans and specifications, the general and supplemental conditions, the bid price proposal, and the agreement form. In design-build, however, fully developed plans and specifications do not exist when the contract is awarded. Instead, the design-build contract is initially governed by the owner's RFP, which may include some general and supplemental conditions along with the agreement form, and the design-build team's submitted proposal in response to that RFP. The proposal typically includes a **Technical Proposal** (which

---

D-B DICTIONARY

Technical Proposal
The portion of a design-build proposal that contains the non-price components, such as the design. Other non-price components, such as the proposed schedule, quality assurance plan, or overall management approach, may also be included.

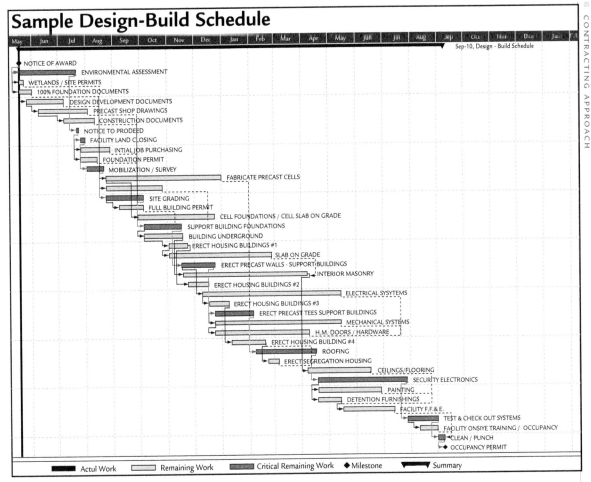

Figure 3.11
Integrated design-build schedule

in turn includes qualitative information such as the conceptual design, the schedule, and a description of the team and management approach) and often a **Price Proposal**. Although an RFP may include some design and prescriptive specifications as part of the criteria, the design-builder typically submits a proposal based primarily upon minimal design and performance requirements. The plans and specifications are only fully developed *after* the design-build contract is awarded. (See Figure 3.12.) You will learn more about the RFP process in Chapter 4.

After the contract is awarded, the details of the design are developed and incorporated into the contract documentation. This model is drastically different from the design-bid-build contract model, setting the stage for a very different approach to the execution of the contract and imposing a significant management challenge.

Another important characteristic of the design-build contract process relates to the development of estimates. Traditionally, in design-bid-build, calculations of project costs are based on completed plans and specs. In this situation, the details of the project design are easily

D-B DICTIONARY

Price Proposal
The portion of the design-build proposal that includes the price for the design and the construction of the project. The price may or may not be a part of the RFP request.

Figure 3.12
Design-build governing documents

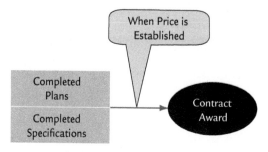

Design-Bid-Build Governing Documents

Design-Build Governing Documents

quantified. However, in design-build, where we lack a complete set of drawings, we are often required to develop and submit firm, fixed prices within an uncertain context. This requires a completely different way of thinking about developing project costs. You will be introduced to this way of thinking in Chapter 6.

Design-Build Contract Issues

Although it is not the intention of this book to address design-build contracting per se, there are a few unique features associated with design-build that deserve special attention. For example, consider the following few issues:

:: **Standard of Care and Warranties:** Traditionally, under a separate contracts model, the designer is professionally bound to uphold a generally accepted standard of care relative to the design delivered to the owner. This means the designer is obligated to provide a design that meets prevailing industry standards. If the designer fails to meet this standard, he will be held accountable to the owner under a negligence theory of liability.[7] No warranty is provided against defects in the design. Meanwhile, the contractor is required to warrant the construction work and all associated materials and workmanship against defects. However, in design-build, from a liability standpoint, the line between the two services partially blur, making it difficult to segregate the design services from the construction work. The use of performance-based requirements may obligate design-builders to warrant both the construction work and the design work against defects. In other words, "in the event that the Contract Documents specify that portions of the Work be performed in accordance with a specific performance standard, the design services shall be performed so as to achieve such standards . . . and the design-builder warrants the design shall be performed in accordance with the Contract Documents." [8] Unless the contract itself creates different standards for the design and construction aspects of the project, the design-builder is required to both provide design services that achieve the contracted performance standards and warrant the construction work.

:: **Ownership of the Documents:** When it comes to ownership of the contract documents, some unique questions arise in design-build. To begin with, who owns the initial conceptual design work that is submitted as part of the RFP process? In the same vein, what rules protect the work of the design-build teams that are not awarded the contract? Also, what rights does the owner have to make use of those ideas and under what circumstances? Furthermore, if the owner receives several different design ideas from several different design-build teams during the solicitation process, what keeps the owner from sharing the ideas of one team with another team? These and other issues related to the ownership of design must be addressed in the contract so everyone knows exactly what the boundaries are.

:: **Owner Involvement in Design:** One of the more challenging aspects of design-build relates to the terms under which the owner is involved in developing and reviewing the design after the contract is awarded. This can have a significant impact on the overall flow of the project. Obviously, most owners will want to be engaged at some level. However, some owners take a hands-off approach, while others demand extensive involvement in the development of design elements. The latter can be problematic if it begins to negatively influence either the schedule or the budget. A wise owner will keep in mind that the details of the design are the obligation of the design-builder; as long as the design meets the performance criteria spelled out in the RFP, the design-builder should not be restricted from moving forward in a timely fashion. In this regard, the owner's right is to review for compliance to the RFP. Unless the procedures are clearly spelled out in the contract, this issue can lead to some real frustrations down the road. It is very important that everyone is in agreement regarding this matter.

These are just a few contract issues to keep in mind when contemplating a design-build project. The DBIA, AIA, AGC, and Engineer's Joint Contract Documents Committee (EJCDC) all have standard design-build contract forms that address these unique design-build considerations. However, the forms provided by these organizations vary in the way they treat the issues discussed in this chapter. All parties associated with design-build contracts need to be aware of these differences. To provide assistance, DBIA publishes the *DBIA Contracting Guide*, which provides

> The DBIA standard design-build contract form is often identified as the most balanced design-build contract by practicing design-builders. In 2008, the DBIA convened a special committee to review, update, and improve the standard DBIA contract documents to better address the integrated nature of design-build and make them more adaptable to owner and design-builder needs. These revised contracts became available in Spring 2009.

owners, design-builders, and other participants in the design-build process with guidance on how to most effectively address the critical contract issues between owners and design-builders.[9] This publication discusses several of these issues and compares the different treatment of them in the standard forms available. Another way to educate yourself and avoid some of the more troublesome design-build contract issues is to take a look at the case studies presented in *Design-Build Lessons Learned* by Michael C. Loulakis, published annually since 1995 by A/E/C Training Technologies, LLC. The books may be found at http://*www.aectraining.com* or by visiting the book store at *http://www.dbia.org*.

Relational Contracting

The design-build approach only works in an environment of collaboration, interdisciplinary integration, and trust. These characteristics are not exactly the hallmarks of the design-bid-build approach, and traditional contracting methods have reflected this. As design-build continues to evolve and become more collaborative, it is likely that we will see the transactional nature of standard segregated services contracts take on a more relational orientation. You have already learned about Sutter Health's use of an Integrated Form of Agreement (IFOA) in Chapter 1. According to an article in the November 2007 edition of *Engineering News-Record*:

> *The IFOA is a team contract, with shared risk and reward, that fosters an all-for-one and one-for-all spirit. The contract is signed by the major players, including but not limited to the owner, architect, and contractor. Thus, the team sinks or swims together. It abandons, as a document, the idea of preparation for litigation. Instead, it is based on focusing the team to concentrate on how to deliver a successful project.*[10]

The tone of the IFOA document demonstrates a distinct shift from the traditional "command and control" type of contract used in design-bid-build to one that encourages increased collaboration and integration. David Long, of Sutter Health, states it this way, "It's a very basic concept, but a paradigm so difficult to shift to because of a century of construction contracting"[11] Sutter Health has implemented the IFOA in conjunction with practices known as Lean Construction or Lean Project Delivery. However, many of the principles applied in the IFOA complement the goals of design-build as well. We will likely begin to see more creative contract approaches attempted over the next several years as we try to adapt from the traditionally adversarial relationship of design-bid-build to the cooperative alignment needed for successful design-build.

High Performance Contracts

As complex and risky as design-build might seem, it is a contracting method that can deliver project results far superior to that of traditional design-bid-build contracting, primarily because it is more flexible and responsive to changing market dynamics. However, that success depends on an owner who is prepared for the challenges of design-build. Diana Hoag, the federal contracting officer who implemented, wrote, and administered the design-build contract for the Department of Defense's Pentagon Renovation Program that you read about in Chapter 1, stated it this way:

> From a business perspective, design-build projects are by their nature different from, and more complex than, design-bid-build. Three interdependent elements of the design-build acquisition strategy will irrevocably influence how the parties will act and react during project execution. These elements are the use of performance-based specifications, an effectively executed source selection, and meaningful contract incentives. When structured with skill and a little luck, the "sweet spot" where these three elements interact synergistically works as an engine to align the interests of the parties. When the acquisition strategy is structured with little or no consideration, the results can range from less than optimal performance to, in the worst case, conflicting motivations that put the parties at odds.[12]

The concept of performance is essential to design-build. A high performance contract, according to the Design-Build Institute of America, is any contract that results in lower costs, faster performance, higher quality, less litigation, and more satisfied customers.[13] Design-build contracts should clearly convey the owner's performance targets along with creative incentives that will spur the team on to delivering a top notch project.

As the contracting party, the owner makes the rules by which the game is played. If an owner is consistently disappointed in the performance realized from contracts, blaming the contractors becomes less viable over time as contractor after contractor renders the same less-than-satisfactory level of performance. It's time for a re-examination of the rules.

Diana Hoag
Senior Consultant
Xcelsi Group, LLC
Dayton, Ohio

A NEW PARADIGM

At this point, you are probably beginning to understand just how different design-build really is. Although I have tried to explain these fundamental differences in relatively simple terms, the nuances of some design-build concepts can be hard to grasp. Some of these concepts will be addressed in more detail later in the book. For now, the bottom line is this: the only thing in design-build that you can expect to resemble design-bid-build is the construction and construction methods themselves. A design-bid-build crane looks just like a design-build crane; we finish concrete in design-build the same way we finish concrete in design-bid-build. But other than the construction methods, there is really very little about the design-build process that looks like the traditional design-bid-build model.

To take on a design-build project without understanding these fundamental differences and unique characteristics is quite risky—and yet it is a mistake that several first-time (and sometimes second-, third- and fourth-time) contractors, architects, engineers, and owners make. While design-build projects will not necessarily fail when unknowledgeable construction and design practitioners embark on one without understanding the implications of the design-build model, the project might be exposed to unnecessary risks—risks that can easily be mitigated through training and education (the purpose of this book).

By now it should be clear to you that design-build is not just a different contracting methodology; it is a completely different way of thinking and doing business. If you want to take full advantage of the benefits of design-build, try approaching it as if you have no idea how it works—and as this book reveals more details of how to do design-build most effectively, even seasoned industry veterans will learn something new!

CHAPTER HIGHLIGHTS

1. The number one mistake that contractors, architects, and engineers make when transitioning to design-build is failing to recognize the significant differences between the design-build process and the design-bid-build process. The two methods have virtually nothing in common.

2. The nine primary areas that distinguish design-build from more traditional approaches are: 1) solicitation and basis of award; 2) team approach; 3) the design-build entity; 4) risk management; 5) performance requirements; 6) who controls the details of the design; 7) the design-cost relationship; 8) the integrated design-build schedule; and 9) contracting approach.

3. The primary solicitation method for design-build is the Request for Proposal (RFP). The basis of award is typically best value as opposed to low bid. The best value approach considers quantitative aspects like price, along with qualitative aspects such as design, schedule, past performance, and the management plan.

4. In design-build, a single contract between the owner and the design-build entity cause the designer and contractor to work together as a team to deliver an integrated, comprehensive solution for the owner.

5. Contractor involvement starts early in design-build, during the planning, preliminary design, and design development phases. This early involvement provides the greatest opportunity to influence the design at the lowest possible cost.

6. The design-build team works proactively to attack problems and mitigate problems as soon as possible. This proactive approach means that design-build teams enhance their projects as standard practice.

7. The design-build entity is generally structured in two different ways—either as a single, multidisciplined firm or as a temporary teaming of two separate design and construction firms.

8. Approximately 56% of all design-build projects are delivered as contractor-led, with the contractor holding the prime contract with the owner. On average, 12% of design-build projects, usually engineering-intensive projects, are designer-led. Approximately 27% of

all design-build service providers are integrated design-build firms, in which the designers and contractors are all in-house.

9. The design-builder impliedly warrants the plans and specifications to the owner. Any loss associated with insufficiencies or defects in the plans and specifications are borne by the design-builder. The owner is only responsible for owner-initiated change orders that alter a project's scope.

10. The key to successful design-build projects is an owner who can allow the design-build team a great deal of flexibility and creativity. The owner can outline specific performance requirements in the RFP, but the design-build approach should entail far less prescriptive specifications than in design-bid-build.

11. The owner begins the process of developing an RFP by preparing the very general functional descriptions for the project, followed by the more results-oriented performance requirements associated with the various project systems. Upon contract award, the design-build team will translate the performance requirements into more technical prescriptive specifications as the details of the design are developed.

12. Performance requirements must be linked to verifiable performance criteria to ensure that the completed facility will perform as required. The design-build team controls the details of the design; this gives the team the flexibility needed to keep the design, cost, and schedule in balance.

13. In design-build, the design is developed in response to the price. In other words, the design-build team must manage the design to fit the price and schedule they have been contracted to deliver.

14. Because the design-build project begins before the design is completed, the design-build team must use conceptual estimating to estimate the cost of the project.

15. The design-build contract is initially governed by the owner's RFP and the design-build team's submitted proposal (which consists of a technical proposal and, often, a price proposal). Details of the design that are developed after the contract has been awarded are eventually incorporated in the contract documentation.

16. Design-build contracts are becoming more relational in nature. The contracts must also address specific issues related to the uniqueness of design-build. Some of these issues include the standard of care and warranties, ownership of the documents, and owner involvement in design.

RECOMMENDED READING

:: *Design-Build Lessons Learned: Case Studies From 2004*, by Michael C. Loulakis, Wickwire Gavin, P.C. and A/E/C Training Technologies, LLC, 2004.

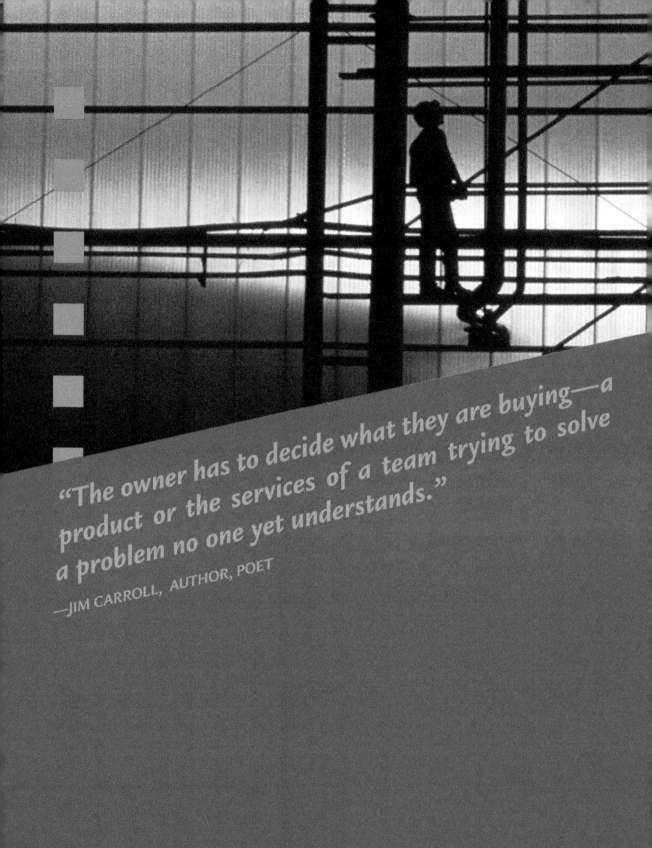

"The owner has to decide what they are buying—a product or the services of a team trying to solve a problem no one yet understands."
—JIM CARROLL, AUTHOR, POET

4
BUYING DESIGN-BUILD SERVICES

After choosing a project delivery method for their design and construction project, an owner then has to decide how to select and acquire the design and construction services needed. You have already learned that with design-bid-build, design services are purchased before construction services, in a linear, independent fashion. In design-build these services are purchased together as a package and coordinating design and construction efforts leads to a more integrated outcome.

So, once the decision is made to use design-build, the next step is to make sure the acquisition plan is set up correctly to take full advantage of the design-build process.

ACQUISITION STRATEGIES

Before any owner embarks on a design and construction project, a lot of planning and decision making must take place first. The results of several strategic decisions contribute to and support the overall business plan of the company, organization, or agency. Whether private, nonprofit, or public, each organization has goals and objectives, and the decision to buy, rent, build, or remodel is at the heart of its capital improvement scheme. Once an owner decides to build a new facility or remodel an existing one, the organization is on course to engage the design and construction process—ready or not, and it is important that decision makers have done their homework before acquiring design-build services.

Every acquisition starts with the formation of some type of planning team charged with determining the best procurement approach. The team may be formal or informal in nature. It may consist of just a few people making all of the decisions, or a large group consisting of contracting personnel, technical advisors, legal counsel, and other experts necessary to ensure a comprehensive consideration of all of the logistical, economic, and risk factors involved. The size and makeup of the

Approximately $1.1 trillion is spent each year on design and construction in the United States. Most of it is still procured under a segregated services model. Your $25,000 car has a better chance of being produced under an integrated design and construction process than does your $250,000 million construction project. However, it is predicted that design-build will overtake design-bid-build as the preferred project delivery method for non-residential construction by 2010. (Reference Figure 2.8 in Chapter 2.)

Figure 4.1

Acquisition planning pyramid

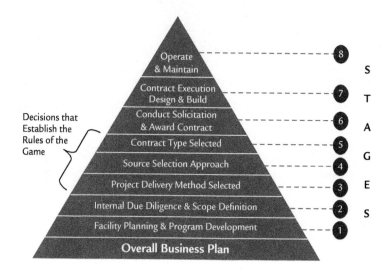

team should be tailored to reflect the specific requirements of the project. In essence, this team will establish the game plan that will govern how the solicitation, selection process, and eventual contract award will transpire.

The acquisition planning pyramid shown in Figure 4.1 has eight stages, all resting on the organization's business plan. Each ascending stage entails a number of important assessments that will frame the owner's expectations and influence when the designers and the contractors become involved in the project. In some instances the design-build team will be employed at the absolute earliest preplanning stages (stages 1 and 2) and in other instances the design-build team will not be engaged until later, during the solicitation and contract award stage (stages 6 and 7), having no influence on the early decision making. The results of decisions made at each stage of the acquisition pyramid form the company's acquisition strategy. The acquisition strategy, however it plays out, will profoundly influence the project results.

Go Slow to Go Fast

The first two stages in the acquisition planning pyramid represent the foundational information for any facilities project. The work that occurs during these data-gathering stages is critical and can make all the difference between satisfaction and disappointment when it comes to the final outcome of the project. I call this early planning the *"go slow to go fast"* approach. The process requires taking or making the time to capture the right kinds of information and knowledge so it can be conveyed adequately to the design-build team. The more the team understands the purpose of what a client or company does and why they do it, the better they can help come up with a solution that serves that purpose at the highest level. Speed at this stage may not be the best course, because the decisions made here will have an impact on the entire organization and its operations, and therefore its success. This type of deliberate investigation takes real effort and usually does not happen by accident. However, it is in the owner's best interest to give this process the attention it deserves—especially in design-build.

> As the leader of a design-build team I've found that it's important to dig pretty deep to get at the heart of your client's primary purpose for their project. You are going to have to make a lot of decisions and you can no longer make them from just a contractor's perspective or the architect's perspective. You really have to be an advocate for the owner and clearly understand what they do and what they want, and why they want it. It's getting to that "why" that is really insightful in terms of their overall mission. Rarely is the project simply about the obvious program and deliverables. There's usually some larger purpose and it's well worth your time to do a little digging to see what the important motivations are. It really helps you see the bigger picture even when individual players on your team can't. It helps immensely to steer the entire team and process in the right direction. The owner's mission becomes my North Star and gives me the benchmark that I need to make good decisions about design, cost, schedule, quality, performance, and any other factor that serves the goals and objectives of the entire team and all of the project participants. Otherwise I am left to my own narrow perspective, which just isn't enough when you are trying to determine the best, most comprehensive solutions from multiple perspectives—which is what doing design-build right is all about.

Joe Flores
Senior Project Manager
Director, Human Resources
The Beck Group
Corporate Headquarters
Dallas, Texas

Mining for the Gold

Presuming the overall facility planning and program development processes have been developed adequately and support the organization's business plan, goals, and objectives, then the next step is to adequately develop the project scope and definition. In design-bid-build, the project scope is defined by the prescriptive plans and specifications. However, as you learned in Chapter 3, in design-build the project scope needs to be defined in terms of function and performance requirements, and the process to develop them takes some effort.

One way to approach this challenging task is to treat it like a reconnaissance mission, in which the owner digs deep within the organization to find and extract the right kinds of information from the operational front lines. The owner needs to "mine the gold" available from their own "expert" resources. (See Figure 4.2.) Many owners bypass this in-depth reconnaissance step because it seems too cumbersome and they do not want to risk having the session become nothing more than a complaint fest. But there are ways to manage the process efficiently, and with proper facilitation, the sessions can be both productive and informative. Although this may seem like an impossible task when applied to an entire company, organization, or agency, it is quite manageable when performed in smaller groups at the department or area level.

Figure 4.2

Mining for the gold

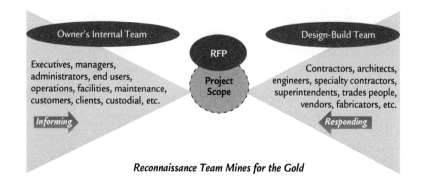

Reconnaissance Team Mines for the Gold

This reconnaissance team should be formed as part of the owner's overall acquisition planning strategy and should include representatives from all levels of the organization—from the custodial staff to upper management. This representative group of individuals is charged with investigating and soliciting specific information about the functional performance of the facility from the people who know best. The reconnaissance team gathers information about what works and what does not work from a performance viewpoint—and more importantly, why they think something works or does not work.

The goal is to capture the proper information, data, and input needed from all levels of the owner's organization. All too often, project needs are defined only by the company's or organization's senior management and executive leadership ranks, and the "down in the trenches" input from custodial and maintenance staff, technicians, operators, facility personnel, administrative support, line managers, customers, clients, end users, and other stakeholders never emerges until the project is complete. Then it is determined that some element of the designed and constructed solution will not work—even though it meets all of the requirements originally spelled out in the RFP. This happens too often, disappointing everyone involved—and it could have easily been prevented.

Once the information is collected, the small group of representatives can consolidate their findings and share them with the acquisition planning team. The acquisition team can then develop a well-informed project scope and definition in terms of performance requirements that will help the design-build teams best serve the organization and meet expectations.

One of the advantages of design-build is that even if the owner does not adequately perform this very important reconnaissance process, the design-build team may be able to correct the oversight by conducting it themselves—especially if they are engaged early. Unfortunately, in a competitive design-build situation, this opportunity is usually not available until after the contract award. However, even then, it is important that the design-build team verify the functional and performance information provided in the RFP as best they can by spending time with the appropriate people within the owner's organization before the team gets too far into the design process. Having information from as many stakeholders as possible ensures the best chance at delivering a project that meets the performance expectations at every level of the organization.

Listening to Learn

MENTORING MOMENT

Meloni McDaniel
Coordinator of Federal
Business Development
The Beck Group
Dallas, Texas
Education – University of
Oklahoma, graduated 2005
Major: Architecture with
a minor in Construction
Science Management

My advice to design and construction students in school is to be open-minded enough to listen and learn from other people's opinions. Take every opportunity to work with the other disciplines in your department or college, getting to know how they think and work. Some students dread such interactions and intentionally avoid them but I loved it. When you're working with architecture students, they've been trained differently; they speak differently, and they think differently. It's important to take in their perspective. Think about what they are saying and just completely turn off your own perspective for a moment so you can actually learn from them.

Likewise, if you're on the architecture side, think about things from the contractor's perspective and his/her experiences and what s/he has seen in the field and how things actually work and go together. That's the benefit of design-build. You have both sides coming together to make decisions from a broader perspective, allowing the project to come together with a really neat, unique design, get the quality that is needed, and stay in budget, all at the same time. You can really give the owner a better product because you're working together as one team. The different perspectives should not be in competition—they should complement one another to deliver all aspects of the project to meet the client's needs. The real challenge for a student is that university curricula aren't necessarily set up to have this multidisciplinary interaction happen automatically, so you have to work on it yourself. Join different clubs, go on field trips together, or if you're lucky, take some interdisciplinary classes. I ended up taking a construction management minor to go along with my architecture major. That's really how I fell in love with design-build. I love working with both sides and fortunately, working for an integrated firm gives me that opportunity every day.

Whether the owner "mines the gold" or waits for the design-build team to get on board, it must be done. To go forward without completing this critical step jeopardizes the success of the project. In accordance with the old cliché, "garbage in, garbage out," the owner has control over the quality of this information. In reality, some owners intentionally keep some internal groups out of the loop for their own reasons. However, if the owner fails to fully and accurately describe and communicate the project scope in a comprehensive fashion to the design-build team, then it will be difficult for the team to deliver a facility, structure, or building that meets the needs of the people using or occupying it. Having more information reduces project risk. As stated by Frank Cushman and Michael Loulakis in the *Design-Build Contracting Handbook* (2001):

> *The more specifically the scope of work and priorities are set out in the project definition, the better chance that the risk associated with that scope of work can be minimized.*[1]

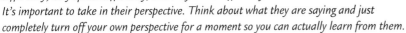

TAKEAWAY TIPS

Reconnaissance Team

Consider going the extra mile when involving people in the reconnaissance mission. Make a special event of it by serving lunch and welcoming people's opinions about the project. It is hard for people to "buy in" until they have an opportunity to "weigh in." The entire acquisition process will go much more smoothly when individuals within the organization feel like they have been heard and are now part of the solution. When this happens, the design-build process is done right.

Rex D. Huffman, DBIA
Vice President
Gibbs and Register, Inc
Winter Garden, Florida

Most owners do their very best to provide you with all of the information that you need to make good design and construction decisions on their behalf. But unfortunately, there are often gaps in that information that need to be filled before you move forward with a full-blown proposal or project solution. After contract award, we spend a great deal of time going back to verify and reconcile the project objectives with all of the various stakeholders associated with the project. Sometimes this means requesting special meetings with different departments or groups within the owner's organization and it does take some extra time up front but slowing down a little bit in the beginning helps us go much faster in the end. We want to do things right the first time and make sure everyone is satisfied. We often find differing agendas for the project and we see it as part of our job to reconcile those interests before we get started. Failing to go this extra mile can cause unnecessary problems down the road. There's just too much unknown risk if you don't have a good idea of where everyone is coming from. . .and those are risks that a good design-builder can easily flesh out and manage if they take the time to do so.

D-B DICTIONARY

Source Selection –
A formal process employed in the procurement of design and construction services. It entails making decisions about the form of solicitation, the evaluation and selection process, and the contracting method.

In certain circumstances, even public owners can choose their design-build team through direct selection. It is called an emergency award and has been used on several occasions—for example, on highway projects after earthquakes in California, or for select projects after hurricane Katrina. In these circumstances there is not time to organize a competitive solicitation.

No one wants to get to the end of a project only to learn that the work designed and constructed will not function or perform as needed. It is important to understand that the design-build process starts deep within the owner's organization—long before the RFP is ever released.

SOURCE SELECTION

Once all of the program and scope development work is complete and the owner decides to move forward using design-build as the project delivery method, there are still important acquisition decisions to make before the solicitation can begin. The first decision has to do with **source selection** and the second decision considers the contracting approach itself. Both of these decisions will directly influence the potential performance and effectiveness of the design-build process.

Source selection has several components. The first entails the solicitation itself. In design-build you have already learned that the process begins with an RFP. Next the owner must decide exactly how to procure, evaluate, and select the design-build team. Private owners can choose their design-build team directly or competitively. However, public owners, such as federal or state agencies, are typically forced to use competitive methods.

Price, qualifications, or a combination of both, forms the fundamental criteria for all procurement options. The three common competitive procurement choices are 1) Low-bid Procurement, 2) Qualifications-Based Selection (QBS), or 3) Best Value Procurement. These three options establish the basis for how the design-build team will be evaluated and selected.

Low-Bid Procurement

Low-bid procurement is the method that most of us are familiar with and the one that would be used by an owner whose primary criteria is initial low cost. As we know, competing contractors submit a lump sum or guaranteed maximum price for a project based upon the same design and set of documents (prescriptive plans and specifications). The contractor with the lowest responsive bid wins the competition. Although qualifications may be used to prequalify firms before being allowed to submit a bid, qualifications or past performance are not evaluation factors in determining who wins the job. Historically this has been the traditional method used for selecting contractors under a design-bid-build scenario.

The low-bid procurement model has also been applied to design-build. The **low-bid design-build** approach is any selection method that has price as the predominant evaluation factor, even if other criteria are considered. When an owner uses price as the primary evaluation criteria for selecting their design-build team, they have usually included a substantial amount of prescribed design in the RFP through a process called **bridging**, in which the owner hires an independent designer to develop prescribed areas. (Bridging will be discussed in detail later in the chapter.) The design-builders are expected to complete the partial design as part of their response, and submit a price for construction. Generally, as shown in Figure 4.3, the more design in the RFP, the more likely it is that the basis of award will be low-bid. The Design-Build Institute of America (DBIA) considers any solicitation having more than 35% design included as part of the RFP to be a predominantly "price-based" solicitation. Many experienced design-builders refer to this approach as **draw-build** instead of design-build and are reluctant to pursue such projects because they have increased risk and limited opportunity to impact the design and add value.

D-B DICTIONARY

Low-bid Design-Build – Any design-build procurement process that stipulates price as the predominant evaluation criteria and selection determinant.

Bridging – A process in which the owner directly hires a designer (bridging architect or engineer), independent of the design-build team, to develop the project design prescriptively as part of the requirements of the RFP.

Draw-Build A variation of design-build in which the owner has completed the design to such an advanced degree that the design-builder's role is limited to completion of the detailed construction drawings and specifications and construction, thus diminishing the opportunity to impact the design and add value.

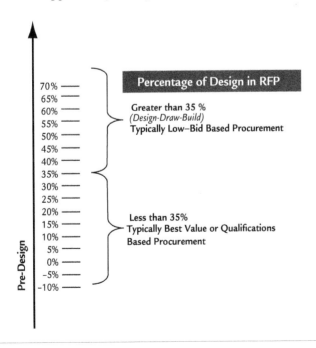

Figure 4.3
Percentage of design in the RFP.

> A 1943 federally-enacted law known as the Brooks Act prohibits the low-bid method from being used to select architects and engineers for publicly-funded projects in the traditional design-bid-build model. However, in design-build, where both design and construction are being procured together, the low-bid method can be used, even on public projects, although it is not recommended.

In some instances, public procurement laws or agency regulations will not allow for anything but a low-bid design-build procurement process, which defeats the purpose of using design-build in the first place. Many public agencies and individuals, in particular, are reluctant to shift from the traditional low-bid approach.

Qualifications-Based Selection

Qualifications-based selection (QBS) compares one competing firm or team to another based solely upon their qualifications; price is not a factor. In the QBS procurement method, the Request for Qualifications (RFQ) is the solicitation instrument. In design-build the types of criteria that are often evaluated in the RFQ include:

:: Design-build experience
:: Past performance
:: Financial strength of the company
:: Qualifications of the individual team members
:: Evidence of design and construction excellence
:: Any other qualitative type criteria an owner might wish to evaluate

In a straightforward QBS competition, the criteria evaluated might not even be project-specific; evaluators may simply consider the firm's reputation, experience, and qualifications. Historically, the use of this procurement method has been limited to design services, at least in the public sector. However, there is increasing pressure being placed on legislative bodies to enact laws that would allow QBS to be used to select design-builders as well. To date, only five states, Florida, Arizona, Nebraska, Wyoming and Alabama, allow QBS as one of the competitive procurement options when choosing design-build teams for public projects.

In order to compete in a QBS design-build competition, the design-build teams must put together a **qualifications package** in response to an owner's RFQ. The RFQ describes the project in enough detail to let potential proposers determine if they wish to compete. The type of information provided in the typical RFQ includes:

:: Owner name, address, contact information
:: Announcement or advertisement regarding the design-build solicitation
:: Competition timetable, key dates, and submittal requirements including deliverables
:: Description of the project and scope including the building type and size, project budget, and project schedule
:: Identification and makeup of the selection jury
:: The evaluation criteria and the weighting of that criteria
:: How the teams will be evaluated and scored
:: Information regarding interviews and/or presentations if applicable

D-B DICTIONARY

Qualifications Package – A proposal developed by a design-build team in response to a Request for Qualifications that identifies and highlights the unique qualifications of the design-build entity in regard to their experience, past performance, and capabilities that relate to the specific project in question.

Although designers are accustomed to competing in this fashion, many contractors are not, and it takes some effort to figure out the best way to differentiate their design-build team from all of the others competing for the same project. However, more construction firms are now seeking negotiated, instead of low-bid, work, and they have experience putting together qualifications-based proposal packages—at least from a construction perspective. The added challenge in design-build is to combine their construction qualifications with their team's design qualifications to deliver a comprehensive, integrated response.

After receiving all of the qualifications packages, the owner evaluates those packages based upon the specific criteria they identified in the RFQ and may or may not include an interview or presentation as part of the assessment process. The owner then selects the firm with the highest score and awards the contract. Project price is not a factor in determining who wins.

Progressive Design-Build

In recent years a hybrid version of the QBS method has emerged and seems to be growing in popularity. Practitioners and clients alike refer to this as **progressive design-build**, an approach that combines the concepts of QBS with a distinct two-stage contract component that is similar to the CM-at-Risk project delivery method that you learned about in Chapter 2. (See Figure 4.4.) This method reflects the evolutionary nature of the design-build approach.

The progressive design-build model uses an RFQ to solicit the qualifications and performance history of the design-build teams. The owner uses the same evaluation process as in the QBS process to select the design-build team. The contract would consist of a Part A and a Part B phase. Only the Part A portion of the contract would be executed initially. Under Part A, the selected design-build team would proceed to develop the design to the point where a lump-sum or guaranteed maximum price for the project could be established—usually to the 60% to 70% complete range. The team then develops and submits the price to the owner for

D-B DICTIONARY

Progressive Design-Build – A qualifications-based design-build procurement method in which the design-builder is selected based primarily on qualifications; the design-builder and the client enter into a progressive two-part contract that is implemented in stages.

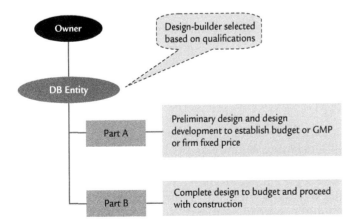

Figure 4.4
Progressive design-build

> The design-build RFQ process should not be confused with the prequalification process in design-bid-build that uses a Statement of Qualifications (SOQ) document. The RFQ process ranks teams by scoring their qualifications numerically or in some other fashion. The SOQ, on the other hand, prequalifies a contractor as having met minimum requirements or not. A contractor who meets the requirements can bid on the job; one who does not, cannot. The RFQ process scores teams from highest to lowest and then shortlists the top teams (usually three) to continue in the competition. In a straightforward QBS procurement, the highest scoring team wins.

approval. If the owner and the design-builder agree on the price, then the Part B portion of the contract would be executed. Part B would trigger completion of the design and construction for the agreed-upon price. If for some reason the owner and the design-builder cannot agree or negotiate an acceptable price for the total project after completing Part A of the contract, the owner can decline to execute Part B. In this event, the owner could take the partially-completed design and negotiate with another qualified design-build team. Or they could use the design as the basis for a traditional design-bid-build or CM-at-Risk procurement.

Best Value

Best value is a unique procurement technique because it considers both the technical solution and price in the evaluation process. Although the best value approach may be used with design-bid-build for construction proposals only, it is particularly well suited to design-build because it evaluates both the design and the construction. The best value method often initiates with the QBS method (using an RFQ) that selects a limited number of design-builders to proceed with the full-blown RFP process. (You will learn this process in the next section.)

The RFP document is the primary solicitation instrument in design-build. It is released to the competing proposers by the owner and should fully describe the procurement process and selection method, outline exactly how each of the design-build teams are to prepare their proposals, and identify how teams should submit each deliverable. The types of information typically included in the RFP are similar to that in an RFQ, but also includes these additional items:

:: General information about the owner and points of contact; instructions to proposers; scheduled preproposal meetings and special requirements; nature, scope and location of the project; any unusual project circumstances; mandatory or regulatory requirements; and overall project description, budget, and schedule.

:: Solicitation information, including submission requirements and critical deadlines; the overall RFP timeline; selection procedures, criteria, and weighting; evaluation method (interview and/or presentation format); evaluation team makeup; and deliberation and award timeframe.

:: Known information about the site including surveys showing property boundaries and ancillary conditions, topography, soils, and utility information; benchmarking controls; and any restrictive covenants or regulatory constraints.

:: Information regarding the specific project parameters and requirements, including a summary of the overall project program, goals, and objectives, functional requirements, performance requirements, site plans, preliminary layouts, technical specifications, codes, standards, and warranties, and information pertaining to the entitlements and permitting status.

- Information regarding the contract terms, including roles of the parties; general and supplemental conditions; and special requirements such as women, minority and/or small business participation. (A copy of the actual contract is often included in the RFP.)
- Information regarding the proposal submittals, including types of required drawings and specifications; presentation materials; the organization of the design-build entity; the project team makeup and overall management strategy; information regarding the project schedule, quality control and quality assurance programs; and schedule and cost control plans.
- Information regarding how the price proposal portion of the solicitation should be formatted and conveyed.

There are two evaluation components in best value procurement. One includes technical and/or management elements establishing the "technical proposal portion" of the best value solicitation. The technical component can include any number of criteria—for example the project schedule, overall project approach, quality control plans, management plans, safety plans, value engineering, sustainability features, low maintainability, durability, and of course the conceptual design—anything but price. The other evaluation criterion is simply the project price for both the design and the construction. The winning design-builder is the firm or team that has the best combination score for both the technical/management component and the price.

Contractors who have spent their careers competing on price only might have a difficult time making the transition to a best value competition. In best value, it takes more than just having the cheapest price to win; many more subjective, qualitative factors help determine the winner. The process entails intensive proposal preparation, and team interviews and formal presentations are also common. Fortunately, many construction firms today are accustomed to competing on factors other than price and obtain a large portion of their work based upon their qualifications, reputation, past performance, and project management skills. These companies generally welcome an opportunity to compete based on non-price factors and if they cannot be selected on qualifications only, most would prefer to compete on a best value basis rather than on price alone.

Although any of these procurement methods can be used in a design-build solicitation, the best value approach seems to be the most prominent.

HOW THE RFQ – RFP PROCESS WORKS

In an ideal design-build world, all design-builders and contractors would be selected based upon their qualifications first and foremost, just as architects and engineers are. However, as an industry, we have not yet reached that stage. So the next best procurement option we have available, which makes the most of what the design-build process offers, is a **two-phase procurement** option that combines a qualifications phase with a best value phase. (See Figure 4.5.)

D-B DICTIONARY

Two-Phase Procurement – A process that selects a team by evaluating qualifications to create a short list, and concludes with a best value assessment of a complete design-build proposal from each of the shortlisted teams. Also called Two-Stage Procurement.

In design-bid-build it does not take much effort on the owner's part to open the bids and determine which contractor has the lowest price. However, in design-build the review and evaluation process is far more intensive and there is more work involved for the owner. That is just one of the reasons why the shortlist should be limited to only the top three design-build teams.

D-B DICTIONARY

Shortlisting – A method used in design-build procurement to narrow the field of potential RFP responders to only the highest-ranked three or four teams based on qualifications submitted in response to a Request for Qualifications.

Stipend or **Honorarium** – A stipulated amount of money sometimes paid to the unsuccessful proposers by the project owner in a design-build competition. The money is intended to help defray proposal development costs.

Qualifications Phase

The two-phase procurement process begins with a qualifications ranking procedure conducted through the RFQ process. The RFQ is distributed to a list of potential design-build contractors or firms; each one submits a qualifications package. The package usually contains boilerplate material like the company's history, mission statement and core values, team credentials, experience, staffing capabilities, awards and honors, company financials, references, and the like. It closely resembles a marketing package and is relatively inexpensive and easy to assemble. The owner evaluates the qualifications packages, as in the QBS process. But in this case, evaluating qualifications is not the end of the competition. Instead of selecting a winner, the owner selects the top three, four, or five competitors—based upon their scores relative to the qualitative factors that were evaluated. This narrowing of the field is called **shortlisting**. To be shortlisted is the first goal of the design-build team because only those firms that made the cut get to go onto the second step of the competition—the best value phase conducted through the RFP process.

In the shortlisting process, more is not necessarily better. According to one successful, experienced design-builder, one of the first things he looks at, when trying to determine whether to pursue a particular project or not, is the number of candidates the owner decides to shortlist to. He had a simple philosophy on the matter: "If there are three we talk, if there are four we balk, and if there are five we walk." This statement represents the opinion of a number of design-builders.

Best Value Phase

Developing a proposal in response to the RFP is a more extensive process and requires a significant investment of both time and money. It is not unusual for proposal preparation on large projects to take weeks or even months to develop and can cost well over $100,000 or even $200,000. That is why, if you are a design-build contractor, and the owner does not shortlist to narrow the competitive field, you might not want to compete, because your odds of winning are greatly reduced. On the other hand, proposals for smaller projects might only take a week or two to develop and can cost less than $10,000, so submitting as part of a larger candidate field entails less financial risk. In either case, competing for projects in a design-build environment is not for the faint of heart or pocketbook.

Because the proposal development process is so involved and expensive, experienced owners will often tender a stated amount of money, usually called a **stipend** or **honorarium**, to the unsuccessful proposers to help offset the costs to compete. Owners have good reasons to offer these. The federal government's Office of Management and Budget Circular No. A-11 (2006) puts it this way:

> Providing stipends to contractors to cover some or all of proposal costs can provide an effective financial incentive to increase competition.... Stipends to non-successful offerors help defray, but rarely come close to fully covering the costs that offerors expend in responding to RFPs. However, providing

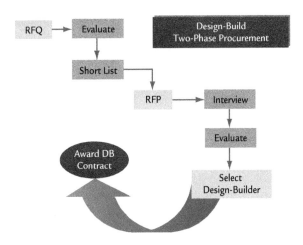

Figure 4.5
Two-phase design-build procurement

a stipend strongly encourages the very best companies to put forth their very best proposals. The government may decide to require permission to use design plans as a precondition to receiving a stipend. Experience in construction contracts has shown that when an optional stipend is given to a non-successful offeror in exchange for the right to use the design plans, the stipend is generally readily accepted. The availability of a stipend and the terms governing its use must be identified in the RFP. [2]

Other reasons listed in the OMB Circular A-11 for offering stipends to non-successful design-build proposers are:

1. Proposal development is very costly
2. Signals the intent that owners are serious about carrying the project forward
3. Improves the quality of firms which are submitting
4. Encourages proposers to give full effort[3]

PROJECT DESIGN AS A COMPETITIVE FACTOR

Unlike a design-bid-build project in which all of the competitors are competing based on the same design, the design-build competitors will each develop their own conceptual design in response to the criteria identified by the owner in the RFP. The RFP may have already included a site plan and preliminary drawings and design parameters, in which case the design-builder will use this information to develop its solution. Or the RFP might have nothing more than written descriptions of the design intent and some performance criteria; it does not contain any completed design. And for an experienced design-builder, the less design provided, the better. The design-build team will then get an opportunity to shine, bringing its unique talents, skills, ingenuity, and innovative ideas to bear. This reveals the true nature of a design-build competition: it will reveal several different design concepts that will meet the owner's stipulated performance

D-B DICTIONARY

Conceptual Design – A preliminary stage of the design process that addresses the owner's initial program and focuses on the broader, overall design parameters such as total square footage, number of stories, and building mass and configuration. Conceptual design typically includes diagrammatical sketches and drawings without detail, outline specifications, and other documentation that illustrates the scale and relationship of the project.

Co-Location – A space or office arrangement that has contractor and designer members of the design-build team stationed in close proximity to one another on a temporary basis in order to work closely together on a specific project.

criteria and project parameters. The best value design-build solicitation encourages the "best design" and price combination—whether the owner provides the fixed budget up front or the teams develop their design and price together. Even though at the proposal stage, the design-build team will typically only be submitting a **conceptual design** or schematic design, it is important to keep the proposed design in balance with the owner's price targets for the project. Therefore, it is highly recommended that the contractor's conceptual estimator and the designers developing the concepts move to closer physical proximity even during the proposal stage. You will learn more about **co-location** and how it actually works in Chapter 7, but for now, it is most important to understand that design-builders must coordinate and manage the relationship between design and cost from the very beginning of the procurement process.

Remember, price is the other component of a best value procurement and because the design is incomplete at the time of the proposal submission, the price is also developed conceptually. Depending upon what the owner asks for in the RFP, the conceptual estimate may be developed as a budget, a guaranteed maximum price (GMP), or a lump-sum fixed price, regardless of the completeness of the design at the time of the proposal. Obviously, the less complete the design, the more contingency will be included in the price—especially if the owner expects a lump sum fixed price at the conceptual design stage. Conceptual estimating is both an art and a science and you will learn more about this unique process and contingencies in Chapter 6.

Jeffrey Cupka, AIA
Design Principal
Ryan Companies US, Inc.
Minneapolis, Minnesota

I consider the design as the primary differentiator in a Design-Build competition. Many times our customers have stated that all the proposing teams looked like "they came off the same boat" and had the same script in interviews. The only difference many times is the design itself—the creativity to respond to the customer needs and the RFP while guaranteeing a price. After all, design and costs are integrally intertwined in a true DB process. We cannot talk about one at a time or we are certainly going to miss the target and churn about with redesigns and repricing exercises similar to the traditional Plan and Bid [design-bid-build] process. The process of designing while concurrently pricing the designs is a skill that is very valuable and unique to the DB approach, but really not that difficult if it becomes a standard procedure. However, even though costs related to the building are important, often traditional contractors tend to focus too much on the scope and costs of the project when presenting to a client and they end up looking and sounding just like all of their competitors. We strive to create exhibits that are attention-getting, creative, and aesthetically extraordinary. We also try to talk about our design processes as much as possible, such as BIM, Integrated Services, 4-D animations, walk-throughs, etc., to illustrate how we can better serve our client's needs from both a design and build perspective. It is truly an opportunity to show a difference and highlight one of the big advantages of design-build.

PIONEER PERSPECTIVES

Formalizing the Process

As a high school student in Alameda, California in the early 50s, I had a keen interest in ships, and more particularly, in the design of small boats and yachts. That interest eventually morphed into an interest in designing buildings and I ended up completing a BS degree in architecture at Georgia Tech, where fellow Tech alumnus, George Heery, FAIA often served as a visiting architectural design critic. Apparently George liked my designs enough to offer me a position after graduation in his then four-person firm in Atlanta, a growing distribution hub, thanks to its being in the center of a new interstate highway web. It attracted many new plants to its surrounding cities and counties. These plant owners were in a great hurry to build and occupy their new manufacturing facilities and distribution centers. George's firm, Heery & Heery, responded to this need with several very intense charrettes to meet the owners' accelerated construction schedules. These early efforts were very much crash production schedules involving night and weekend work. Later, the firm developed a simple form of fast-track design and construction with separate bid packages for grading & foundations, structural systems and long-lead mechanical and electrical items.

Edward C. Wundram,
AIA Emeritus, CSI,
DBIA
Senior Project
Construction Manager
Cumming
LACCD-Los Angeles
Southwest College
Los Angeles, California

Heery & Heery's reputation for fast-track design and construction of manufacturing facilities caught the attention of the Lockheed Corporation, which had just received a government contract to design and build the new C-5A troop transport plane for the United States Air Force. Lockheed's need was for a very large engineering design facility (and eventual classroom facility) to house hundreds of engineers to design the new plane and they wanted to house the first group of engineers at their Marietta, Georgia plant in just five months! Lockheed's facility department had done research on the building type they needed and suggested we look at the results of a School Construction Systems Development (SCSD) program in California that might give us some ideas about how to get this facility designed, built, and up and running in such a short period of time.

The SCSD program, housed at Stanford University's School of Education was a program to develop specific building systems and products for the K-12 educational building market. It consisted of a consortium of California school districts that had agreed to pool their building budgets to amass a large enough target to allow ind ustries to develop unique products for the educational sector of the market. Examples of products that came out of this program were modular structural systems specific to school buildings, rooftop multi-zone HVAC units, demountable walls systems, ceiling/lighting/air distribution systems and modular classroom and laboratory furniture. The program's methodology was to define the size and budget for a single building component, say a column/roof structural system, and to publish a very detailed "performance specification" for the system. The specifications were neutral regarding the materials used in the component; for example, structural systems could be precast concrete, steel, glue-laminated wood beams or composites of several materials. It was this program and it components parts that Heery & Heery and its client, Lockheed Corporation, used to design and build their engineering building on time and on schedule.

(continued)

PIONEER PERSPECTIVES (CONTINUED)

In 1971, the Educational Facilities Laboratory (EFL) of the Ford Foundation asked Heery if it could send someone to Portland, Oregon to apply the SCSD principals of "performance-based" systems building programs to a district-wide school building renovation effort. The goal of the program was to convert a large number of the district's well built and well maintained stock of 1920s–1930s eras, K-8 buildings into modern 5-8 middle schools. I accepted the challenge and moved to Portland to direct the school district's new systems building program. Unfortunately, the bond issue necessary to support the program failed and only one prototype classroom was completed. But by now I was pretty well sold on the concept of a performance-based approach to design and construction.

After a stint at the startup Heery-Farrow, Ltd. in London, I returned to Portland in 1977 and I took a position with Dielschneider Associates, a small program management and consultancy firm. I was there when the City of Portland requested Dielschneider to develop a building program for a new city hall annex to be eventually known as The Portland Building. It was to house city departments then located in nineteen separate buildings throughout the city. As a component of this effort, Dielschneider Associates were also asked to recommend a building budget and a design and construction schedule. When our report was almost complete, the city's Director of Administration, Earl Bradfish, asked me if I could write a letter to the building contractors interested in the project and ask them if they could submit a building design at the same time they submitted their price. Remember, this was a period of very high inflation in the building industry. I admitted that I understood his need to tie down the building's cost early in the project schedule, but it would likely be more complicated than a single letter.

It just so happened that I was serving as the chairman of the Portland Chapter AIA Design Awards Committee around this same time. One afternoon, the chapter executive director suggested that I might be interested in a newly received National AIA publication entitled "Design-Build-Bid, a 1975 Task Force Report." The Task Force was charged with developing a strategy to counter the influence of building contractors that offered design and construction in a single package to private industrial clients, (often referred to at the time as "package-dealers"). The architectural quality of these buildings was judged to be of very low quality, and the AIA wanted to convince building owners to engage architects independently of their building contractors. The essence of the report was that, while independent and separate architectural services were recommended, if building owners were to continue to use design-build-bid methods to acquire their facilities, there were some specific procedures to follow if AIA members were to participate. These recommendations were: 1) prepare a complete and well-written program of requirements; 2) appoint a qualified jury that included at least one design professional; 3) provide for the payment of honoraria to the participants commensurate with the design effort required; and 4) employ a professional competition advisor. These procedures were very similar to those of the AIA's Design Competition Guidelines. While the report did not encourage the use of design-build-bid procurement methods, a minority of the Task Force suggested that design-build competitions provided considerable benefits for owners and the method should be taught in architectural schools.

It ended up that this Task Force Report and the procedures described in it served as the basis of my recommendations to the Portland City Council for their new city hall annex. The city accepted my recommendations and instructed Dielschneider Associates to proceed with the development of the solicitation documents for their project. At the time, Oregon public contracting procedures did not contemplate design-build contracts or any deviation from the traditional design-bid-build method, but because the Portland City Council sat as their own public contracts review board, the approach was approved.

The mayor of Portland, Neil Goldsmith, when it was explained that design-build procurement would guarantee that the project would be designed and built on budget, asked if, at the same time, we could "design the building to put Portland on the architectural map of the country?" I explained that building cost was not necessarily a hindrance to architectural design excellence and there was a chance his goal could be met, with some specific changes to the design-build request for proposals (RFP). In order to attract the leading-edge architectural designers to our project, it would be necessary to put a nationally-known and respected architect on the selection jury. With the city encouragement, I convinced Philip Johnson, FAIA to serve as a special advisor to the city and the jury. His involvement was, no doubt, the reason the relatively low-budget project was able to attract an array of national and international architects of the caliber necessary to achieve the mayor's goal of architectural excellence. The end result of the 1978 design-build competition was The Portland Building designed by Michael Graves, FAIA, Architect and Emery Roth & Sons, Architects and built by a joint-venture of contractors Pavarini Construction Co. and Hoffman Construction Co. The building was completed and occupied in 1982.

Subsequent to the success of The Portland Building project, I opened my own one-man office in Portland and specialized in the organization and administration of design competitions and design-build programs for significant public structures. Design-build projects that followed included: The Tacoma Dome, Tacoma, WA; Harold Washington Library Center, Chicago, IL; The Hawai'i Convention Center, Honolulu, HI and many others. In 2003, I was able to expand the design-build RFP method to include the cost of operating and maintaining a building over a long period of time. Referred to as a design-build-operate-maintain (DBOM) contract, the method was used at the University of Washington in Seattle to acquire a new physical science research laboratory with a guaranteed cost to design, build and operate a 122,000 GSF, six-story building on the University's campus for thirty-years after completion.

My association with the Design-Build Institute of America (DBIA) began around 1993 when I noticed a short announcement in Engineering News Record Magazine (ENR) from Preston Haskell that he and a few other design-build contractors were forming an association to encourage and promote design-build contracting in America. Unfortunately, the initial dues were $10,000 for each firm, far too rich for my small firm. However, I thought that with the three or four design-build projects that we had successfully administered to completion, my firm, The Design-Build Professional Group, deserved to be a member, but with some consideration for our size and limited cash flow. I wrote to Preston listing the firm's design-build credentials, and argued for a new class of membership for design professionals and others, with an appropriate dues structure. My letter must have made the founding fathers realize that it takes many separate disciplines to make design-build contracting work. The following year, DBIA announced several new membership categories, and invited me to join the association. My first national convention was in Atlanta in 1994, followed up with a DBIA Workshop in Denver six-months later. By this time I was the chairman of the process committee and as the committee's first act, we recommended that DBIA initiate a Manual of Practice to guide new members and the design and construction industry in the practices and procedures of design-build procurement and contracting. I volunteered to write the first section, Design-Build RFQ/RFP Guide for Public Sector Projects. In 1995, I was awarded the first DBIA Leadership Award, along with Mark Shambaugh.

Later, I served as an advisor to the Board of Directors, representing public owners, and was Co-Chair of the Manual of Practice Task Force in 1996–97 and was made Vice-Chair of the Policy Committee in 1998–2003. In 2002, I had the pleasure and distinction to be invited to be a charter member of DBIA's first class of Designated Design-Build Professionals. I continue my membership today as an individual representative of a public owner.

HOW THE PROPOSAL IS SCORED

Clearly, preparing a design-build proposal is more complex than preparing an estimate for a design-bid-build project. Likewise, the owner must evaluate a great deal more than price before declaring a winner. About the only similarity between the two selection processes is that design-build proposals are usually due to the owner at a prescribed date and time just like with a design-bid-build competition. However, unlike design-bid-build, there is no public opening of the proposals. Instead, a jury or panel, typically identified in the RFP, reviews all of the proposals.

The proposals are usually packaged in three-ring binders and may include several exhibits in the form of plans, renderings, or even scaled models. It can take the panel more than two weeks to complete their review, depending on the size of the project. In addition to the proposal evaluation, it is common for owners to conduct individual team interviews or presentations.

When presentations are a part of the evaluation process, each short-listed design-build team is given a specified amount of time to present and explain their proposal to the jury—usually around forty-five minutes. The typical design-build presentation team is made up of one to three members from the design group and between one to four members from the construction group. The usual participants in a design-build presentation are the project director, project manager, lead architect, project architect, estimator, one or two engineers, perhaps a special consultant, and last but not least the project superintendent. The owner is always

> The days of the "rip 'um and read 'um" low-bid competition are numbered. It takes a great deal more than low price to compete and win in today's competitive market. Similarly, project managers with "construction only" orientations will have to learn how to respond to issues other than costs and schedules.

Weighted Criteria Matrix - Technical Components

Design-Build Team	Design Solution (Weight 5)		Green Building Techniques (Weight 3)		Quality Management Plan (Weight 4)		Financial Capability (Weight 3)		TOTAL POINTS
Superior Design-Build	3	15	4	12	3	12	2	6	45
Master Builders Incorporated	4	20	4	12	3	12	3	9	53
Design-Build Associates	3	15	3	9	3	12	4	12	48

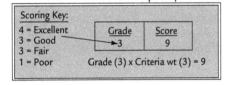

Scoring Key:
4 = Excellent
3 = Good
3 = Fair
1 = Poor

Grade	Score
3	9

Grade (3) x Criteria wt (3) = 9

Best Value Matrix – Technical Score + Price

Design-Build Team	Technical Score (60 Pts. Max)	Price Proposal	Price Score (40 Pts. Max)	TOTAL SCORE (100 Pts. Max)
Superior Design-Build	45	$2,647,000	38	85
Master Builders Incorporated	53	$2,564,000	39	92
Design-Build Associates	48	$2,528,000	40	88

(*Master Builders Inc. wins even though they did not have the lowest price – they did have highest technical score)

Figure 4.6
Design-build scoring using a weighted criteria matrix

interested in seeing and hearing the person who would be running their job on a day-to-day basis. After the team presents for their allotted time period, the jury usually takes an additional twenty to thirty minutes to ask questions and obtain necessary clarification.

Throughout this evaluation process, the owner will use a scoring system that reflects the type of selection technique they are using. One of the most common techniques used in design-build procurement is called a **weighted criteria matrix** as illustrated in Figure 4.6.

In the matrix, all the elements the jury wants to evaluate, as well as the weights assigned to each one, appear in the top row. The assigned weights represent the value that the owner places on each criterion. In Figure 4.6, "Design Solution" has a weight of 5 while "Financial Capability" has a weight of 3. If the evaluation criteria and their weights are disclosed in the RFP,

D-B DICTIONARY

Weighted Criteria Matrix – A tool used to assess a best value selection in which evaluation criteria and weights are displayed and scores are distributed to determine the winner in a design-build competition.

TAKEAWAY TIPS

All the World's a Stage"

As your company pursues more and more design-build projects, consider training your technical people, including your engineers and estimators, in "platform and presentation skills." If they come from the design-bid-build world, they might not be accustomed to getting up in front of a panel of owners and owner representatives to explain their solutions. Include your superintendents and other senior field personnel; they, too, will need to make presentations regularly and are critical players on a successful DB team.

they can help teams put together a winning proposal because they know where to focus their time and energy.

All jury members evaluate every proposal according to the criteria and assigns a numeric score that represents their evaluation of each one. In Figure 4.6, a score of 4 is given for excellent, 3 for good, 2 for fair, and 1 for poor. The numeric score for each criterion is multiplied by that item's weight to calculate an applicant's total points for each criterion. An owner can include any number of criteria in the evaluation; some RFPs list as many as twenty different factors. This same type of criteria matrix is also used in scoring qualification packages in a QBS competition.

When owners use the best value procurement method, price becomes part of the scoring mix as well. In Figure 4.6, the lower matrix is an extension of the top matrix and indicates how the owner has assigned overall points for the technical/management components and the price component. In this example the technical/management components are worth 60 total points and the price component is worth 40 points—in other words the technical elements of the proposal are more important than the price to this owner. The full 40 points are awarded to the lowest bidder, in this case to Design-Build Associates. The other proposers are awarded points inversely proportionate to the 40 point potential. In other words, the higher your price, the fewer points you will receive. After assigning points for the technical and price criteria, the owner adds up the two scores and the design-builder with the highest number of total points wins the competition.

In determining which RFPs they want to respond to, design-builders should take care to ascertain the order in which owners will evaluate the two proposal components. This may seem like a minor consideration, but because the low-bid mentality is so pervasive in the construction industry, evaluating the price component before the technical component could negatively influence the technical scores. To avoid this, owners should include a stipulation in their RFPs that requires the design-build teams to submit their technical proposal and their price proposal in separate envelopes. This safeguard is referred to as the **two-envelope proposal**. Furthermore, it is important that the jurors not open the price envelope until after they have scored the technical components. As unbiased as jurors would like to be, human nature, compounded by a history of low-bid procurement, can cause any juror to allow price to influence their scoring of nonprice factors. It would be in the design-builder's best interest to avoid competing for projects where the RFP calls for the technical solution and price proposal to be submitted in the same package.

OTHER EVALUATION METHODS

Although the Weighted Criteria Matrix is one of the most common forms of evaluation used for design-build, it certainly is not the only method. There are four other frequently-recognized evaluation techniques. They are:

D-B DICTIONARY

Two-Envelope Proposal – A proposal process that requires the technical components of a design-build proposal be submitted separately (as in separate envelope) from the price component of the same proposal, with the price component remaining sealed until all of the technical factors have been evaluated and scored.

1. Fixed Price/Best Design

In this commonly-used method, the owner discloses the project budget in the RFP as a fixed price and only evaluates the technical/management components using a Criteria Matrix. The design-builder still must conduct a conceptual estimate to verify their price and confirm that they have not designed a solution that exceeds the budget, but the evaluation scores apply to the technical criteria only. The team that scores the most points without going over the budget wins the contract award. In this evaluation method, creativity, innovation, and true value engineering can take center stage. Although there are some minor variations on this method, it is one of the more successful best value approaches because price becomes a nonissue. The biggest consideration is whether the owner has established a realistic budget in the first place. But even when the owner's budget does not cover their required scope of work, the design-build process will allow the teams to offer proposals showing the client how much program they can build for their dollars and help them find a solution that does meet their budget.

2. Adjusted Low-Bid

This method is similar to the weighted criteria evaluation process in that technical components are evaluated and scored using a criteria matrix. Once the technical scores are known, the price envelopes are opened. However, this is where the approach varies significantly from the weighted criteria method. In the adjusted low-bid method, the design-builder's technical score ranges from 0 to 100 and is expressed as a decimal. For example, a technical score of 87 is expressed as 0.87 and the price component is not assigned any points. Instead the design-builder's price is divided by the technical score resulting in an adjusted "low bid." The design-build team that has the lowest adjusted bid wins the contract award. However, the adjusted bid is used for selection purposes only, and the design-builder's original price remains the actual contract price. This method is illustrated in Figure 4.7. Many Departments of Transportation use variations of the adjusted low-bid method for their design-build procurements.

It is not uncommon for owner-developed project budgets to miss their mark for the project scope represented in the RFP. This is primarily due to the timing of the budget development relative to the release of the RFP for bidding. Often, these budgets are established many months and even years before the RFP is distributed to the design and construction community and the numbers are not updated. Labor and material prices can fluctuate significantly with economic conditions over time and the accuracy of the budget can be drastically affected.

DB Team	Technical Score	Price Proposal	Adjusted Bid
Superior Design-Build	0.79	$2,145,000	$2,715,190
Master Builders Inc.	0.87	$2,229,630	$2,562,793
Design-Build Associates	0.92	$2,301,500	$2,501,630

Design-Build Associates wins this contract award by having the lowest adjusted bid even though their original proposal price was the highest, it has been offset by their high technical score. The adjusted bid is for selection purposes only and the contract price will remain the same as the proposal price at $2,301,500.

Figure 4.7
Adjusted low-bid evaluation method

D-B DICTIONARY

Discussions –
One-on-one meetings between the proposal reviewers and the design-build team regarding deficiencies, mistakes, errors, misinterpretations, and needed clarifications in the technical aspects of its submission.

Technical Leveling –
A technique used in the equivalent design – low-bid evaluation method in which the owner's evaluation team reviews the technical components of the design-build proposals and, in discussions with each design-build team, asks for adjustments in their proposal. Technical aspects of competing design-build proposals are critiqued and proposers respond with adjustments to create technical equivalency across all proposals.

Competitive Range –
A term that describes those proposals that are reviewed and are found to contain no "fatal" errors or deficiencies and therefore deemed responsive in certain low-bid types of design-build competitions.

Best and Final Offer (BAFO) –
A term used to describe the final revised proposal submissions in an Equivalent Design / Low-bid design-build competition. Also referred to as *Final Revised Proposal.*

---

3. Meets Criteria/Low Bid

This method is most closely related to the traditional design-bid-build method. Here, the jury evaluates the technical/management components of the proposal to determine whether they meet a "minimum" criteria standard. Rather than scoring and ranking the competitors, the jury simply checks boxes indicating whether the proposal meets the required criteria or not. Of the proposals that meet all criteria, the contract is awarded to the lowest bidder. However, this method is not designed to capture significant advantages that one firm may have over another. For example, one proposer might include 250 beds in their design for a new dormitory and the other may include 255 beds. But if the requirement was 250 beds, both proposers meet the criteria and receive a checked box. The firm offering the 255-bed design gets no credit for designing in an extra five beds, even if their overall proposal equates to a lower cost per bed overall. This type of procurement method provides no incentives for "adding value" which is one of the benefits of design-build procurement. Because of this, the Design-Build Institute of America finds this the least desirable procurement method.

4. Equivalent Design/Low Bid

Although this is classified as a best value procurement method, equivalent design/low bid, as the name implies, is actually a low-bid competition. In this method the jurors do not assign proposals points or score their technical components. Instead they conduct formal critiques, or **discussions**, with each of the design-build teams. The purpose of the discussions is to correct errors or mistakes in the proposals, expose and communicate deficiencies, and obtain clarifications on any portion of the submittal that might not be clear. These discussions attempt to establish an "equivalency" across all designs so they represent a more consistent submission from each of the proposers and in essence create a level playing field. In fact, the technique is often referred to as **technical leveling.** In essence, the reviewers help each design-build team adjust its technical proposal to fall within what is commonly called the **competitive range.** The technical components are not competitive factors and low price becomes the primary, if not sole, selection criterion. Once the discussions have been completed, each design-build team is given a deadline by which they are to respond with revised designs and adjusted prices (up or down) according to what they learned in the critiques. This practice is commonly referred to as a **Best and Final Offer** (**BAFO**). (The Federal government now refers to the BAFO as the **Final Revised Proposal**). The revised designs are then evaluated to make sure that they still comply with the requirements of the RFP, and the price envelopes are opened. The design-builder with the lowest price wins the contract award.

This evaluation method is not favored by many design-builders because the line between technical leveling and **technical transfusion**, the transferring of information and ideas from one competing team to another, is a fine one—no matter how well-intentioned the reviewers are. This is especially true if multiple iterations of discussions are allowed. However there are several public agencies that mandate this method and many design-builders are accustomed to the process and take it in stride as they develop their proposals. And some agencies have improved on the BAFO approach by retaining the technical factors scoring, even after revisions, keeping the best value concept intact.

Although design-build is meant to operate as a best value or qualifications based procurement, not all methods appear to meet the spirit or intent of the approach. Much design-build work is currently procured using methods like Meets Criteria/Low Bid and Equivalent Design/Low Bid where price is really the sole discriminator, especially with government and public agencies. In these instances it appears that the owners are more interested in using design-build as a risk transfer vehicle than in benefiting from the value that the design-build process can deliver. However, some agencies are constrained to do so by their governing procurement policies.

> **D-B DICTIONARY**
>
> **Technical Transfusion** – Disclosing technical information from one design-build proposal during discussions that results in the improvement of a competing proposal.

My experience with design-build as a project delivery method has been both as a federal "owner" with the Air Force and now as a general contractor. From both perspectives, I highly prefer design-build over traditional design-bid-build. The primary reason I prefer design-build is that it offers a much higher opportunity to meet the owner's goals for cost, schedule, and quality. In traditional "hard bid" or "rip and read," fully-designed projects, the designer is influenced to give the owner as much of what they want as possible without a full understanding of the cost impacts of actually building the project. In contrast, during a design-build project, the designer, general contractor, and owner can make design tradeoffs for such things as structural approach, material costs, etc. to keep project goals reachable.

Another reason that I prefer design-build is that it enhances overall project delivery schedules. This is principally because the design can be accomplished in "packages" and allows the design and construction to be overlapped, thereby shortening the overall schedule. There's no need to wait for a "100% design" to break ground; the civil design package can be advanced to allow the site work to begin while the structural design package is underway and the MEP and interior outfit design packages are just starting. This allows for key items like structural steel to be ordered long before the total design is complete which really helps keep the project on target.

And lastly, design-build by its very nature when done right sets up and continues a very collaborative working relationship between the owner, general contractor, and designer. Once everyone experiences a mutually successful project, it's quite likely that they will all want to work together again—and the more they work together, the better they are able to deliver superior results. It's a win-win for everyone.

Pat Burns, PE, (Brig. Gen., United States Air Force, retired.)
Vice President
Federal Contracting Group
Mortenson Construction
Alexandria, Virginia

Project Enhancement Approach

There are also many hybrid evaluation methods that combine aspects of the various design-build approaches and evaluation methods. One of the more interesting hybrid methods, which I call the **project enhancement approach**, is one that starts out with a highly-prescribed design in the RFP, generally counterproductive in design-build. However, the design-build teams are not restricted by that design and can change it—in some cases

D-B DICTIONARY

Project Enhancement Approach –
A hybrid evaluation method in which design-builders compete on the basis of the enhancements that they propose to the bridging design included in the RFP and overall proposal price. The total cost is divided by the total enhancement points, called quality points. The team with the lowest cost per quality point wins the competition.

Enhancement Criterion –
An evaluation item associated with a design-build RFP involving some component of a project design that the owner wishes to see improved or enhanced.

---

they can discard it altogether if they choose. The bridged design is used as a benchmark to establish the minimum basic program parameters (such as the building's footprint, overall height, number of rooms, number of specified spaces, etc.). Because of the time required to essentially predesign the building in advance of the design-build team, this is an expensive way for the owner to communicate this basic information.

The design-builders, in essence, are asked to compete based upon their ability to "enhance" the RFP design—even if that means starting from scratch. What makes the approach challenging is that the RFP also comes with a "fixed price" budget based upon the RFP design.

The owner's evaluation criteria, also provided in the RFP, provides direction for the design-build teams by stating where the client would like to see the enhancements. For example, on a dormitory project, the number of beds might be an **enhancement criterion**. If the RFP design stipulates 700 beds, an enhanced design would include more than 700 beds for the same price, or less. Other enhancement criteria might include energy efficiency, sound transmission, building class, LEED certification, life cycle expectancy, and even aesthetics.

One set of circumstances that prompts an enhanced design approach is when the owner has a certain limited budget, and hires a traditional architecture firm to partially design a project that meets their basic program requirements (though not necessarily performance requirements) and cost limitations. The owners may decide that perhaps they could get more for their money if they choose to use design-build procurement rather than finishing the design in a traditional fashion. Next the owner goes through the preliminary design and makes a list of all of the areas where they would like to see more—more quantity, quality, space, durability, aesthetics, value, and so on—all for the same price or less.

In this unique scenario, the scoring technique divides the overall project price by the point accumulation of the enhancements:

Proposal Price / Total Quality Points = Price per Quality Point

The final tally determines the lowest "cost per unit of quality" and the team with the lowest price per quality point wins the competition. The sample Evaluation Summary shown in Figure 4.8 indicates that Premier Design-Build Associates would end up winning the competition under this approach.

Although Premier was not the lowest bidder in this example, the owner realizes a number of benefits:

1. The enhancements have added value to their project above and beyond their original RFP design
2. They have received this added value at a significant cost savings—over $1 million less than their original budget.
3. They can now take the savings and use it elsewhere, or negotiate to add additional enhancements to their project

Figure 4.9 shows technical evaluation criteria taken from a design-build RFP document for a mixed-use dormitory project. This score sheet

DB Team	Proposal Price	Quality Points	Price per Quality Pt
Best and Better Joint Venture	$40,000,000	896	$44,643
Premier DB Associates	$38,890,600	914	$42,550
Class Act Design-Build	$37,420,000	788	$47,487

Not-to-Exceed Project Budget = $40,000,000

Figure 4.8
Project enhancement evaluation summary

Technical Evaluation Criteria

Each proposal will undergo a detailed technical review by the Evaluation Team, which will assign the quality point value to each item indicated in the Technical Evaluation Criteria. The criteria to be used for evaluation of Technical Proposals will be as listed below. Each criterion will be scored based on a 10-point scale, 10 being the highest score, and weighted relative to the multipliers indicated for each criterion.

The Quality Point Value determined by the Evaluation Team is final, and no appeal of its decision will be considered.

Technical Evaluation Criteria	Maximum Quality Points	Multiplier	Total Possible Score
1. Design Bed Space Enhancement	10	4	40
2. Structural Enhancement	10	2	20
3. Exterior Finish Enhancement	10	2	20
4. Interior Finish Enhancement	10	3	30
5. Mechanical System Enhancement	10	3	30
6. Design Enhancement	10	3	30
7. Consideration of operational and maintenance factors; mechanical and energy efficiency.	10	2	20
8. LEED Certification (Points given only for Certification of one LEED category)	10		10
Bronze (4)		1	
Silver (6)		1	
Gold (8)		1	
Platinum (10)		1	
Grand Total Points	80		200

Figure 4.9
Project enhancement score sheet

makes it clear where the owner expects to see enhancements. In this particular RFP, the following three-step evaluation process was outlined:

1. The first step of the analysis procedure will be the determination of Quality Point Scores assigned to each Technical Proposal by the Evaluation Team. At a public opening, the Evaluation Team's Quality Point Value assigned to each proposal will be announced.
2. The second step will be the public opening of sealed Cost Proposal Packages, matching Cost Proposals with the Quality Point Values assigned, and reading of those Cost Proposals.
3. The third step will be the computation of each proposer's cost/unit quality, arrived at by dividing the Cost Proposal by the Unit Quality Point Score. The cost/unit quality formula will determine the apparent finalist, and the second, third, fourth, and fifth place awards.

On this particular dormitory project, the winning design-build team of Clark Construction and Niles Bolton Associates did modify the RFP design in order to meet the owner's budget. They actually reduced the overall number of dormitory buildings indicated in the RFP from eleven to nine, but still met the required bed count and added a number of other enhancements to the project as well. In addition, they were able to deliver five buildings in year one of the project instead of the required four.

The project enhancement approach is an interesting concept, but the same result could likely be achieved without spending the money for bridging documents up front or expending the time to produce them. In other examples of this method, the original RFP design has been completely discarded by every shortlisted design-build team. Each team significantly improved upon the original design for exactly the same budget and, in many instances, for less. Given design-build's propensity for reducing waste and increasing efficiencies, paying for extensive prescriptive design up front to communicate program requirements and project scope seems superfluous.

Negotiated Design-Build

Not all design-build procurement is competitive. In private industry, owners can select their design-build team through a negotiated selection process. In these instances, the owner may be looking for a design-builder to work with them from the very early stages of their project planning. It is not unusual for the design-build team to be on board during the programming and feasibility stages, and carry through to the design and construction phases. In some cases the owner has not even figured out where they will build, or what kind of facility they actually need—or if they need one at all. They may only know that they have a business challenge before them, and it involves facilities. It could be a problem with capacity, or production, or product research. And once they decide to use design-build, then they want to get a team involved as quickly as possible to help them solve their problem. In most private enterprise, "time is money" and getting the proper experts engaged early is an important goal. The case study Icing on the Cake in this chapter is an excellent example of a direct selection, negotiated design-build scenario where the owner brought the design-build team on board before they even knew exactly what is was that they were going to do.

How Negotiated Selection Works

When negotiated selection is used, the process for selecting the design-build team is fairly straightforward. After the owner has defined the project in broad business terms, she will identify one or more design-builders that she potentially wants to work with. These decisions are usually made based upon the design-builder's reputation, experience, past performance, previous relationship with the client, design capabilities, and the like. The shortlisted design-build teams are interviewed and evaluated against a set of criteria developed by the owner, possibly using a weighted criteria matrix. The owner can also request that the design-build teams submit a qualifications package and do a presentation to highlight past projects, capabilities, and experience.

After interviews and presentations, the owner will select the team they want to work with. The owner may first solicit the design-builder's services to continue with project feasibility work, due diligence, and other preplanning and preconstruction services before proceeding to the design phase of the project. The services the design-builders supply depend on how early they are involved in the project. Their services are usually rendered on a fee-for-service or consulting basis. However, once it is clear that the project will proceed, the final terms of the contract can be negotiated. After these terms have been agreed upon, the contract is typically executed in two parts. Part 1 authorizes the team to develop the design to the stage where it can be adequately priced to establish a project budget. If the parties agree on the design and budget, then Part 2 of the contract is executed to complete the design and proceed to construction. The two-part contract, as in progressive design-build, allows several checkpoints at which times the owner can decide whether to proceed with the project or not.

It is common for private owners to engage the services of their design-builder on a negotiated basis, especially if they have a long-standing relationship. Once a client has had success with the design-build process and with a particular design-build team, there is usually not a need to look elsewhere. To produce high-value results for a client, the design-build team needs to know who the client is, what they do, and why they need to do it. The more the parties work together, the more efficiently and effectively they will be able to produce the successful results the client needs and expects.

Integrated design-build firms that also perform development services are most likely to participate in a negotiated design-build selection.

CASE STUDY

ICING ON THE CAKE

THE FACTS
Project Name:
Krispy Kreme Doughnuts
Project Type:
Manufacturing and Distribution Facility
Location:
Effingham, Illinois
Description:
A 187,000-square-foot bakery mix manufacturing plant using advanced material handling and distribution technologies
Cost:
$35 million, including process equipment
Duration:
16 months
Completion Date:
Fully operational May 2002

THE TEAM
Owner:
Krispy Kreme Doughnut Corporation
Design Builder:
The Haskell Company

Source Haskell

THE LEAD-IN

Nothing like a hot cup of coffee and a freshly-baked doughnut to start your day off right. Ahhh. . .but not just any doughnut. A sign flashing "Hot Doughnuts Now" is the magical lure for Krispy Kreme devotees who want to witness the making of their doughnuts as much as they want to taste them—and they can do so right there in the store as they make their purchases. As a matter of fact, the opening of a new Krispy Kreme store has become a cult, a community event, often with lines of eager customers circling the block to get some of those delicious doughnuts. Many of us enjoy this special treat, but do not give much thought as to how the doughnuts come to be at a location near us.

THE STORY

In 1937 the founder of Krispy Kreme, Vernon Rudolph, started making doughnuts from a secret German recipe to sell through a few Winston-Salem, North Carolina, grocery stores. Interest in the Krispy Kreme doughnuts increased to the point that he began selling them to locals through a window in his small manufacturing facility. As the Krispy Kreme reputation grew, the operation expanded and a new plant was built to make doughnut mix for distribution to Krispy Kreme stores in the area. Growth continued at a steady pace in the Southeast until the

1990s, during which time expansion became exponential and widespread throughout the U.S. The original plant and building were expanded incrementally as business grew, but the cobbled together facility was outdated, inefficient and fast reaching the point where it would not be capable of meeting the demand of the rapidly added new stores.

The operations staff was focused on getting the most capacity possible from this single plant and working hard on the ever-widening distribution network. They projected less than two years until they could no longer meet the needs with this facility, which had no more room on site for expansion. Needing an immediate plan to add capacity, they hired a logistics consultant who recommended that they begin design and construction of a new plant in western Pennsylvania. From their own experience in distribution, the Krispy Kreme operations staff was not convinced that this conclusion and its rationale were correct. The clock kept ticking and their dilemma kept deepening as they began a search for some other outside entity that could help them with the capacity and business issues while they concentrated on continuing to maximize capacity at the Winston-Salem plant. Outsourcing production of the mix was a possibility, but one that would only be considered as a last resort due to special secret mixes, processes and exacting quality control.

Through common membership in a food industry business organization, Krispy Kreme staff found Haskell, and invited the design-builder to meet with them to discuss their problem and possible solutions. Haskell presented their extensive food industry experience and their capabilities to assist Krispy Kreme with their capacity issues. Based on that presentation and the similar services Haskell had recently provided for Starbucks, Krispy Kreme requested that Haskell propose a menu of "front end" services to help Krispy Kreme define needs, plan for short and long range growth and to make the business case for approval by Krispy Kreme executives.

Leveraging its in-house specialty design services and real estate units, Haskell proposed a supply chain/business case analysis to determine optimum location(s) for additional capacity, site selection services following the supply chain identification of general geographic area(s), facility programming to include defining the manufacturing process, equipment, material flow, building spaces and site requirements and an overall conceptual project budget. Haskell added to its team a professional supply chain consultant and a baking industry process consultant. Authorization was given to proceed with this increment of services, keeping open the idea that Krispy Kreme would have the future option to continue working with Haskell for design-build, process equipment procurement and installation and potentially financial services.

From the start, of course, the Krispy Kreme operations staff had to become an integral part of the interactive Haskell team. Much of the operational information required for the initial work had to come from the client, or from jointly made assumptions where actual data was not available. The supply chain analysis identified costs of raw material supply, current and future mix demand based on projected store locations, inbound/outbound transportation, labor, land, site development, building,

On the Double

Krispy Kreme Doughnuts applies advanced material handling and distribution technologies at its new Central Illinois facility.

BY LAURIE GORTON

"We dream big here at Krispy Kreme," said Fred W. Mitchell, senior vice-president, manufacturing and distribution, Krispy Kreme Doughnut Corp., Winston-Salem, N.C., as he described the fast startup of the company's new bakery mix manufacturing plant at Effingham, Ill. "And this facility will help us realize those dreams."

At $35 million and 187,000 sq. ft. Effingham is a big plant, a major boost to bakery mix capacity not only for Krispy Kreme but for the mix industry nationwide. But even more telling is the way Scott A. Livengood, Krispy Kreme chairman, president and c.e.o., described the plant in terms of the company's future. "This facility provides much needed mix production capacity," he said shortly after its May 2002 startup. "It is also a symbol of our growth in recent years and our optimism about our prospects over the long term."

Just six months after startup, the Effingham plant supplies 35% to 60% of system-wide needs, according to Mr. Mitchell. Total volume this year will be 3.8 million 50-lb

Effingham's lab tests the proofing and frying performance of samples from every batch, using equipment identical to that in its stores.

Source Haskell

equipment, value of potential governmental incentives and included a payback analysis. It considered not only the optimum location and capacity for the immediate manufacturing facility but also, based on the rapid growth nationwide, projected a location for a potential third facility. Concurrent with the supply chain analysis, Haskell proceeded with the project programming, which indicated an initial 187,000 sq. ft. plant on a site of approximately 40 acres, both allowing for future expansion. Idealized conceptual site and building footprint templates were developed as part of the programming effort for use as templates in the upcoming site search. Major subcontractors, members of Haskell's Preferred Partners Program, were consulted throughout the early conceptual site, building and process planning work.

The supply chain analysis recommended that a new manufacturing/distribution facility be located in the corridor between St. Louis, Missouri and Indianapolis, Indiana, and that an additional plant follow shortly thereafter in southern California. Immediately, keeping the client's name confidential, Haskell's real estate professionals began identifying potential sites for the new facility in this area that would meet the requirements of the idealized footprint templates. From the initial total of twenty-five sites, the Haskell team quickly narrowed the list to twelve and then to five prime sites. It required the knowledge of the real estate professionals, the ability of the architects, engineers and construction managers to visualize site development and input from the client to create this short list—none singularly could have done an adequate job. The Haskell team and client (incognito) then visited these five sites and met with community officials. From these visits, the list was narrowed to two, one admittedly only a "straw dog" to assure a good negotiating position for potential governmental incentives. At that point Krispy Kreme was identified as the client, and undertook incentive negotiations, assisted by Haskell. The selected site in Effingham, Illinois, was a virtual duplicate of the idealized site configuration identified through the programming effort.

Next, Haskell was authorized to proceed with design development and to provide a Guaranteed Maximum Price (GMP) for building and process equipment. During this phase, major subcontractors were brought on to the team and worked closely with the design professionals, process consultant, construction staff and, of course, the client in developing further project details. Preliminary subcontractor proposals for major trades were received and reviewed with Krispy Kreme. In May 2001 a GMP was agreed upon, and, to assure dry-in before winter,

Haskell immediately began construction of building the shell on a fast-track basis (not simply trying to do conventional activities faster, but doing things in new ways and sequences to meet the needs of this specific project) to assure dry-in before the rapidly approaching winter. The schedule included a three-month "shake-down" period to assure consistent product quality through the new process, which incorporated the latest equipment and technology.

During the course of final design, Haskell offered its furnishing selection and procurement services, working professionally on a fee basis. Krispy Kreme indicated that they had recently leased a new corporate headquarters, were happy with the furniture vendor they engaged there, and had asked that vendor for a proposal for the new manufacturing facility. When Krispy Kreme received the furniture proposal, it was accompanied by multiple thick sample books for color and finish selection. They were unsure of how to deal with the furniture finishes to assure that they would be coordinated with the building interior finishes being selected by Haskell. In addition, they had no sense as to the quality of the furnishings proposed nor the fairness of the pricing. As a nominal add-on service, Haskell's interior designers reviewed the proposal, checked pricing (which proved to be fair) and made suggestions relative to quality. In addition, they prepared coordinated building and furniture material and finish recommendations and reviewed them jointly with Krispy Kreme and the furniture supplier. Thus, the furniture supplier became an additional resource and member of the project team. The grand opening of the new facility, with ceremony and festivities, occurred just as originally scheduled in May 2002, an event that included the whole community, Krispy Kreme executives and the complete project team with donuts and coffee for everyone.

THE CLOSE

Although a "design-build" project, this engagement extended services not only linearly forward into the client's early business case analysis and project definition, but also expanded services laterally into furnishings, process design, and equipment installation, all in the interest of helping the client solve a problem—not simply accomplishing the design and construction of a building.

From the Owner

"We fired up the equipment in April, less than a year after ground breaking in May 2001. In fact, 97% of everything we put in here worked "right out of the box" and in just six months, this plant was supplying 55%-60% of our system wide needs. That and the assistance from Haskell are why we got such a good start. The Haskell Team provided complete architectural, engineering, construction, real estate, and facility management services on a single-responsibility basis. Haskell was with us every step of the way."[4]

Fred W. Mitchell, Senior Vice President, manufacturing and distribution, Krispy Kreme Doughnut Corporation (Baking & Snack, December 2002).

AMOUNT OF DESIGN IN THE RFP

One of the factors that influences the ability of the design-build team to add value to the project is how much design the owner prescribes or mandates in the RFP. In an ideal design-build project, the owner describes their wants, needs, and desires in terms of performance, rather than products, parts, and pieces, and lets the integrated design-build team, with their unique experience and expertise, propose a solution that optimizes the project outcomes for function, time, cost, and quality. However, some owners feel a need to control the process by supplementing the RFP with some prescribed design. Tradition has conditioned us all to be skeptical of any process other than the linear design-bid-build procurement model. Owners must radically change their thinking to grasp the concept that less design is better in the RFP.

In order to proceed with a design-build procurement, the owner must decide how much design they are going to prescribe and specify to the design-builder up front, and how much they are going to trust the design-build team and trust in the design-build process. This fundamental decision is a critical one.

There are three approaches that an owner can choose regarding the amount of design to include in their RFP, shown in Figure 4.10. Each option reflects the amount of trust an owner is willing to place in the design-build approach.

1. Direct Design-Build (-10% to 10% design)

Direct Design-Build, in which the owner specifies very little, if any, of the design, is a design-builder's dream come true. In essence, no design decisions have been made and the design-build team is brought on board early—often at the programming stage and sometimes even earlier than that. The design-builder may be involved in the feasibility stage, site and environmental analysis, facility and capital budget planning, pro forma development, risk analysis, and many more of the up-front services involved with planning, designing, and building new facilities. In these instances, many design-build firms supply what Chapter 2 referred to as design-build plus.

Figure 4.10
RFP design strategies

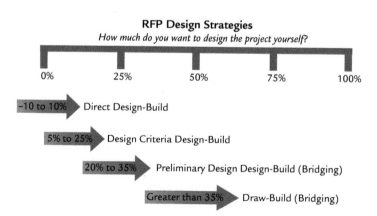

The "plus" can be anything from preproject planning tasks, such as arranging financing for the project, and lease agreements, to postproject activities like operating and maintaining the facility. This option is not always available for public projects but it can bring tremendous assets and expertise to a company trying to make major fiscal and facility decisions.

2   Design Criteria Design-Build (5% to 25%)

Short of a direct selection approach, this is the best method for performing competitive design-build as it was meant to be. In this scenario the owner writes the program and defines the scope, but allows the design-build team to provide the complete design solution. The owner provides just enough project criteria and performance data for firms to develop a conceptual design and costs estimates. The design-build team is given full freedom to solve the owner's problem at the highest and best value within the given requirements for performance, quality, schedule, and budget. When using this method, an owner creates the widest range of opportunity to receive multiple creative design solutions, which is what a design-build solicitation is all about. And in many cases that is exactly what the client needs.

3   Preliminary Design Design-Build (20% to 35%)

Preliminary design design-build, often referred to as **bridging**, is popular with owners trying design-build for the first time. In this method, the owner contracts with an architect or engineer directly to develop the initial design for her project before the RFP. (See Figure 4.11.) Although this design, known as the bridging design, is preliminary, it often contains a fair amount of detail. It is then included in the RFP and the design-build proposers are expected to use it as the basis of their proposals. At this stage the owner's bridging architect's contract for services expires and she is usually released from the project. Significantly, the bridging architect has no liability relative to the design that they have initiated, even though it has become part of the contract documents. The design-build architect, as part of the design-build team, must then complete the design and finish the drawings, assuming all associated professional liability.

Many owners believe that by using the bridging method that they have achieved maximum risk shift while still controlling the design, but this is

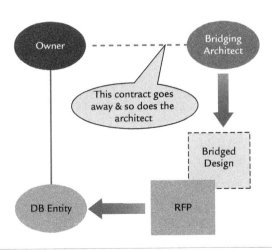

Figure 4.11
Bridged design-build

not necessarily so. To the extent that the bridging documents prescribe and specify design details, the design risks may shift back to the owner. Figure 4.12 illustrates how risks shift back to the owner with increased design in the RFP—the less design, the less risk associated with that design.

It is an important concept to grasp because many owners' believe that design risks always shift to the design-builder; many design-builders incorrectly believe the same thing. But that is not true if the owner still controls the details of the design, as is true with bridging. If there should be a problem with the project stemming from the initial design prepared by the now-absentee architect, then the owner might be left with the responsibility for that error, omission, or flaw—and not the design-builder. In a review of *Design-Build Lessons Learned: Case Studies from 2003* by Michael Loulakis, the author confirms the following:

> *In addition to the philosophical and policy issues related to bridging, one major legal question repeatedly arises relative to this subject—who bears responsibility for problems arising out of the owner's design? Most owners take the position that their design is preliminary, should not be relied upon, and that the design-builder, as the designer of record, bears these risks. As expected, most design-builders take issue with this approach, arguing that they have the right to rely upon the accuracy of what the owner puts into the RFP package, and if there are mistakes in the owner's preliminary design, the design-builder should be entitled to payment for correcting those mistakes. Only a handful of cases have addressed this issue—and virtually all have sided with the design-builder's point of view that the owner should bear the risk of defects in the bridged design documents.*[5]

The ultimate outcome in these situations may depend primarily on how much prescribed design is in the RFP and how restrictive it is. In some cases it is minimal—a site plan and maybe a few diagrammatical floor plans to describe design intent or preferences, still leaving the design-build team flexibility when it comes to the details of the design. In other instances owners specify 40, 50, 60, or even 70% of the design in the RFP and calling it design-build. The only thing left to do is finish the drafting of the plans. This draw-build approach defeats the purpose of design-build and usually ends up being nothing more than a low-bid competition. And regrettably, owners are unwittingly exposed to risks that they do not understand. In many cases they would be better off to

Figure 4.12
Design-risk shifting

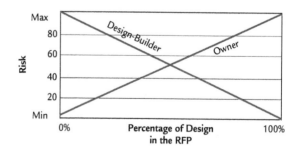

use design-bid-build because at least then they would have a professional designer under contract to share the exposure.

Granted, there may be cases in which the owner wants to control the details of the design. For example, the owner may want to specify a particular piece of equipment that has performed well in the past and that their facilities people can maintain. This is acceptable, as long as the owner understands that the liability for that design decision may fall back on their shoulders, just as if it were design-bid-build. Furthermore, prescribing that design may restrict the creativity of the design-build team or diminish the opportunity to achieve an even better solution or higher performance.

In general, an owner using design-build should focus on controlling the outcome through their project performance criteria and requirements, and refrain from prescribing design in the RFP as much as possible. Owners experienced in true design-build have learned that if they let the process proceed as it was intended, they usually end up with better results—enhanced design, lower costs, better quality, and faster schedules.

Many top-quality design-build firms will not propose on any bridging procurement because they view it as low-bid based. They know it will have many of the undesirable attributes of a low-bid procurement, and that is not what they do. In addition, their ability to win the project due to their creative design approach or past design-build experience is diminished or lost altogether, and they are expected to assume the risk of a design that they did not create. In their eyes, this is worse than design-bid-build.

Robert J. Hartung, DBIA
Founder/President
Alternative Delivery Solutions LLC
Laguna Niguel, California

SELECTING A CONTRACT TYPE

The third critical decision the owner makes in the procurement planning process is selecting the contract type that is going to best serve their project objectives. You learned in Chapter 3 that a design-build contract contains unique terms and conditions. However, there are a number of issues that a design-build contract handles in the same way as a design-bid-build contract. One of those issues is the payment method. Most of us are familiar with the standard four-contract payment structures—lump sum, cost-plus, cost-plus with a guaranteed maximum price (GMP), or unit price. To review:

- **Lump Sum** – The design-builder agrees to complete the design and construction services spelled out in the contract for a fixed lump sum price, regardless of how much it actually costs to perform these services.
- **Cost-Plus** – The design-builder is reimbursed for actual cost incurred to perform the design and construction services plus an agreed upon fixed fee or percentage.

:: **Guaranteed Maximum Price (GMP)** – This method combines features of both the cost-plus and lump sun contract methods. The design-builder is reimbursed for actual costs incurred to perform the design and construction services, plus an agreed-upon fixed fee or percentage, up to a guaranteed maximum price.

:: **Unit Price** – The design-builder quotes a stipulated unit price for quantities of time, equipment, or completed work items. The final proposal price and compensation depends on the number of units used for the project.

While any of these payment methods can be used with design-build, some are more fitting than others. Design-build at its best focuses on delivering performance. And if an owner wants to achieve the highest performance possible, then the contract payment model should tie in with performance. The Design-Build Institute of America's *Design-Build Contracting Guide* suggests:

> *Because each design-build project is unique—impacted by prior relationships among the parties, financing arrangements and specific design requirements—there is often more than one means of appropriately achieving contracting goals. Parties should be careful about attempting to "cookie-cut" their design-build contracts or blindly following the recommended language contained in model design-build contracts. Effective contracting is accomplished when the parties have given specific thought to their contracting goals and then tailored the contract to meet the unique needs of the project and the design-build team.* [6]

At the heart of the high-performance contract is the notion of incentives. In reality, every contract has incentives—it just depends on what it is the contract is incentivizing. For example, in a lump sum, fixed price contract, there is an implied incentive for the contractor to control costs—which at first glance may seem appropriate. However, if they fail to control costs, or if circumstances beyond their control start to impede their progress and impact their profitability, then their focus will naturally turn toward cost avoidance as a means of cost control—not necessarily a good thing for the owner.

> In design-build it is not unusual for the contract's payment method to adjust in relation to the design progress. For example, the contract could start out as a GMP while design is still in the schematic and preliminary design development stages, but as the design progresses toward construction documents, the contract can convert to a lump sum payment method.

From the very first day the builder signs a lump sum, low-bid contract he has an incentive to cut cost wherever he can, while still meeting the minimum requirements of the contract. After all, he is burdened with all of this uncertainty and risk that is beyond his control—he's fighting weather, craft availability, daylight hours, delayed suppliers, etc. Let's face it—his attention isn't exactly focused on delivering the owner the highest quality, performance, or best value—it can't be. He is tied into this fixed price while trying to manage these dynamic, ever-changing conditions and situations. It took me a while as an owner to figure out that this isn't the best position to be in, and there were other ways to set up the contract that would incentivize the type of performance that I was looking for.

Craig Unger, DBIA
CEO and President
Unger Security
Solutions, LLC
Knoxville, Maryland

In fact, incentives are always a part of the contract, whether they are explicit or not. And if an owner desires high performance, then she has to create the right kind of incentives to elicit the behaviors, and therefore the performance, desired.

Incentivizing for Performance

There is an alternative to the traditional contract payment methods that focuses on performance and has been used successfully on design-build projects. The **target cost contract** sets a target price as a goal for project performance and holds both the design-builder and the owner responsible for cost overruns and savings. Like a GMP contract, it pays the design-builder on a cost-reimbursable basis plus a fee, and provides guaranteed maximum price protection for the owner. However, it is unique in that both the design-builder and the owner agree up front to recognize the dynamic nature of the design and construction process and share in the risks associated with that dynamic. There are three components to a target cost contract:

:: **Target Cost** – The target cost is a realistic estimate of project design and construction costs that will attain the performance stipulated in the contract.

:: **Target Profit** – The target profit is linked to the achievement of the target cost. Profit percentage may be tied to a sliding scale to incentivize cost savings.

:: **Sharing Arrangement** – The sharing arrangement for a target cost contract operates just like the sharing arrangement associated with a standard GMP contract, except the owner and the design-builder share both the contract cost savings and the contract cost overruns. The sharing split is expressed as a ratio (50/50, 60/40, etc.).

> **D-B DICTIONARY**
>
> **Target Cost Contract** – An incentivized cost-reimbursement contract that sets a target price as a goal for project performance and holds both the design-builder and the owner responsible for cost overruns and savings.

How a Target Cost Contract Works

Once the design-build team is selected, before the design is complete, the owner and the design-builder collaborate to establish a realistic target cost and a target profit for the project. They also agree upon a split ratio that will apply to both the cost savings and the cost overruns. However, they will establish a cap or ceiling percentage for the owner's side of the cost overruns. For example, presuming a 10% cap, if cost overruns are less than 10% and the sharing arrangement is 50/50, then the design-builder and owner share equally in the burden of the overrun. However, if the total cost overruns exceed 10%, then the design-builder must accept sole responsibility for the total amount over the cap. (Actual costs are measured against targeted costs to determine the variance percentage.)

On the savings side of the equation, there are no caps. The design-build team has the incentives it needs to be as innovative, resourceful, and efficient as possible in delivering the project design and construction as long as they meet or exceed the performance requirements set forth

Figure 4.13
Contract cost-risk diagram

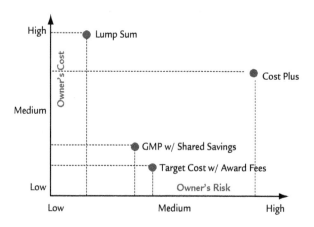

in the RFP. The design-builder must deliver the project performance regardless of the actual cost.

Once all of these ground rules are established and agreed upon, the big challenge is in properly monitoring the contract performance relative to the cost targets. As with any cost-plus/GMP type arrangement, full disclosure and accountability are necessary. The administrative and overhead costs associated with this effort must be taken into consideration; every invoice, payroll record, and receipt associated with the project becomes part of the payment application process. Although this is a challenge for both parties, the results are usually well worth it. As shown in Figure 4.13, incentivized contracts usually result in lower overall cost to the owner even though they are paying more money to the design-builder in shared savings. The concept of incentives and the target cost contract appears to reinforce the notion that "you've got to spend money to make money."

> By agreement, target costs may be adjusted up or down throughout the life of the contract. Owner-initiated scope changes, inflation, or major design variations are all reasons why the contract targets might need to change.

Applying Award Fees

Although the target cost contract does motivate the design-builder to be as efficient as possible in order to meet or exceed the price targets agreed upon, the contract still focuses on cost control. And even though the joint gain sharing or pain sharing aspect of the arrangement does provide incentives for both parties to help control costs and improve efficiencies, it still does not address the noncost features that would lead to project optimization across multiple performance criteria.

However, there is another contract mechanism that can be applied to the target cost model that can significantly influence the overall project performance. **Award fees** are incentives above and beyond the normal cost sharing opportunity that is available with the target cost contract. These incentives are specifically designed to motivate high performance in areas such as schedule, safety, workmanship, durability, sustainability, and responsiveness. An example of an owner's *Incentive Award Fee Evaluation Form* is shown in Figure 4.14. In this form, none of

> **D-B DICTIONARY**
>
> **Award Fee –**
> An incentive applied to target cost contracts that triggers a financial reward when the design-builder meets or exceeds performance standards set forth at the onset of the contract. They are typically used to incentivize noncost performance items such as safety, schedule, communication, and customer satisfaction.

Contract Number JX00c-340
Modification One (1)

FEDERAL BUREAU OF PRISONS-CONTRACTOR INCENTIVE AWARD FEE EVALUATION FORM				
Date of Review	Reviewer:			
Performance Factors	Marginal	Meets Contract Expectations	Exceed Owner's Expectations	Superior Performance
Workmanship New Work Only)				
Provides a superior quality of craftsmanship				
Enforces an approved Quality Control Program				
Promptly corrects noted deficiencies				
Timeliness of Completion				
Establishes and maintains and acceptable schedule within the framework of the contract requirements				
Maintains minimal negative float in the schedule				
Develops and follows an acceptable submittal status				
Provides time and cost impact analyses meeting the contract requirements				
Complete inventory of existing materials				
Responsiveness				
Resolves issues promptly without delay to the progress of the work				
Submits responses to Modification Proposal Request in a timely manner				
Provides complete and accurate Modification Proposals				
A release of claim is provided for every Modification				
Anticipates and deals promptly with issues and obstacles				
Attitude of Offices and Field Personnel				
Maintains a professional working relationship with FBOP staff				
Maintains a high level of cooperation				
Communication				
Exhibits a high quality of oral communications				
Practices openness and honesty during negotiations				
Provides clear and concise written communications				
Avoids unnecessary paperwork				
Safety				
Complies with safety standards and codes				
Employs and monitors approved Safety Program				
Proactive responds to safety issues				
Effective controls use of hazardous materials				
Maintains a clean and orderly site				
Zero Lost Time Accidents				
Overall Commitment to Project				
Contractor exhibits a commitment to the principals of the honesty and fairness				
Maximum utilization of DBE and SDBE resources				
Provide Government maximum value for dollars spent				
Overall Performance Rating for this Period				

Figure 4.14
Incentive award fee evaluation form. Courtesy of Craig Unger.

these evaluation criteria deal with project cost. They are all qualitative in nature but add greatly to the overall project's performance and value.

How Award Fees Work

The owner and the design-builder negotiate the terms of the award fee agreement at the onset of the project, along with the other contract conditions. First they determine a fair and reasonable amount of money to be set aside in an award fee pool. The amount set aside usually sends a clear message to the design-build team as to the importance of these noncost issues. For example, at the Pentagon Renovation Project in Washington D.C., the award fee pool is substantial:

> *The award fee pool in a typical PenRen contract is established at twice the normal profit rate for construction contracts in the Washington D.C. area. This is consistent with PenRen's determination to emphasize the non-cost performance issues that are usually at the heart of owner dissatisfaction, and it clearly communicates to the contractor that the more subjective performance aspects must be addressed to earn substantive profit.* [7]

After the pool is established, the terms of the evaluation process are negotiated, including the distribution of dollars among the various criteria and how the design-build team's performance will be measured and assessed. The parties must also determine at what intervals they will make these assessments and the conditions of payment. For example, in some instances, the owner may require that a minimum score be achieved before any amount of fee is earned. In other instances, the award fees might be earned over and above the design-builder's profit and underrun bonuses, or the various fees and bonuses may be linked. These details represent some of the issues that must be worked out before the plan goes forward.

It takes creativity to structure an incentivized design-build contract, but even diehard opponents have come to understand the power of the properly placed and executed enticement. And even though it may seem that the design-build team should perform at these higher levels without incentives, it is most likely the competitive spirit that rises to such challenges; effort levels tend to rise when something more than the usual reward is at stake. But the advantage of award fees is that they address human issues like communication and collaboration, which increase the opportunity for integrated solutions. And integrated solutions lead to optimized results that can benefit the owner and the end users long after the project is completed. Even though the design-build team may realize additional profit than was originally planned, the overall project may not end up costing a single dime more than the originally forecast amount—especially if the award fees are also tied in with the savings incentives associated with the target cost contract.

However, the subjective nature of award fee evaluations makes it paramount that all parties operate in an environment of trust and cooperation. The performance expectations must be realistic and

---

> Award fee incentives only work in situations where performance-based requirements are applied and the design-build team has some flexibility in the details of the design and the construction process. They will not work in a prescriptive plans and specifications situation.

> Target cost contracts and incentive award fees are relatively new contracting techniques and not well known in the industry. The Design-Build Institute of America offers a full-day public course to owners and practitioners who want to learn more about these high performance contracting methodologies and how they work. You can learn more about the course by going to the DBIA website (http://www.dbia.org).

achievable, and the award fees themselves must be substantial enough to make it worth the team's efforts. And the process requires that the owner act in a spirit of fair-mindedness and provide proper feedback and guidance regarding performance expectations. This is not a process that can operate successfully without proper oversight and attention. However, if these conditions are met, award fees have proven to be an extraordinary mechanism for achieving specific, targeted performance objectives.

YOU GET WHAT YOU PLAN FOR

We have all heard the expression, "You get what you pay for." Well in project procurement, the expression "You get what you plan for" might be more fitting—especially in design-build. If you do the necessary due diligence and think through your source selection and contract options, you could end up getting more value and performance than you paid for.

Acquiring design-build services requires proper preparation and an effective strategy. The overall success of the project will be directly correlated to the quality of the acquisition planning and appropriateness of the source selection and procurement model. Owners who take the time and do the proper due diligence up front and develop well-conceived RFQs and RFPs will attract the very best design-builders available. If they fail to do so, they may miss the opportunity to work with the exact teams that could most effectively deliver their needs.

But the entire burden for project success does not lie with the owner alone. Once those solicitation documents are released, it is up to the design-build teams to properly prepare and strategize, not only to win the project but also to deliver the superior results that the owner wants and the team can be proud of. In the next chapter you will learn what it takes to do that.

CHAPTER HIGHLIGHTS

1. Every acquisition starts with the formation of a planning team charged with determining the best procurement approach. The team will establish the game plan that will govern how the solicitation, selection process, and eventual contract award will transpire.

2. There are eight stages in the acquisition planning pyramid. The first level above the foundation of the overall business plan is facility planning and program development. This data and information gathering phase is crucial to the success of the project.

3. The second step is to develop the project scope and definition. This requires a reconnaissance team with representatives from all levels of the organization charged with gathering specific information about the functional performance of the facility.

4. The third step is source selection, which begins with an RFP. The owner must then decide how they will procure, evaluate, and select their

design-build team. Three common competitive procurement choices are: 1) Low-bid Procurement, 2) Qualifications-Based Selection (QBS), and 3) Best Value Procurement.

5. In low-bid procurement the contractor submits a lump sum or guaranteed maximum price for a project and the contractor with the lowest bid wins the contract. When applied to design-build, low-bid requires design-builders to submit a partial design along with their price for construction.

6. In QBS, teams are selected based upon qualifications only and not on price. The solicitation instrument is the Request for Qualifications (RFQ). The design-build team responds to the owner's RFQ with a qualifications package. A hybrid of QBS is progressive design-build, which has a two-stage contract similar to CM-at-risk.

7. Best value procurement considers the technical solution and the price. An RFQ usually starts this process, followed by an RFP that should outline the procurement process and selection method, team proposal guidelines, and identify the deliverables to be submitted.

8. Contracts are awarded in design-build through a series of evaluations beginning with a review of the proposal by a panel or jury, possible interviews with the owner, and presentations of the project proposal to the panel by key members of the design-build team.

9. During this evaluation process, the owner uses a scoring system that shows the value the owner places on all the particular criteria of a project. Two common approaches are fixed price/best design and the use of a weighted criteria matrix. Other forms of evaluation are adjusted low-bid, meets criteria/low-bid, and equivalent design/low-bid.

10. One unique hybrid evaluation method is the project enhancement approach that starts with a significantly prescribed design in the RFP, which the team can use as a benchmark to establish the minimum basic program parameters. Enhancement criteria allow the team to compete to "enhance" the original RFP design. Scoring divides the overall project price by the point accumulation for the enhancements.

11. Private project owners have the option of employing teams through a negotiated selection process. This process allows the team to work with the owner from the earliest phases of programming and feasibility.

12. An RFP for design-build can have varying amounts of prescribed design: direct design-build (-10% to 10% design), design criteria design-build (5% to 25% design), or preliminary design design-build (20% to 35% design).

13. The four standard contract payment structures that can all be used on design-build are lump sum, cost-plus, cost-plus with guaranteed maximum price (GMP), or unit price.

14. One alternative structure that is successful is the target cost contract, which has three basic components: target cost, target profit, and sharing arrangement. The sharing arrangement involves an agreed-upon split ratio that will be applied to the cost savings and cost overruns.

15 Owners may offer award fees to further motivate high performance on the qualitative aspects of a project. The process starts with an agreed-upon amount set aside in the award fee pool. Next the terms of the evaluation process are negotiated. The parties also determine the intervals for assessment and the conditions of payment.

RECOMMENDED READING

:: ***Design-Build Contracting Handbook***, by Frank Cushman and Michael Loulakis, 2001, Aspen Publishers.

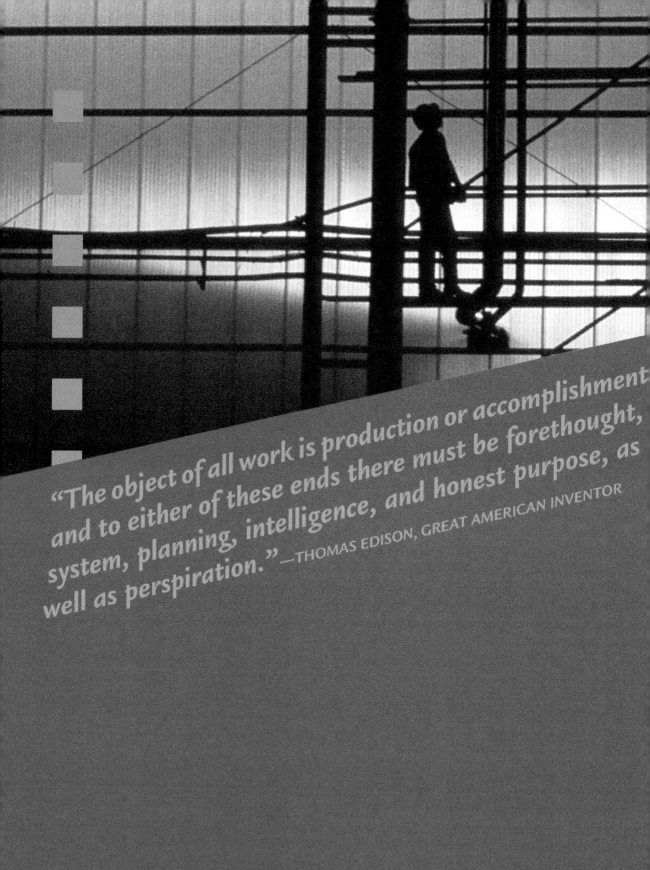

"The object of all work is production or accomplishment and to either of these ends there must be forethought, system, planning, intelligence, and honest purpose, as well as perspiration." —THOMAS EDISON, GREAT AMERICAN INVENTOR

5 GETTING READY TO COMPETE

Unless your work is primarily secured through the negotiated selection process or the direct design-build model, you will probably be competing for your design-build projects. By now you should understand that winning a project in design-build requires a lot more than just having the lowest price. The preparation and effort required to even begin competing is significant and must involve the entire team. Strategy and creativity are essential, with the scope of services involved often extending beyond the traditional services provided by a hard-bid contractor or the traditional architect or engineer.

In Chapter 4, you learned how the design-build procurement and solicitation process differs from other processes from the owner's perspective. Now it is time to consider these differences from the perspective of the contractors, architects, engineers, and subcontractors. Just as the owner should have a logical, organized plan before going forward with a project, a design-build team needs to approach a competition for a contract in a logical, organized fashion, so that it can produce the best possible response to the RFP and increase its odds of winning the contract award. Even though the short-listing process reduces the number of teams contending for the project, responding to an RFP can still be an expensive proposition, both in time and dollars. It only makes sense to do whatever you can to increase your chances of success.

In getting ready to compete in design-build, you first need to understand what you are getting yourself into and how the process works. Second, you need to learn as much as you can about the client and the project. Third, you need to pick the right people for your team. And finally, you need to make sure you have a good plan for developing the proposal and a strategy for winning.

■ KNOW THE OWNER

If you own a home or have ever dabbled in real estate at any level, you already know the three most important real estate rules: 1) location, 2) location, and 3) location. In design-build competitions, we have three similarly simple rules: 1) know the owner, 2) know the owner, and of course, 3) know the owner. This is one of the most important tips that I can share with you as you prepare to compete in design-build.

Unlike in a low-bid competition, where price is the only criterion and comparisons are made objectively on an "apples to apples" basis, in design-build you need to respond to many subjective criteria. In a best value, design-build competition, the owner must make sense of "apples to oranges to bananas" comparisons, with every design-build team interpreting and responding to the RFP differently. If your team happens to come up with an apples approach for an owner who prefers oranges, you may lose out even though your proposal met the overall performance requirement of being an "edible fruit." In other words, you better know going in that the client prefers oranges. Of course, the problem is that such idiosyncratic preferences are not necessarily discernable from the RFP alone.

Reading between the Lines

Clients communicate their preferences in many different ways, so you have to take your cues from multiple sources and learn to read between the lines. Sometimes you can learn the most from what is *not* being said. It is important that you find a way to get inside the world of your client and the end users that your client serves. You need to understand how they think, including their major concerns and greatest fears about the project. It may even be useful to learn about the client's personal interests and hobbies.

It is also important to do your homework when it comes to the entire evaluation jury. Who are these individuals? What part of the company or organization do they represent? What department or division are they accountable for? What are their biggest challenges, fears, or concerns? How do their views differ from those of the other jury members?

For instance, on a recent university project, a design-build firm was able to learn which universities (nine in total) the owners visited when researching ideas for their criteria documents. The design-builder spoke with the facilities and maintenance personnel at each of those nine universities and investigated the qualities they considered most desirable and undesirable in a new facility. All of this information was incorporated into the team's proposal. This information gave the design-builder a better understanding of what the owner was looking for. Just as important, in presenting their findings to the client, the design-build team demonstrated its initiative and determination to understand the entire project. The design-build team also investigated the primary accountabilities for each of the voting members on the selection jury and directed specific portions of their findings to those individuals. You might not be surprised to learn that this particular design-build firm won the project.

In an ideal world, you would always be able to undertake this type of preparation when competing in design-build. However, you will not usually have significant access to information about the client until after the contract award. Thus, vital information might be hard to get. So where do you start?

You start with the RFP itself. A well-written RFP should provide you with all of the pertinent project information as well as a clear scope definition, and program and performance expectations. However, it will not provide you with the more in-depth, client-centered information that you need in order to be as responsive and competitive as possible. This takes some behind-the-scenes investigating. So as you prepare to compete, you should investigate other sources of information that will provide some insight regarding the client.

Most every company or organization today has a website that reveals an assortment of useful information such as its history, corporate or organizational vision, mission, and core values. The website might also tell you about the way the company is structured, how it does business and the names of some clients. Get your hands on the company's written literature and promotional brochures as well, including copies of its annual reports, if possible. Try to talk with the people that the company does business with—its clients, customers, and end users—to learn more about the company's culture, and the particular challenges the company faces. You may also discover something about your client by learning about the types of community or philanthropic activities the company is involved in. The bottom line is this: in order to add value to your project proposal beyond what the client is asking for, you must have more information than is readily available in the standard RFP. You must *know* your owners and their businesses.

> As discussed in Chapter 4, we always hope that the client has taken the time to gather information about the interests and goals of his or her own organization. But you cannot be sure this has happened, and you probably will not know until after you get the project and can make some inquiries inside the organization yourself.

Your ultimate goal as a design-builder is to demonstrate how you intend to become a **business enabler** for your client through your design-build proposal. In other words, you need to show the client how your proposed solution will help them more effectively achieve their goals—whatever they may be. If you are competing in design-build and plan to deliver just the bricks and sticks or concrete and steel of the project, then your chances of winning may be minimized—especially if you are competing against an experienced design-builder.

Some companies choose to strategically target certain projects well in advance of the actual solicitation and get this part of the investigative procedure out of the way early. Obviously, the time and resources that you put into this investigative effort must be weighed against your desire for the project and the rewards associated with it.

D-B DICTIONARY

Business enabler – An individual or a team whose contribution helps clients improve their overall business performance by helping them achieve their goals and objectives in a more efficient manner.

Preemptive Marketing

Experienced design-builders understand that preparation and planning are the keys to winning the most desirable projects. Most design and construction firms have people within their companies whose primary focus is pursuing and securing new work. These folks typically operate under the banner of "Business Development," "Sales and Marketing," or

Jill Wilson
Vice President, Marketing
Gray Construction
Lexington, Kentucky

We make every effort to know as much as we can about our new customers and also keep up-to-date with key existing customers. We have various team members on board who are strategic thinkers when it comes to working with our customers on how they run their businesses. We often offer process improvements that make their production, or whatever they are doing, run more efficiently and effectively. These individuals are not just on our design side—they are also project managers and site managers. We've learned that if we can bring "extra" value to a project, then we truly have a customer-for-life. Design-build allows you to do that every time.

> The competitive portion of the design-build process is very different from design-bid-build; not everyone is up to the task. You might be thinking, "Why would I want to go through all of this trouble just to go after a project where I have to carry so much of the risks?" Or you might be thinking, "Wow, I actually get to use my initiative and creativity to position myself to win this project, and to gain the insight I need to manage and control the risks that I assume."

similar titles. Although it is not unusual, even in the design-bid-build world, for these individuals to be tracking a project a year or two before it goes to market, a number of design-builders today are taking the concept of "know the owner" to a whole new level. These innovators are doing the investigative portion of the preparation long in advance of the RFP ever hitting the streets.

Some design-builders specifically target the clients they want to work for–whether these are new potential clients, or clients with whom they already have a relationship. Once they have identified these individual firms, agencies, or organizations, they intentionally set out to learn everything they can about them, their project goals, and their plans for future expansion. It is not unusual for organizations to have a long-range development plan that is publicly accessible or available by request. You can bet that a successful design-build business development team will be extremely familiar with the development plans of all current and potential clients.

However, the strategic design-build business development team does not stop there. Knowing that the projects that they are most interested in will not actually be procured for a number of years actually gives them significant time and opportunity to research the owner's needs, interests, and desires at a very deep level, well in advance of the RFP development or release. I refer to this preparation strategy as **preemptive marketing**. When conducting preemptive marketing, the design-build team intentionally hones its expertise for a specific project, with a specific client in mind, long before there even is a project. Among other things, a team engaged in preemptive marketing should:

:: Initiate a series of **information interviews** with the potential client to gain as much insight as possible about the future project.
:: Visit similar projects that exemplify state-of-the art facilities around the country.
:: Find out in advance what has worked and what has not worked on the same type of projects in terms of design and function, as well as overall construction approach, materials, methods, etc.

:: Speak to other owners, end users, maintenance personnel, and other knowledgeable individuals associated with similar projects to ask their opinions about what improvements they would make if designing or building the facility all over again.

Obviously, such an effort takes a commitment of both time and money. Not every design-builder can go to such lengths in pursuit of preferred projects and clients. However, it is important for you to know that some can and will. They are planning and preparing specifically to out-perform their competitors in both the proposal and project delivery stage. They have no intentions of going through the motions of a design-build competition without walking away the winner. As you prepare for your own design-build pursuits, that is something to keep in mind.

Procurement Workshops

Design-builders are not the only ones getting a jump-start on preproposal planning. Some savvy owners conduct **procurement workshops**, open to all interested design-build contractors, to discuss potential projects before the RFP ever comes online. In other words, they openly give design-builders an opportunity to get to know them better. Such owners sometimes conduct informal "public design-build brainstorming" sessions around upcoming projects to discuss their ideas about the project, as well as special challenges, constraints, risks, potential solutions, and other considerations. In this open-dialogue arena, the owners get valuable feedback from the design-builders regarding their ideas about the potential project and concerns regarding the owner's approach to the procurement process. No one shares secrets at this early stage, when neither RFP documents nor drawings exist. At this point, it is just a conversation about potential projects and the owner's goals for these projects.

Instead of this somewhat informal, conversational way to tap into the expertise of the design-builder community, some owners prefer the formality of releasing a draft RFP to their short-listed design-builders before the final RFP is issued. In doing so, the owner is soliciting design-builder comments and suggestions regarding the procurement and the project. These comments can be delivered in a written format or in private or public meetings as requested by the owner. Again, the owner is not soliciting solutions at this stage, but is instead simply asking for feedback from the designer-builder perspective. The owners sift through the information gleaned through this process and make appropriate adjustments to the final RFP before releasing it to the short-listed proposers. This proactive approach on the owner's part helps alleviate everyone's concerns and usually ends up saving all parties time, money, and anxiety.

If handled well, this pre-RFP dialogue does not hinder competition. Just the opposite is true. It encourages the better design-builders, who appreciate the opportunity to learn enough to influence the RFP process. Their involvement, in turn, helps head off issues that would keep them from competing for the project at all. Although you initially might think

---

D-B DICTIONARY

Preemptive Marketing – An intentional and directed effort to target specific projects and clients in advance of an RFP release or procurement effort on the part of the client. The directed effort entails specific project preparation and due diligence activities to gain information, knowledge, and insight about the client and the project. Some companies refer to this as "pre-winning a project."

Information Interviews – Conversations conducted in an interview format, with the goal of gaining specific insight, knowledge, and understanding.

Procurement Workshop – An organized (usually public) briefing session in advance of an official RFP release, where design-builders are invited by the owner to voice their opinions about an upcoming project.

TAKEAWAY TIPS

Surrogate Owner

Consider conducting information interviews with a number of surrogate owners of similar businesses to gain insight into your client's business and their project expectations. In a competitive RFP process, you cannot get access to the real owner until after the competition is over and you have won the contract. But that should not keep you from learning everything you can about your client's major concerns and special challenges.

D-B DICTIONARY

Surrogate Owner – A substitute or alternate owner who is in the same or similar business as a project owner; business development teams sometimes interview surrogate owners in order to gain information and insight into the needs, goals, and objectives of their actual client.

---

these techniques give one design-builder an unfair advantage over another, several owners who have used these procedures report very good results and favorable comments from the design-builder community.

Show Off Your Best Qualities

One of the beauties of design-build is that it allows you to compete on something other than price. Simply being the low bidder will not be sufficient. In a best-value, performance-oriented competition, you have an opportunity to "strut your stuff," so to speak. In fact, you *must* if you want to be successful. In his book, *Work Like You are Showing Off*, Joe Calloway puts it this way:

> *Showing off is about bringing the best you have to any situation. It is about excellence and exceeding expectations. Showing off is about the joy, jazz, and kick of being better tomorrow than you were today. Showing off is a mindset that leads to success.*[1]

The effort required to win design-build projects is far greater than in design-bid-build. In design-build you have to bring your best ideas, creativity, innovation, experience, and skills to bear on every aspect of the project—the design, the overall approach, the construction means and methods, the phasing, the buy-out, the schedule, the quality control plan, the safety program, and any other areas where value may be enhanced. On top of that, you will need a good dose of energy, passion, and enthusiasm.

AUTHOR EDITORIAL

MEDIOCRE NO MORE...

Early in my career, I spent about ten years working in the "hard bid" environment, where, essentially, the contractor who made the biggest (unintentional) error in his or her initial estimate won the job. The winning contractor would then be forced to find flaws and inconsistencies in the architect's plans and specs. They did this so they could mitigate their potential loss by issuing change orders to remedy these flaws after they got the contract award. In fact, finding flaws was not hard, because the architects were also being forced to reduce their fees, which, in turn, often resulted in poor quality construction documents. To make matters worse, once you got started on the project, you were likely to find that your "low-bid" subcontractors ended up getting the job the same way you did.

So there we all were—fighting, usually with each other, for our economic lives. After the end of a project, many of us would end up feeling emotionally and professionally bruised. As a result, we were often less than enthusiastic about the project in which we had just invested months and sometimes even years of blood, sweat, and tears. It is a terrible feeling to give everything you have to a project that ultimately satisfies no one—not you, not the architect, not the end user and certainly not the owner. You find yourself going home each evening a little more exhausted than you did the day before, and going to work each day with a little less energy.

I will never forget being on the roof of one of my "low price" design-bid-build projects, and hearing my roofing subcontractor sighing deeply as he installed some flashing. When I asked what the problem was, he said "This flashing detail is going to leak at the first rain." I asked him why he was installing it anyway. He explained that he was following the detail shown on the plans. He said he had already mentioned the problem to the owner's representative, but was told to build it the way the architect designed it—after all that is what architects get paid for. I then asked if he could recommend a better approach, one that wouldn't leak. He proceeded to recommend a way to make the joint watertight.

At that point, I instructed the roofer to stop what he was doing. I immediately called the architect and explained the situation. I was stunned by the architect's response. He was immediately skeptical, saying something like: "Why are you recommending this change and what's in it for you?" He suggested that I was just looking for ways to make him look bad and put more money in my pocket. When I suggested that we get the owner out on site right away and figure out a way to fix the problem, he said no. Instead, I was instructed to install the flashing detail as designed.

For the life of me, I could not get him to see that I just wanted to do a better job; I wanted to build a detail that would not to leak. You would think he would want to thank me, but instead he offered only doubt and distrust. After all, I was a contractor, and you can't trust those contractors. I don't know about you, but as a young construction manager, it wasn't my goal to go out into the world and do my mediocre best. I doubt that is the goal of the architects and engineers that work in the industry either. And I have yet to meet an owner who wants to build average facilities that cost too much and do not really work. But my experience with the roofing subcontractor told me that mediocrity was exactly what owners risk getting when they depend on the low-bid, adversarial, design-bid-build approach for delivering their projects.

I tired of trying to convince owners and architects that I truly cared about their projects and building things right the first time. I realized that, if I could compete on the basis of something other than price, I could prove that to them. Design-build allowed me to do just that. I can report that it is a completely different experience when you are able to bring your best contribution to the table for the benefit of the project and satisfaction of everyone involved.

SELECTING THE DESIGN-BUILD TEAM

One of my favorite quotes comes from the movie *Miracle,* which dramatizes the 1980 United States hockey team's triumphant Olympic victory against the Soviet Union. In one scene, Coach Herb Brooks is trying to select the final members of the 1980 Olympic team. As Brooks is making the cuts, his assistant coach becomes concerned when he bypasses some

> **D-B DICTIONARY**
>
> **Signature Architect** – An architect of renowned stature with significant name.
>
> **Proposal Response Team** – The group of contractors, designers, and consultants who put together the proposal package in response to a project RFP. These same individuals usually stay with the project after contract award if they win the job.
>
> **Project Director (Project Executive)** – A person, usually from the business development or sales side of the organization, who selects the design-build projects that the firm will pursue. He or she usually initiates the proposal development process and continues to participate in a leadership role on the project after contract award.

of the top star players in favor of some of the less talented players. Brooks explains his choices to his assistant coach by declaring, "I don't need the best players, I need the right players."[2]

Whether it is a hockey team or a design-build team, you need the right players. And quite frankly, some of the personality traits that go along with "being a star" do not always improve the performance of a team. In the end, a good team will most likely make a better decision than an individual. The Collaborative Process Institute put it this way:

> *Teams make better choices than individuals. Teams, which are diverse groups of individuals organized for a common purpose, are good at achieving optimal outcomes for three reasons. First, they enlarge the set of possible solutions; second, they have more capabilities than one individual; third, they are more likely to identify the best solution from among the possible options. Creative thinking and extraordinary outcomes are more likely to occur in a team setting.*[3]

Of course you want to put the most talented people you can find on your team. But you must consider other factors as well.

One of the most important factors to consider during the planning and preparation phase of the project pursuit is that not everyone in your organization may be cut out for design-build—especially at the beginning of the process. A traditional plan and spec contractor will have a hard time responding to a performance-based RFP. Such a contractor is not accustomed to being involved in the design process, having to develop the design, and pricing a project before the design is complete. On the design side, a talented and renowned **signature architect** is likely to have a hard time accepting the ideas of other designers, let alone the ideas of a contractor.

It is a risky proposition to put the wrong people on your **proposal response team** and even riskier having them on your project after the contract award. (Chapter 7 will discuss this in more detail.) The bottom line is that not all contractors, engineers, or architects are suited to this kind of competition; they will be frustrated by the process, and may even obstruct the team's performance. On the other hand, some contractors, engineers, and architects are eager for the opportunity to compete on a best-value basis. As a **Project Director**, you must discern who should be on your design-build team and who should not. By selecting the best-suited individuals from both inside and outside your organization, you will increase your odds of winning the job. (Chapter 8 discusses the actual mechanics of a functioning team.)

Proper team selection is a critical step on your way to successful design-build. Unfortunately, not all design-build companies have given this vital task the attention it deserves. It is much easier to simply take whoever is available to work on the proposal and put them together to create the team. This happens more than you might imagine. But keep in mind that most RFPs require that the key people from the proposal team (such as the project manager, lead designer, estimator, project engineer,

Crunching the Numbers

When I came into the work force in 2005, I knew sustainability was quickly becoming the norm rather than the exception, so I began studying for the LEED Accredited Professional (LEED AP) exam. At the time, my superiors questioned the validity of the accreditation, and seemed almost to scoff, as if I just wanted more badges on my sash. But after I passed the LEED exam in 2006, I became only the fifth LEED AP in the company, and one of the first in my work division. Then, in early 2007, we were competing on a design-build LEED Certified Student Housing project and I was placed on the team as the LEED AP for the project. In addition to being responsible for design and construction LEED credits, I was also assigned to serve as one of two lead designers on the team. This was an extraordinary opportunity for me.

When I first decided that I wanted to be a design-builder, my goal was to work on the design of a project and then to follow that design into the field and watch it become a reality. As the design phase of the project came to a close, the only missing piece for me was the construction experience. So I leveraged my intimate knowledge of the design, along with my LEED AP status, to press for an opportunity to be assigned to the field in a construction role. My request was granted. I was one of three assistant project managers on the job. It was exciting and very scary at the same time. All kinds of questions went through my head. "What if I am not as productive in construction as I am in design? What if I am the weakest link on that side of the team? What if I don't speak enough "construction" and look like an idiot?"

I can tell you, it is quite humbling when you get to see firsthand how much of what you design is not actually buildable. But it is also an awesome learning experience. In the end, I think the construction guys respected the fact that I was willing to take the criticism and jump in and get my hands dirty. At the same time, they learned the advantage of having a design-minded professional on the construction side of the team, and the time savings and other benefits that can bring.

Through this experience, my confidence in my abilities has grown. I never expected to have such opportunities only three and a half years out of school. It is amazing how one seemingly small training request, such as becoming LEED AP, can open a door a crack. Once that happens, you can pry it further and further every day. Today, there are over eighty individuals within my company who have achieved LEED AP status, and it would not take much effort at all to convince a manager that pursuing such a credential would be valuable to the company. But back when I originally wanted to pursue this goal, I would have never been able to achieve it without being persistent. Since then I have attended the DBIA Designation Boot Camp and achieved my DBIA Designated Design-Build Professional status—the result of another hard-fought request. I have also earned my architecture license.

The bottom line is this: you have to generate your own career. You cannot count on anyone else to figure it out for you. If you want to be a true design-builder, you must seek training across various disciplines, in whatever way you can. You will not learn it by osmosis. You need to jump in there with both feet, humbling yourself, acknowledging that you don't know it all, and being willing to take criticism, learn from it and move ahead. You have to be somewhat fearless and, although your managers may come to think of you as a pain in the neck, that's okay as long as you continue to learn, and grow, and contribute. After all, if you are going to be so daring as to demand company time and money to pursue accreditations and licensure, you darn well better be ready to come up with decisions, ideas, and problem-solving abilities that prove the investment was worthwhile.

MENTORING MOMENT

Jordan Moffett, AIA, DBIA, LEED AP
Architect, Assistant Project Manager
The Haskell Company
Jacksonville, Florida
Education: California Polytechnic State University, San Luis Obispo, graduated 2005
Major: Architecture with Construction Management minor

Project directors are often at the vice president level within a company. They serve as part of the business development or sales side of the organization, often making first contact with potential clients. Some project directors stay closely connected to the project from proposal to project closeout, while others only participate in the solicitation and proposal portion, handing off the project to a project manager after contract award. It varies from company to company.

superintendent, key consultants and specialty contractors) form the core personnel on the project team.

These are just some of the characteristics to look for when putting together your design-build team:

- Experience and technical expertise
- Initiative, creativity and innovation
- Personality, attitude, and commitment
- Adaptability, responsiveness, and discipline
- Communication, collaboration, and trustworthiness
- Sensitivity to design, cost, and schedule
- Platform and presentation skills
- Leadership

Neil Johnson, LEED AP
Team Leader
Ryan Companies US, Inc.
Minneapolis, Minnesota

By using Design Build delivery, we find efficiencies in many areas; our sub-contractors help us do that. We engage several of our design-build subcontractors to help us select products that fit our predetermined budget and meet a performance-based criteria, thus giving us the quality we require for the budget we can afford. It is important to note that subcontract awards are awarded to the best-value bidder in lieu of the low-price bidder. Quality, budget, and schedule are all factors in making an award decision. Sometimes all three criteria can be provided by the low bidder, but it is not uncommon to award to the second or third bidder because they are providing the best value for the budget.

Blending Cultures

D-B DICTIONARY

Cultural Fit – Compatibility between the personality of firms or individuals working together. Compatibility is essential in terms of business practices, attitudes, and behaviors.

Team Alignment – The compatibility of personal or corporate values, beliefs, and principles, as well as project goals and objectives.

Unless you are a multi-disciplined integrated company, capable of delivering both the design and the construction for the project, you will have to look outside your own firm to complete your team. Even an integrated firm may have to look beyond its own walls to find people with the right skills for some projects.

As you might expect, selecting such outside partners is one of the biggest challenges you will face in creating the right design-build team. When shopping around, you will of course verify capabilities and reputation, but it is also critically important to check for **cultural fit** (personal compatibility) and **team alignment** (the compatibility of personal or corporate values).

Cultural fit refers to the personalities of the two firms. Some firms are very structured, while others are more relaxed. Significant differences will cause trouble down the road. Likewise, it is important that the two firms, share similar values, principles, and beliefs. They must fundamentally agree on the overall project goals and objectives. (See Figure 5.1.)

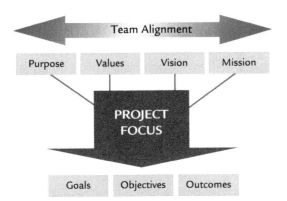

Figure 5.1
Team alignment

Team alignment across these core issues will either make or break the relationship.

Checking for compatibility and alignment does not stop at the company level. It is also important that you select the right individuals from within the firm you partner with. Quoting again from the Collaborative Process Institute:

> In building a team, pre-qualify firms and select people. It is tempting to select team members by picking the organization on the basis of size, experience, financial strength, fee, etc. However, a team is not made up of organizations; it is made up of people. While some organizations foster collaboration more than others, personal chemistry, individual capabilities, and teamwork skills are, in general, better determinants of team performance than organizational factors.[4]

One of the fastest ways to get a read on cultural fit and team alignment is to develop a **teaming profile**, or formal team assessment, before signing any **teaming agreements** that define roles, expectations, and responsibilities. This is especially true if firms and individuals are working together for the first time. Once sufficient alignment is reached and the decision is made to partner with a firm, the fun begins. The team should establish its project goals and objectives, set its targets, aim to win, and fire away.

Although integrated design-build firms may benefit from having their contractors and designers working under the same roof, they still face challenges when it comes to team alignment. Fundamentally, contractors and architects think differently and view the various aspects of project performance from different perspectives. Thus, it is important to keep in mind that differing perspectives do not necessarily mean opposing values.

For example, a common perception among architects is that, because contractors value budgets and schedules, they do not care about design. The same can be said for contractors, who believe that

---

D-B DICTIONARY

Teaming Profile – A formal assessment of a team that focuses on compatibility among members, team alignment, and cultural fit. Can include both individual and team assessments. The process of creating a teaming profile is often accompanied by a proposal briefing session.

Teaming Agreement – A vehicle for structuring the relationship between the various team members associated with a design-build project. The agreement establishes and defines the roles, responsibilities, and expected performance of the team members.

architects do not care about project budgets. The entire team *must* care about design, budgets, schedules, and quality, in a holistic, comprehensive fashion. Our traditional, segregated roles have forced us into "my part/your part" dichotomy that has no place in design-build. Team alignment is only possible among team members who value multiple perspectives.

Peter Beck
Managing Director
The Beck Group
Dallas, Texas

Even in an integrated firm, cultural friction occurs whenever you put one or the other discipline in charge of a project. But one person must be in charge of the entire design-build process, and this can be difficult. Architects don't want to relinquish the ultimate design responsibility to a contractor, and contractors don't want to relinquish ultimate cost responsibility to an architect. Many people who have practiced in a traditional capacity for fifteen or twenty years, whether it be construction or architecture, are going to place the importance of their discipline much higher than integration—at least until they realize just how rewarding the results of that integration can be. If it is difficult in an integrated firm, just think about what has to happen when you are talking about two firms that are working together for the first time.

TAKEAWAY TIPS

Mixing and Matching

Consider gathering information regarding your employee's personal interests and hobbies, on a volunteer basis of course, along with professional credentials and experience and then creating a database that can be linked to project pursuits. You certainly want to match your most technically qualified personnel to your projects, but at the same time, if you can also match personal interests to the project, you will improve your chances of winning. For example, people who have interests in water activities will bring an enthusiasm to a marine project that someone without that perspective might lack. Someone who collects art or is an artist herself will have a natural connection with a museum owner. An employee who plays in a local orchestra on weekends will bring a passion to a concert hall project that someone who lacks an interest in music might not be able to equal. You can't always play matchmaker in this way, but when a person's technical experience and personal interest coincide, the spontaneous enthusiasm and connection to the client is almost tangible.

Stepping Outside Your Comfort Zone

For some people, the notion of contractors, architects, engineers, and owners working side-by-side is foreign. As a matter of fact, some teams never gel, and therefore never take advantage of the synergy that arises when you approach a project from multiple perspectives in an integrated fashion. It is not uncommon to encounter design-build teams whose members relate to each other as if they were involved in a traditional design-bid-build project, with little or no engagement between design and construction. On such teams, the individuals continue to operate from their segregated services orientations. This is unfortunate and potentially quite risky. However, it can be difficult to suddenly set aside years of traditional, often adversarial, behavior, and adopt an integrated, fully cooperative approach instead. You may have to go way outside your comfort zone in order to put together the right team, with the skills and talents needed to win the competition and deliver the project. But by doing so, you will significantly reduce your risks and take the necessary steps to assure project success. And although it might be awkward, at first, to sort through the differing perspectives, personalities, and egos at work on a design-build team, the effort may very well result in an award-winning, breakthrough project for you and your team. For example, see the case study "Dancing on the Edge."

GO OR NO GO

It is best if you can track a project over a long period of time and plan your strategy for winning it, but in many cases you will not learn about the prospective project until the owner releases the RFP. Only then will you begin to weigh your interest and start to evaluate the opportunity that lies before you. From the beginning it is important to understand that the quality of the RFP and your ability to deliver a quality project in accordance with the owner's expectations are directly related.

Unfortunately, not all RFPs are created equal. If you are a design-build contractor, architect, engineer, or specialty contractor you better be able to recognize a good RFP when you see one—and more importantly, you better be able to recognize a bad one.

In a design-build competition, design is often a primary factor in determining who wins the project. Even when the owner has made other factors the dominant selection criteria, the design will always influence the final decision. Good design will definitely increase your odds of winning, but, at the same time, that design must be developed in accordance with a budget and schedule. It is important to partner with designers and consultants who understand the design-build process; otherwise the learning curve can be brutal for the whole team, adding unnecessary risk to the pursuit.

The main difficulty that public-sector owners have with the design-build delivery process is their need to define what they want in terms of scope, quality, and performance requirements without having the benefit of going through a full design period. However, it is important to understand that the quality of the response from design-build contractors will be directly proportionate to the quality of the RFP; a good RFP should always clearly communicate what is required from the proposers, and the format in which it should be provided. The RFP should also have a clear and defined process for proposer inquiries and owner responses. It is also very important that the client defines what flexibility the proposers have in providing their technical solution; what is allowed and what is not acceptable and to give proposers adequate time to respond and prepare their proposal. A copy of the final design-build contract should be included in the RFP, with a fair risk allocation and reasonable liabilities. And finally, the RFP should clearly communicate how the proposal evaluation process will be conducted, and what selection system will be used. Also, I highly recommend that the RFP provides stipends of up to 0.2% of estimated project cost to the unsuccessful firms, especially if it requires a significant amount of engineering and design effort to provide technical solutions. This is the best way to attract the very best design-build firms.

Chuck Williams, PE, DBIA, PMP
President
ODC Synergy
Overland Park, Kansas

CASE STUDY
DANCING ON THE EDGE

THE FACTS
Project Name:
Caltrans (California Department of Transportation) District 7 Headquarters
Project Type:
Office Complex
Location:
Los Angeles, California
Description:
1.2 million square-foot complex consisting of a thirteen-story tower with 756,000 square feet of office space, along with a cafeteria, daycare center, wellness center, credit union and four levels of subterranean parking for 1,142 cars.
Cost:
$172.3 million
Duration:
Twenty-nine months
Completion Date:
Phased completion, August 2004 to December 2004

THE TEAM
Owner:
State of California
Design Build Entity:
Main and First Design Build Associates Inc.
Executive & Design Architect:
Morphosis Architects
Contractor Managing Partner:
Clark Construction

Victor Muschetto Photography Courtesy of Clark Construction Group

THE LEAD-IN

Design-build is like dancing. It requires a great deal of flexibility, adaptive collaboration, and some fancy footwork. The ultimate goal is to accommodate the needs and interests of an entire team of stakeholders and project participants. So what happens when the design-build dance partners are an edgy and progressive architectural firm known for its innovative, iconic buildings and urban environments, and a century-old, conservative general contractor? Well, in design-build, when the entire team is committed to making the process work, such a match-up can mean magic—even in the face of extraordinary challenges.

THE STORY

In the mid-1990s, the State of California adopted design-build legislation to allow for a best-value selection process instead of the traditional design-bid-build process. While some California state projects were procured under a bridging format, the California Department of Transportation (Caltrans) project was procured under a pure design-build method, thereby transferring all design and engineering risk (as well as code compliance, jurisdictional approvals, program compliance, etc.) to the design-build entity. In transferring such significant risks on such an important project, the state demonstrated its faith that the design-build team would handle these risks fairly, equitably, and responsibly. To ensure that it got its money's worth, the state set up a competitive selection process that included proposal enhancements designed to encourage teams to offer even more than was required by the project's RFP documents.

At the time of the design-build solicitation for the new Caltrans District 7 Headquarters, the State of California was experimenting with a new "Design Excellence" initiative. This was part of a statewide effort to see whether design-build could achieve what design-bid-build typically failed to produce—aesthetically interesting buildings that are completed quickly and with modest budgets. The Caltrans project was the largest project to be undertaken under this initiative. After an initial Request for Qualifications (RFQ) process, three design-build teams were short-listed to compete in a three-month RFP process during the fall of 2001.

The Design Excellence initiative spurred interest from renowned architects around the globe. The architects represented on the three short-listed teams included Thom Mayne from Morphosis Architects, Rem Koolhaas who founded the Office of Metropolitan Architecture, and EMBT Architects, the firm founded by the Spanish architects Enric Miralles and Benedetta Tagliabue. In the end, it was the design-build team of Main and First Design/Build Associates Inc., consisting of Clark Construction and Morphosis Architects, that won the contract. It was time for the "dance" to begin.

This was Thom Mayne's first design-build project. He was initially skeptical; not surprisingly, the initial efforts at "dancing" were awkward. Clark Construction, which was at risk for both the design and the construction, immediately tried to impress upon the renowned architect that Clark would control the process. As you might guess, this did not sit well with Thom Mayne, who insisted from day one that he would never put his design pencil down until the project was complete.

It did not take long to figure out that arguing about who was going to lead this dance was not productive. The project leader for Clark, Marc Kersey, decided that the best way to manage risks and make this unique relationship work was to take a different tact. So rather than wasting time with posturing and positioning, he decided to move the Clark team into Morphosis's design studio, lock, stock, and barrel. By spending everyday elbow-to-elbow with the architects, his construction team could provide immediate cost and schedule information throughout the design process. A relationship that started off on shaky ground soon transformed into a trusting, synergistic partnership that paid big dividends in the end. Thom Mayne would later say that "the design-build relationship produces more transparency between the design and the building processes, making it possible for the architect to have a more integrated role in the daily decision-making."

The scale of the project was a challenge in itself. It required packing 1.2 million square feet of office space and parking facilities into a 140,000-square-foot site. Given these constraints, combined with the project's modest initial budget of $165 million, it seemed that the only option was a conventional office block. Instead, to force the state to evaluate how highly it valued architecture, Morphosis presented Caltrans with a basic design package with several potential upgrades-much like when buying a new car. Ultimately, the Main and First team employed

THE TEAM (continued)

Associate Architect:
Gruen Associates

Developer:
Urban Partners

Structural Engineer:
John A. Martin & Associates

Mechanical Electrical Engineer:
Arup

Specialty Contractors:
Model Glass, Raymond Interiors, Herrick Corporation, Conco Cement, Dynalectric, Western Air

Specialty Consultants:
Shoring Engineers

Victor Muschetto Photography Courtesy of Clark Construction Group

several innovative techniques to construct a high profile building while saving the state time and money. The design-build team turned the structure into a technological marvel and demonstrated the best qualities of design-build: its high tolerance for experimentation, its valuation of collaboration and its freedom to move quickly.

The heart of the project is an urban lobby, which is carved right through the building's core, linking the plaza to the main entrance. The lobby ends at a glass-enclosed cafeteria. A metal bridge spanning the atrium connects the offices with natural light spilling down from a light well. This outdoor space is highlighted by a four-story interactive and integrated neon art sculpture designed collaboratively between Morphosis and New York Artist Keith Sonnier, who was a selected member of the design-build team from the outset.

To keep the project on schedule and within budget, the team incorporated a relatively straightforward primary structural system. The basic diagram, with its L-shaped footprint, was very efficient, leaving room in the budget for some of the buildings unique features, such as a 26-foot cantilever above level six. Stretching over a sidewalk on the north side of the building, the cantilever resembles a freeway overpass structure, thus serving as a visible representation of the work done in the Caltrans building. However, the most striking design feature of the building is a skin of perforated horizontal panels that act like mechanical eyelids. This effect is magical at night, when the building dissolves into a slab of glass. These scrim walls are supported by a curtain wall mullion system and are not weatherproof. A one-foot space between the scrim wall and the building's exterior wall acts as a chimney to move heat away from the building.

The team remained quite flexible throughout design and construction, making some significant last-minute changes to the exterior of the building. For example, after visiting the Massachusetts Institute of

Victor Muschetto Photography Courtesy of Clark Construction Group

Technology, Thom Mayne saw a rubber coating called Sarnafil that is traditionally used as a roofing membrane that he wanted to use on the walls of the new building to make them more waterproof. He liked the material, because it would ultimately leak and crack less, and also provide better seismic performance than the originally planned plaster wall. This change required a redesign of the building's slab edges. Without the flexibility of the entire team, this modification would have been problematic so late in the project's development. Wrapping the entire building in this membrane was the ultimate win-win approach: saving time and money, while improving the quality and longevity of the exterior wall system for the mutual benefit of all involved.

In addition to some of the architectural complexities, the design-build team faced a number of other project challenges. At the project's start, access to occupied buildings along the northern frontage of the site were not secured until months into the project, requiring the excavation operations to be sequenced around existing structures. The design-build team re-sequenced these activities without affecting the budget or schedule. Later, towards the end of the design phase, a major redistribution of the tenant groups resulted in a restacking of the tenant floors. Although well into construction, this modification was completed, again without affecting the schedule or budget. Due to the extreme efficiency of the design, the state actually had one entire floor of extra space which it could either use or lease for additional revenue.

Critics of design-build point to quality assurance as the potential weak point in the delivery method. The "fox watching the hen house" metaphor is often tossed around. From the outset of this project, Main and First employed independent third party consultants and testing agencies to perform all contractually required tests and inspections, and also to peer review design and engineering elements throughout

Victor Muschetto Photography Courtesy of Clark Construction Group

the entire process. A QA/QC manager oversaw these requirements and ensured compliance with design criteria and programming, building codes, ADA ordinances, testing criteria, fire-marshal approvals, and the like. However, more importantly, Clark, Morphosis, the consultants, and key subcontractors employed several layers of their own quality assurance throughout the project, realizing the ultimate benefit of managing risk through these means. As an example, no fewer than 50 models of the building were prepared during the design phase, one of which was chamber tested for wind effects. Also, three-dimensional design and engineering was employed to make certain all elements would fit within their designated spaces. Perhaps the most influential quality control assessment was housing the construction project management staff and the design and engineering professionals in the design studio at the beginning of the project; this ensured that real-time constructability input was provided from the early conceptual-design stage all the way through fully developed construction documents.

Some of the more extensive quality assurance measures were taken on the exterior wall systems. No one had ever wrapped a structure in Sarnafil roofing membrane before, so extensive study and research went into each and every component of the exterior wall as the design progressed. Many tests were performed to confirm quality and function. Five exterior wall mock-ups were built, including one full-scale three-story wall that was subjected to wind, rain, and seismic movement. Leak tests were performed on the actual installed exterior systems as construction continued. Just after the final completion in December 2004, Los Angeles experienced record rainfall through the 1st quarter of 2005. Thanks to the design-build team's extensive quality assurance efforts, not one leak has been detected in the entire skin system.

THE CLOSE

Caltrans District 7 Headquarters is an exceptionally successful design-build project because of its contribution to the civic environment, relatively low-cost, architectural innovation and aesthetic power. The project was awarded the Leadership in Energy and Environmental Design (LEED) Silver certification status from the United States Green Building Council. It stands as a monument to what can happen when the right dance partners commit themselves to collaboration. As this project proves, the results can be extraordinary. The substantial overlap of the design phase and the construction phase sets this project apart from other significant state projects. From essentially a standing start, the 1.2 million-square-foot facility was designed and built in just twenty-nine months.

From the Owner:

Thanks to the team's ingenuity and well-managed design-build delivery system, this facility offers the State of California the highest degree of architecture and innovation for the dollar of any similar facility in the state's inventory. This beautiful new facility will be enjoyed by Caltrans, other building tenants, and the city of Los Angeles for decades to come.

Brian Day

State of California

Project Director

[Case study adapted from Clark Construction's submission for the Design Build Institute of America 2005 Design-Build Award Competition, by permission.]

Clearly, it is important to know what you are getting yourself into before you spend weeks of time and many dollars preparing a proposal. In Chapter 4, you learned of the effort that goes into preparing for a design-build procurement. The better you understand what goes into developing a quality RFP, the better you will be able to recognize the flaws, holes, and gaps within it. This allows you to identify, anticipate, analyze, and mitigate the risks associated with the procurement and the project. Once you are aware of the inherent risks, and have conducted enough investigative research to feel comfortable going forward, it is time to begin the formal project pursuit process.

Obviously, you are not going to go after every RFP that comes across your desk; so first you need to make the **go-or-no-go decision**, (also referred to as the **pass or play decision**). You cannot afford to waste your time going after a project that is not worth having. It is

> **D-B DICTIONARY**
>
> Go-or-no-go Decision – An analytical decision about whether to pursue a project or not based upon information provided in the RFQ and RFP documents and other project circumstances and conditions. Also referred to as the **pass or play decision**.

imperative that you have a reliable, systematic method for evaluating an RFQ-RFP package in order to determine whether you want to pursue the project or not. Once you have established a reliable evaluation method, it is important to apply it each and every time you consider a project. I recommend the following ten-step process to guide you from the initial RFQ-RFP evaluation all the way through to contract award.

Step 1 – Analyze the RFQ

You need to ask several practical questions right off the bat to determine if the design-build project is right for your company and your team. Here are just a few of these important questions:

- **Will the competitors be short-listed after the RFQ?** Remember, it is quite expensive to put together a complete proposal; if the owner is not going to narrow the field to only the most qualified, then you might want to pass and wait for the next one.

- **Is the project funded?** This may seem like a silly question, but I can tell you that more than one contractor has had to learn this lesson the hard way. Do your homework. Find out where the funding is coming from and when.

- **Do you have the required project experience and expertise?** If not, can you supplement the team by adding an outside member or consultant to fill the gap? Could you partner with another firm?

- **Do you have a geographic advantage?** Is the project in your back yard? Do you have a good handle on the subcontractor base? If not, would adding an outside team member or partnering with another company help the situation?

- **Is there a political component to this competition?** For example, with a university project, would it make any difference if members of your team were alumni?

- **Do any of the project owners and potential team members have similar interests or participate in the same types of activities?** For example, there may be existing relationships that you are unaware of, such a potential team member who sails at the same yacht club as the owner. These things are not supposed to make a difference, but many subjective factors affect a design-build evaluation. Of all these subjective elements, trust, which begins with familiarity, is key to winning a project and establishing a good working relationship.

- **Is the owner's budget and scope realistic? Do the project requirements match that budget?** This one seems obvious, but failing to assess this could result in a lot of wasted time. If the budget is out of line, the best thing to do is to call the owner and get some clarification. If that does not result in an adjustment, then you need to pass on the project.

:: **Who is on the selection panel and what criteria are being evaluated?** It is extremely important to know who the jurors are and how you will be scored. This will help you determine where to focus your attention and exactly how to present your proposal material. Different audiences require different approaches.

:: **Do you have the right staff available for the job?** One of the biggest mistakes I see contractors make in design-build is to assume that just any project manager, superintendent, or field engineer can do a project. As you will learn a little later in this book, it takes a certain mentality to be successful at design-build.

Step 2 – Respond to the RFQ

Now that you have reviewed the RFQ and decided to go for the project, it is time to develop a **win strategy**. At this point, you should have a good sense of how the game is going to be played and what it will take to win. Some contractors wait until after they are short-listed to focus their attention and efforts, but the competition really starts at the RFQ stage.

Always keep in mind that, at this stage the owner might be reviewing as many as ten to fifteen different RFQs. Thus, you want to make sure you highlight the aspects of the RFQ that emphasize your strengths and your competitor's weaknesses. For example, if you have a local office in the same city as the project whereas your competitors will be sending their teams in from a distance, you can highlight your familiarity with the area and the local subcontractor base. Your competitors will lack established relationships in the area, putting them at a disadvantage when it comes to sourcing labor. Take care to answer all the questions asked in the request. Organize your answers in a way that makes it easy for the jury to follow. Also, make the packet as graphically appealing as possible. A good rule of thumb is to use a graph, chart, or table whenever you can in place of text. Reviewers can grasp a concept illustrated by a graphic supplemented by an explanatory sentence or two much more easily than they can grasp the same concept explained in a paragraph of text.

Step 3 – Make the Short List

The initial goal in a two phase design-build competition is to end up on the short list. Once you do, the real work begins. After you make the cut, find out who else is on the short list. Then keep track of your competitors throughout the RFQ process, because sometimes companies drop out along the way. Knowing your competition should play into your overall win strategy. However, do not focus on "beating" another team; instead, focus on the client's real needs and delivering the best solution you can. If you are too focused on the competition, you might miss what you really need to see to best serve the client and win the award.

> **D-B DICTIONARY**
>
> **Win Strategy –**
> An organized plan that focuses on highlighting the team's strengths, competencies, and capabilities relative to the client's specific goals and objectives for the project.

The Design-Build Institute of America recommends that the short list be limited to the three most qualified teams. Including more than three teams defeats the purpose of the short list concept and will discourage design-builders from proposing at all. Furthermore, the time it takes to properly review a proposal package is significant; most owners cannot afford to invest the time that it would take to evaluate more than three packets. Also, if you have already identified the top contenders, why waste your time on other options?

Step 4 – Analyze the RFP

The RFQ document should give you the project parameters (description, budget, schedule, etc.), but you need more information than that. As you prepare your response to the RFP, you need to assess the technical aspects of the project, digging much deeper into what it will take to put together a rock-solid response. By this time, you should have selected and confirmed all the members of your team. When analyzing the RFP for strategic advantage, it is critical that you identify the three to five most critical issues for the client and focus your team on addressing those issues first.

Also, at this point, I strongly recommend doing a **joint risk assessment** (designers, contractors, major subcontractors, vendors, etc.) to get as much information out on the table as possible. Most contractors are accustomed to doing a project risk analysis from a construction perspective; however, they are not used to inviting outside participants into the room. This joint discussion is critical. Your goal should be to identify every possible risk, with the whole team present. Figure 5.2 illustrates a risk identification technique. (Chapter 6 provides more detail about risk identification, evaluation, and analysis.) Once all the risks are identified and evaluated, the team can strategize and come up with a comprehensive **risk mitigation plan** for the entire project. In other words, the team can agree upon the best ways to lessen the risks.

Sometimes, after all of this analysis, your instincts might tell you that something just does not seem right. In that case, you should probably pass on the project. I can tell you from firsthand experience that ignoring those instincts is usually something you will regret. However, if you have taken a good hard look at the risks from an interdisciplinary perspective and you have a plan to mitigate those risks and everything feels right about the project, then by all means jump in with both feet.

Step 5 – Conduct a Design Charrette

Unless you are an architect or student of architecture, you probably aren't familiar with the concept of a design charrette. As a design-build contractor, you are going to end up learning a great deal about the design process and this is where it often begins. A charrette is a design brainstorming exercise where several ideas are generated in a very short frame. The typical project charrette lasts about one day, although some teams will go to the project site (or somewhere nearby) and completely immerse themselves for several days in finding the best solution. Ultimately, the team comes out of these sessions with an agreed-upon schematic design. Contractors are not usually invited to a traditional design charrette. But in a **design-build charrette**, it is *critical* that they participate. Furthermore, you might even consider inviting a surrogate owner to your charrette to have access to their unique views relative to the design concept. As mentioned earlier in this chapter, a surrogate owner is someone in a business that is similar to the business of the client whose project you are trying to win.

D-B DICTIONARY

Joint Risk Assessment – A multi-disciplined, comprehensive risk identification, evaluation, and analysis session that involves representatives from the entire design-build team.

Risk Mitigation Plan – A structured approach to managing the uncertainty associated with a design-build project through calculated decision-making and management strategies

Design-Build Charrette – A collaborative design session in which people from multiple disciplines gather in the same space to conduct an intensive brainstorming session; the goal of a design-build charrette is to flesh out multiple design options quickly. Ultimately, the multi-disciplined design team will emerge with one design concept around which it will build its response to the RFP.

Risk Analysis (Risk = Probability × Impact)

	Risks	Potential consequences	Probability	Impact	Risk	Mitigation strategy
	Design risks					
A	Incomplete drawings and specifications	Project delay	3	3	9	Thoroughly check drawings for errors or omissions
B	Owner directed changes	Schedule, costs	3	3	9	Develop a clear understanding of owner's objective
C	Meeting design milestones	Not awarded the job	1	5	5	Proper time management and team meetings
D	Incomplete or incorrect soil/geotechnical data	Increased costs, redesign necessary	1	5	5	Hire a pre-qualified expert consultant.
E	Non-compliance to standards & codes	Design and/or construction rework	1	5	5	Integrate Code Reviews into the design process
F	Scope creep and cost	Depletion of design fees and impacts on construction cost/schedule	3	4	12	Clear and transparent communication between the owner and design-build team
G	Design defects affecting constructability	Increased construction cost/schedule	1	4	4	Integrate construction knowledge into the design process through the co-location of design-build team
H	Conceptual design instability	Depletion of design fees, scope creep, and/or budget/schedule effects	3	3	9	Effectively define executive summary of project and integrate owner in design team in order to keep project intentions clear
	Construction risks					
I	Availability of materials	Material procurement delay, material change	3	2	6	Proper material management with emphasis on long lead items.
J	Safety	Worker's injured, litigation, project delay	1	4	4	Injury Illness Prevention Plan (IIPP), consult OSHA
K	Weather	Schedule delay, material damage, increased costs	3	2	6	Follow accurate weather reports, include weather days in schedule
L	Earthwork	Increased need for specialty machinery	3	4	12	Proper site research, add contingency (%)
M	Re-work, defective work	Schedule delay, need to replace labor	1	4	4	Hire qualified subcontractors, maintain documentation
N	Supplier/manufacturer delays	Material procurement delay, schedule impacts	1	4	4	Collaborate with suppliers/manufacturers during design process
O	Workforce availability	Increased labor costs, possible decrease in construction quality	1	4	4	Implement effective QA/QC personnel that remain independent of the construction operations of the company
P	Constructability issues	Decrease in construction quality, possible construction rework	3	4	12	Integrate construction knowledge into the design process through the co-location of design-build team, implement effective QA/QC personnel
Q	Trade quality	Potential construction rework, decreased owner satisfaction	3	4	12	Implement effective QA/QC personnel that remain independent of the construction operations of the company
	Financial Risks					
R	Material price escalation	Increase costs, decrease profits	2	3	6	Lock in prices as soon as the job is secured
S	Gas prices	Increase costs	4	3	12	Purchase local materials, use efficient machinery
T	Lack of owner funding	Schedule delay, project termination	2	5	10	Raise funds from within university, search for alumni donations
U	Damage/theft	Schedule delay	2	3	6	Fence the jobsite, lock all equipment, hire security
	Political Risks					
V	LEED-NC 2.2 rating system change over to LEED-NC 3.0	LEED certification delay, certification fee increase	5	5	25	Train personnel in office and in field on LEED 3.0
W	Transportation delivery restrictions	Schedule delay	5	5	25	Collaborate laborer parking, staging areas, and material delivery with university
	Environmental risks					
X	Unforeseen site/facility contamination	Scope creep, schedule and cost increases	3	5	15	Remain familiar with local environmental abatement contractors and remain clear with owner about original scopes outlined in original contract
Y	Community disturbance	Students and faculty of university distracted by construction operations	5	2	10	Maintain regular working hours between 8am-5pm and use necessary precautions to minimize impacts on the surrounding student community
Z	Endangered species	Schedule delay	5	5	25	Investigate site conditions further before continuing design development
	Planning and approvals					
A*	Permitting process	Schedule delay, design modifications	2	4	8	Investigate permitting process thoroughly and allow for enough time for designers to modify plans according to agency suggestions
B*	Community/university opposition and activism	Schedule delay, design modifications	2	3	6	Implement Public Relations personnel in order to attend to the needs of the student body, faculty, and staff needs
C*	Agency regulations and restrictions	Schedule delay, design modifications	2	3	6	Investigate governing agencies and maintain a close relationship between project team and governing bodies
D*	Endangered species	Schedule delay	5	5	25	Investigate site conditions further before continuing design development
	Operation and maintenance risks					
E*	Operating performances	Construction/Design re-work	2	4	8	Implement QA/QC personnel to make sure that all performance requirements of the owner are met
F*	Liability to end users	Construction/Design re-work	2	4	8	Implement QA/QC personnel to make sure that all public safety and public satisfaction requirements are met
G*	Inadequate infrastructural support (power supply, water, etc.)	Construction/Design re-work	2	4	8	Investigate all existing infrastructure before design initiation

Figure 5.2
Risk identification

> *I am a big believer in on-site design charrettes. They are without a doubt the most invigorating and exciting thing that I do in my job. They are like a microcosm of the design-build process. You get everybody there together, all contributing at the same time. You learn more at a charrette than anywhere, because you are getting hit from all sides and your narrow views get completely expanded. If you are a designer, you are hearing the perspectives of all sides of the equation— from the owner, the contractors, the other design disciplines, and even the end users, if they are there. You can't get that anywhere else. It is very rich and very rewarding and a great way to start off any design-build project.*

Alan Wilson, AIA
Chief Architect
The Haskell Company
Jacksonville, Florida

A lot of contractors may ask, "What do I know about design?" In fact, you know more than you think. The design team needs your knowledge about material availability and cost, constructability, lead times, labor issues, and productivity and fabrication challenges, to name just a few important issues. The superintendent, in particular, can provide vital contributions about methodology and project approach. It is not a bad idea to invite your major subs and vendors as well; their involvement can help fill in the gaps when it comes to some of the design schemes. This is particularly important for specialized projects, such as hospitals or biotech facilities. One of the greatest benefits to a design-build—charrette is that it allows you to bring the whole team on board regarding the design from the very beginning. Giving everyone an early opportunity to voice their concerns and contribute to the overall plan means fewer problems later on.

Step 6 – Produce the Proposal Documents

Once the conceptual design idea has been determined, the design team goes to work developing the preliminary site plan, floor plans, elevations, and probably a section or two. It all depends on what the RFP requires in terms of deliverables. At the same time, the construction team works on both the conceptual estimate and the **cost model**, which will be used to control costs throughout the design process. (You will learn just how important cost models are in design-build in Chapter 6.) At this time, the construction team also prepares the initial project schedules.

It is very important that both sides of the team stay in communication during this period. As a matter of fact, it is highly recommended that the team (especially the designers and estimators) work in the same location during the design-cost development phase. It is not unheard of for a company to require the project estimator to sit at the elbow of the designer during the proposal development stage. Such companies have discovered that their odds of winning the job go way up when they do so. If they are so lucky as to end up with the project, they will be in the excellent position of knowing they are starting off on solid ground when it comes

---

D-B DICTIONARY

Cost Model –
A framework for quantifying all project costs relative to a conceptual design in a systematic fashion. The finished model is used to guide the completion of the design by setting the limitations for costs and quantities originally allocated for the project.

to balancing design with project cost. (As you will learn in Chapter 7, this is a critical project management function in design-build.)

Although developing the proposal is a team effort, designers are required to invest much more time and money during the proposal development process than the contractors. After all, it is the design that makes up the bulk of the proposal. If the owner does offer a stipend or honorarium to the non-winning proposers, make sure that both parties have agreed as to how that stipend will be shared and that the split is equitable. Some contractors end up giving the whole stipend amount to the designers.

Step 7 – Deliver the Proposal

Deliver the final proposal by hand whenever you can. This is another one of those lessons you do not want to learn the hard way. Do you really want to rely on some $24.95 "guaranteed overnight delivery" service to deliver a proposal package that cost you $140,000 to put together? Better to put someone in a car, or even on an airplane if necessary, to protect your investment.

Step 8 – Prepare the Oral Presentation

One of the most distinctive characteristics of a design-build competition is the interview or presentation process. I have had the opportunity to coach a number of winning design-build teams and I can tell you that, no matter how great your design or approach may be, if you cannot communicate it to the jury, it could all be for naught. Unfortunately, this is an area where many teams drop the ball. They leave this crucial step to the last minute, without preparing properly.

It is extremely important to organize your presentation in a logical fashion. The following list walks you through the essential elements of a presentation:

- **Introductions:** Start by introducing your team, and then discuss your overall project approach and management plan. Be sure to highlight the attributes that differentiate you from your competitors.
- **Design Solution:** Next, present your design solution. Start with the big picture items first, such as the site plan, explaining how the project fits in and relates to the whole community.
- **Outside to Inside:** Work your way through the design from the outside in. At this point, it probably makes sense to present the exterior elevations and then move to the floor plans. Somewhere along the way, take care to present the major systems—structural, mechanical, and electrical for example—using cross sections, wall sections, and details.
- **Reveal the Model:** If you are using a physical model (or even a digital one) as part of your design presentation, you will want to wait until after you have thoroughly introduced your design solution before

---

Although the initial charrette typically takes place without the owner's involvement, once you win the contract award, you may choose to go through a modified version all over again. Or you might proceed with a series of review or briefing meetings to reconcile and clarify design, quality, performance, and aesthetic expectations. This way the owner has an opportunity to become familiar with the thought process and logic that went into the initial design concept and has an opportunity to contribute and "buy-in" before proceeding any further.

Do not ever show up at a design-build presentation without your superintendent! Every owner wants to meet the person who is going to be in charge of the day-to-day project operations. In a recent design-build competition presentation, each team's superintendent and project manager was asked to stay behind for an additional thirty-minute interview after the rest of the team was excused. Although unusual, there is nothing that restricts the owner from making these special requests. Owners are beginning to customize their proposal requirements to better assess the capabilities and talents of the design-build teams.

you reveal it. You do not want your jurors gawking at the model while you are trying to explain some key component of your solution.

:: **Estimator:** After the design portion of the presentation, have your estimator summarize the project cost information.

:: **Superintendent:** Next comes a key player in your presentation—the superintendent—who presents the site logistics. Among other things, the superintendent should discuss your plans for onsite traffic and parking, the staging of materials and equipment, plans for deliveries, and plans for keeping the site safe. Of course, the superintendent should also focus on the always important project schedule.

Step 9 – Deliver the Presentation

The ability to stand and deliver your ideas is a critical component of design-build competition; every design and construction professional should be trained in the basic **platform skills** associated with public speaking. While explaining the project elements required of the RFP, the presentation should illustrate the team members' relationships with each other.

Here are four basic rules for delivering a successful design-build presentation:

Presentation Rule One: Stay Focused

Throughout the presentation, no matter what element of the project you are discussing, always tie the material back to the three to five critical issues you identified when analyzing the RFP. For example, if the environment is very important to your client and you are talking about a particular roof treatment, you should explain how eco-friendly the system or product is. Make it clear throughout that you understand the owner's concerns. Stay on-point and try to provide succinct answers to any questions. All too often, a team talks all around an owner's question and yet never answers it. In many cases, although the answer is a simple yes or no, the presenter fails to communicate this clear response.

Presentation Rule Two: No Talking Heads

Nothing is more boring than watching half a dozen people get up at a design-build presentation, one after another, speak their part and then sit down after they are done. This "talking heads syndrome" sometimes indicates a fragmented team. Instead, you want your presentation to model your team's interactive nature. At least two team members should present each topic. For instance, one might hold the presentation board and make supportive comments while the other presents the main points. It is especially effective to have a construction team member present with a design team member. This shows that the team is a **seamless team**,

---

D-B DICTIONARY

Platform Skills – Specific public speaking and presentation techniques and skills designed to improve the effectiveness of professional communications, presentations, and interviews.

Seamless Team – A fully integrated team and that can perform and communicate in a holistic, comprehensive fashion; a seamless team is fully responsive to a project as a single unit.

with a comprehensive understanding of the entire solution, a team that understands how design-build is supposed to work.

Presentation Rule Three: Justify, Justify, Justify:

If you really want to knock it out of the ballpark with your next presentation, you need to justify, justify, justify. In other words, you should have a "why" response for every decision your design-build team make. Of course, the reason why should never be "because it is what we've always done before." So often teams present the details of their design solution like they are presenting a report. They communicate the *what* of their design solution but they fail to communicate the *why*.

This bit of dialog illustrates the problem:

- **Example 1:** "For your new office building we are proposing a concrete structure with a tinted glass curtain wall exterior. We are also proposing a flat roof with a single ply membrane."
- **Example 2:** "Given your concerns for security and durability, we are proposing a concrete structure for your new building. We know that blending in with the rest of your corporate campus is important to a number of your end users, and therefore we are suggesting a tinted glass curtain wall for the exterior skin. And because we know how important it is to you to keep costs down, we are proposing a very economical, but durable flat roof with a single ply membrane."

In the first example, the team tells the owner what they are proposing but not why. In the second example, the team demonstrates just how thoroughly they have considered the owner's concerns. Which team would you choose?

Every justification you make should be tied to those three to five issues that you learned about back when reviewing the RFP. This is also an excellent opportunity to educate your client about value—remember, design-build is all about adding value. You might be surprised to learn that owners cannot always distinguish "value" when they see it because they do not have the knowledge that you do about time, cost, and quality.

Presentation Rule Four: Speak to be Heard

The key to a successful presentation is communication. You cannot do that if your audience does not know what you are talking about. Quite simply, you need to speak in words your listeners can understand. For example, if you want to explain a project schedule to a group of accountants and engineers from an aerospace manufacturing facility, you probably would not think twice about presenting a critical path schedule. However, what if your project is a new elementary school and the jury is made up of K-4 teachers and administrators? In that case, a critical path schedule would be useless. Instead, a simple, storyboard indicating key dates for project start-up, construction operations, and project completion would be far more effective.

Step 10 – Win the Award

And finally, after all of that work and effort, your goal is achieved. You have won the contract award. Once you have the job in hand, it is time to incorporate the most important party into the team—the owner. Up to this point, despite your best efforts to get to know the owner, you have probably had very limited access because you were working within a competitive environment. But now it is time to kick off the project with an **award celebration** followed by the initiation of some form of **partnering program** or **teaming session** that focuses on establishing and maintaining team alignment. Although this program may be informal, it sets the stage for the project's working relationships. Create a positive, cooperative tone right from the start, because once the project gets underway, you will have to shift your focus to managing the design and building the project. (You will learn more about partnering options for design-build in Chapters 7 and 8.)

> **D-B DICTIONARY**
>
> **Award Celebration** – A team celebration that acknowledges the work involved in putting together a successful proposal.
>
> **Partnering Program** or **Teaming Session** – A series of collaborative activities that help build stronger team relationships.

We like to kick off a design-build project with formal teaming meetings conducted by a behavioral specialist—usually a professional team psychologist who understands people and how to mesh the various personalities and individual strengths together. We stay away from the traditional partnering sessions used in the hard-bid environment, where some construction, engineering, or scheduling person comes in and conducts the meetings. We really want to focus on the teaming component—building a strong relationship based on trust. We usually start off with an initial all-day session, maybe two, and then we regroup with the entire team every month or so for maybe two or three hours at a time, all the way through to the end of the project. It takes discipline and commitment to keep it up, but it makes all the difference in the world.

Dave S. Crawford, PE, DBIA
President and COO
Sundt Construction, Inc.
Tempe, Arizona

I base this ten-step approach on the shared experiences and mistakes made by many good design-builders over the years. You may add to or customize these steps to suit your specific organizing principles, but the key is to come up with a reliable plan for evaluating potential project pursuits through the RFQ/RFP process. Consistent review and painstaking analysis will help you sort through the various risk factors, unique project components, and strategic opportunities, so you can direct your attention to making a responsible go-or-no-go decision.

DELIVERING ON THE PROMISE

Texans have an expression, "all hat and no cattle," which describes someone who talks a big story but never delivers the goods. If you want to be successful in design-build, you have to deliver the goods. After all, when you submit your proposal you are really only offering a promise. You are promising to design and build a project that meets the owner's requirements for performance, quality, function, value, budget, and schedule. In awarding you a contract, the owner is putting his faith in your ability to manage a multi-disciplined team, leverage their talents, and provide a solution that will deliver the promised goods and services. You cannot point to flawed plans or other people if things start going wrong. Your team, operating as a single entity, is held accountable for the entire project. Winning the project is just the beginning of the process. Now you must go about managing the process in such a way that you are able to consistently and confidently deliver on your design-build promises time and time again. The next few chapters will focus on just how to do that.

CHAPTER HIGHLIGHTS

1. To compete in design-build you need to: 1) understand what you are getting yourself into and how the process works, 2) learn as much as you can about the client and the project, 3) pick the right people for your team, and 4) make sure you have a good plan for developing the proposal and a strategy for winning.

2. It is essential to know the owner in design-build. One way to learn about the client is through the RFP. You should look at the company's website and written literature to gain more in-depth knowledge about the history, organizational vision, mission, core values, etc. of the potential client.

3. Savvy design-builders often do preemptive marketing by tracking and researching potential projects and clients years in advance. Preparation can include a series of informational interviews with potential clients, visiting similar state-of-the-art projects, finding out what has and has not worked on similar projects, and speaking to owners, end users, and maintenance people to get their opinions on what they would improve if they did the project all over again.

4. Owners sometimes conduct procurement workshops prior to the RFP which allow the design-builders to get to know the owner better; such workshops also allow the owner to gain valuable feedback. More formally, owners can release a draft RFP which allows the owner to adjust the RFP based on initial feedback from design-builders.

5. Performance-based, best-value competitions offer design-build teams an opportunity to differentiate themselves from competitors in a way that design-bid-build never does. Design-builders can use their ideas, creativity, innovation, experience, skills, and talents to win the project.

6   Proper team selection is a critical part of a successful design-build project. Most RFPs require that key people from the proposal team form the core of the project team.

7   Partner firms must share similar values, and have fundamental agreement on the overall project goals and objectives. Team readiness assessments should be conducted before any teaming agreements are signed.

8   Upon receiving an RFP, you should make a go-or-no-go decision based on the quality of the RFP and your ability to deliver the project in accordance with the owner's expectations. If you decide to go after the contract, you need to assemble your team. Designers and contractors should conduct a joint risk assessment and devise an initial risk-mitigation plan.

9   The competition starts at the RFQ stage, with the initial goal of getting on the short list and receiving the RFP. Questions to ask when analyzing an RFP include: Will the competitors be short-listed after the RFQ? Is the project funded? Do you have the required project experience and/or expertise? Do you have a geographic advantage? Is there a "political" component to the competition? Is the owner's budget and scope realistic? Do the project requirements match the budget? Who is on the selection panel and what criteria are being evaluated? Do you have the right staff available to do this job?

10 A design-build charrette is often conducted at the proposal stage. It involves both the contractors and designers of the design-build team and can also include owners, superintendents, major subs, and vendors. The greatest benefit to an interdisciplinary design charrette is getting the whole team on board with the design from the beginning of the project.

11 After the conceptual design is complete, the design team works on the RFP deliverables. These could include the preliminary site plan, floor plans, and elevations. Meanwhile, the construction team should put together the conceptual estimate, build a cost model, and develop the project schedule. The most effective way for the design and construction teams to form a cohesive response to the RFP during is to work under the same roof during this process.

12 Teams need to prepare to present their proposal orally. The presentation typically starts with a review of the overall project approach and management plan, followed by an explanation of the design solution, cost and schedule information, and site logistics. Both physical models and digital "walk through" models are often presented during the interview period.

13 The four basic rules for delivering a successful design-build presentation are: 1) stay focused, 2) no talking heads, 3) justify, justify, justify, and 4) speak to be heard.

14 After winning the contract, the team should host an award celebration and initiate a partnering program that incorporates the whole team, including the owner. After that it is up to the design-build team to deliver the project as promised.

RECOMMENDED READING

- ***Work Like You are Showing Off: The Joy, Jazz, and Kick of Being Better Tomorrow Than You Were Today,*** by Joe Calloway, John Wiley & Sons, 2007.
- ***Selling the Invisible: A Field Guide to Modern Marketing,*** by Harry Beckwith, Warner Books, 1997.

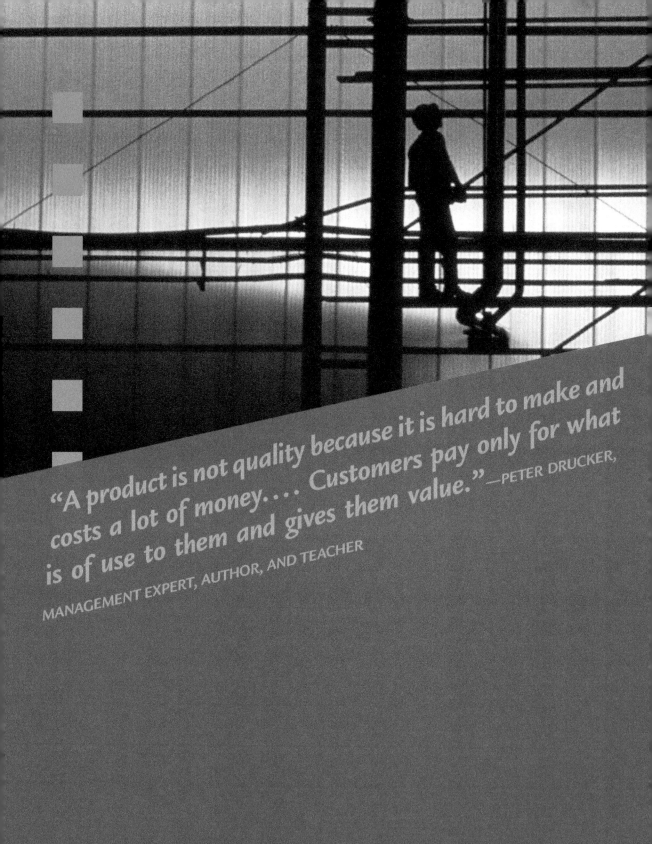

"A product is not quality because it is hard to make and costs a lot of money.... Customers pay only for what is of use to them and gives them value." —PETER DRUCKER, MANAGEMENT EXPERT, AUTHOR, AND TEACHER

6
DEVELOPING DESIGN-BUILD ESTIMATES

No matter how wealthy clients might be, or how unconstrained their budgets, they always want to know, "How much is this project going to cost me?" And they would prefer to have the answer to that all-important question sooner rather than later—especially if they are working with a limited budget, which is usually the case. The design-build project delivery method lets owners determine their project cost earlier in the procurement process than do design-bid-build or CM-at-Risk. Owners don't have to wait until they have already spent a great deal of money on design services to develop complete plans and specs, only to learn that their project is over budget, incurring schedule delays and redesign costs they cannot recover. Even if the redesign brings the project back into line, the waste of time and money is a huge disappointment to everyone involved.

However, in design-build, depending on which procurement method the owner uses (see Figure 6.1), she can work with the design-build team to establish a realistic project budget based on their program and performance criteria, even before the first line of design has been drawn. More importantly, the design-builder can guarantee that the budget will remain intact through the design process and construction.

These advantages are possible because, as you learned in Chapter 2, design-build allows the price to be determined before the design is complete. But switching from the design-bid-build estimating mentality to a design-build estimating mentality can take time and effort, and represents one of design-build's biggest challenges.

The essential question is this: How do you determine a price for a project without having a full set of plans and specifications? It is risky enough to estimate a project when you have a completed design. Surely, it would seem, pricing a project with only a few sketches and limited design criteria would significantly increase the design-builder's risk even more, especially if you must guarantee that price and sign a contract. Clearly, creating a design-build estimate requires an experienced

Figure 6.1
Early knowledge of project cost

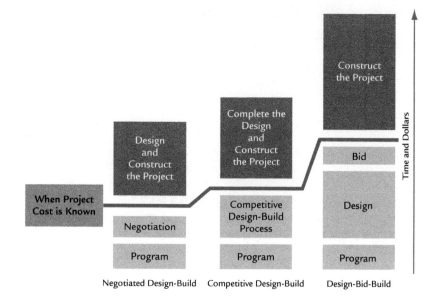

estimator to anticipate all the factors that will impact the project price, but design-builders create estimates this way regularly. See the Hilton San Diego Bayfront Hotel case study in this chapter for an example. Fortunately tools and techniques exist that can reduce risk and assure project success for all parties, once you understand the fundamental concepts of the design-build estimating process.

PRICING THE UNKNOWN

Every day, thousands of contractors across the country sign agreements promising to complete projects or project segments for stipulated amounts of money under lump sum or guaranteed maximum price contracts. But none of these contractors knows exactly what the cost will be until the project is 100% complete and until they have accounted for every invoice, purchase order, and subcontract. So even if the contractor has a full set of plans and specs, as is the case with design-bid-build, she still cannot price the project with complete accuracy because it is impossible to know all the conditions and circumstances that will impact the project costs (weather, labor availability, price fluctuations, etc.) with certainty. The estimate that underlies the contract price is still just a guess—an educated, calculated guess, but a guess all the same.

The uncertainty level of the estimating process relates directly to the level of design completeness at the time we provide the contract price. Sometimes we literally price a job based on a paper napkin sketch and a few handwritten notes. At other times, we might have schematic or conceptual drawings and perhaps a site plan, simple floor plans, and an elevation to guide our estimate, but certainly far less than a full set of drawings. Yet design-builders willingly take on this estimating challenge every day and confidently offer guaranteed prices on projects of every size and type. They have learned that they do not need to have all project details to price a job because they clearly understand how design-build works and the unique symbiotic relationship between project design and project cost.

In hard money estimating, you are just counting up all the widgets that are shown on the plans and specs. But in design-build, the architect hasn't developed the design much beyond the schematic stage, when we frequently have to set a guaranteed price. How do you account for widgets that are not even drawn yet? You have to be able to visualize exactly what it's going to take to build the project and anticipate what will "eventually" be on those drawings. This is where experience comes in and there is absolutely no substitute for it. You've really got to have a thorough understanding of construction methods and how buildings go together before you can be a good design-build estimator. This includes knowledge of code requirements and details of design. You've got to think through, for example: Where is flashing and miscellaneous metal going to be required? Where are vertical shafts that have to be fire rated? Where do the exit corridors have to be and what are the fire codes associated with them? You end up building the project on paper long before the architect has stamped the drawings. It's kind of risky business but there are folks so knowledgeable about construction and construction costs that they can come up with estimates that are darn close to the final numbers in the end. It's quite amazing and sometimes makes you wonder why we have to go through all of that drawing effort in the first place. That's why the concept of "design to budget" makes sense—you literally have designed the project through the estimating process.

Wayne Lindholm
Executive Vice President
Southern California
Hensel Phelps
Construction Co.
Irvine, California

CHOOSING VALUE AND QUALITY OVER LOW PRICE

The design-build concept of price fundamentally differs from that of design-bid-build. Instead of emphasizing price alone, design-build encourages bidders to supply maximum value and quality for their available dollars. Owners have seen, time and again, the effects of the low bid process and the resulting escalation of their final project cost through multiple change orders. Today public and private owners have huge capital improvement needs and little time to go through the slow, painstaking process of a design-bid-build, low-price competition that ends with cost overruns, legal claims, and a project that does not meet their needs. In trying to find a better way, owners are realizing a fundamental design-build advantage: Optimizing their project budget by emphasizing value and quality is far superior to finding the cheapest price.

Before the design begins, owners can declare how much money they have to spend, along with their project needs, and then put the burden on the competing design-build teams to come up with the best solution to their building challenges for the specified dollars allotted—and not a penny more. They are not interested in merely obtaining the lowest price or in receiving budgeted money back. Instead, they want bidders to be as efficient as possible and to add value to make their project better than they asked for in the first place. Because design-build's final results differ so radically from design-bid-build results, it stands to reason that the estimating process necessary to arrive at these results will differ as well.

CONCEPTUAL ESTIMATING IN DESIGN-BUILD

In design-build, we usually must estimate the project's future cost long before we lay out its design details, using the process of conceptual estimating, which you learned about in Chapter 3. According to Jim Crockett, "Conceptual estimating…is the process of establishing a project's cost, often before any graphical representation of a facility has been developed."[1]

In design-bid-build we estimate projects the way most of us were trained, using detailed estimating techniques. All university Construction Management programs teach at least one estimating class, which usually focuses on the traditional low-bid, detailed estimating format and approach. In this method, we take the 100%-complete plans and specifications, count all the parts and pieces, square footages, linear feet, etc. shown in the drawings, and apply unit prices for materials, labor, and equipment. In other words, we do a quantity takeoff in accordance with the plans and prescriptive specifications. The plans tell us "how much and how many" and the prescriptive specifications tell us precisely "what kind" of parts and pieces. Figure 6.2 shows an example.

However, in design-build, the design is not complete at the time of the estimate. Sometimes the design consists of only a few conceptual sketches and a site plan—or less—and ideally it prescribes no design specifications. Instead of specifying parts, pieces, manufacturers, and model numbers, the owner outlines the project scope and describes how she wants their building or project to work. The owner communicates these needs in a set of performance requirements (described in Chapter 3) and perhaps a

Figure 6.2
Detailed estimate process

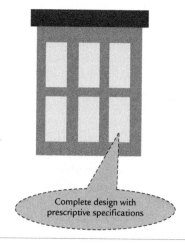

Detailed Estimating

- 100% design complete
- Prescriptive specs
- No design flexibility
- Count parts and pieces and multiply X unit pricing for labor and material
- 6–4060 metal windows with low-E insulated glass
- Material = $ 950/EA and Labor = $250/EA

6 windows @ $1200/EA = $7200

Complete design with prescriptive specifications

> **Typical Conceptual Estimating Data for a Building Project**
> - Gross floor area
> - Foot print area
> - Building perimeter
> - Roof area
> - Exterior wall area
> - Exterior window-glass area
> - Floor-to-floor height above grade
> - Floor-to-ceiling height above grade
> - Number of stories above grade
> - Number of stories below grade

Figure 6.3
Typical conceptual estimating data.

few design criteria set out in the RFP. Because the design details do not yet exist, the design-builder uses these scope and performance requirements to create a design concept, calculate some basic **design parameters** (see Figure 6.3), and apply some **parametric formulas,** formulas derived from past projects, to generate a systems-oriented estimate corresponding to that conceptual design. For example, according to many HVAC contractors, approximately one ton of cooling is required for every 1,000 square feet of floor area. Structural engineers suggest that it takes about eleven pounds of reinforcing steel for every square foot of elevated slab in a concrete building. And the 2008 *R.S. Means Square Foot Costs* book applies a parametric formula of one interior door for every 200 square feet of floor area for a one-story Outpatient Surgery Center with a 15'4" story height and 12,500 square feet of floor area.

Even the organizational framework that we use in design-build estimating is different. Instead of using the traditional sixteen-division (or newly-revised fifty-division) MasterFormat™ to organize our estimate, we use the UniFormat™ organizational framework. The UniFormat structure is an arrangement of construction information based on physical parts of a facility called systems and assemblies. (See Figure 6.4 for a comparison.)

These systems and assemblies are characterized by their function (foundation, exterior closure, etc.) without identifying the products that compose them. Systems and assemblies render a view of a constructed facility different from the view rendered by a breakdown of building materials, products, and activities.[2]

According to the Construction Specifications Institute, UniFormat serves the following needs:

:: Performing economic analysis of construction alternatives, particularly in early design stages;

:: Building and evolving a construction cost model arranged by functional parts of a facility;

:: Arranging descriptions of a project's functional parts to communicate design intent for accomplishing an owner's construction program;

:: Arranging Project Manuals for applications when design-build project delivery is employed;

D-B DICTIONARY

Design Parameters – Information that sets the boundaries of the design and guides the direction and decision making of the design-build team relative to scope, size, cost, quantities and any other types of constraints and limitations created by the owner's program and economic requirements for the project.

Parametric Formulas – Formulas derived from data collected from past projects that help you quantify and/or estimate certain elements of the design before specifying the design details. Some are informal rules of thumb and others are more elaborate. Parametric formulas should be adjusted to account for project specific conditions and requirements.

Uniformat Categories / **Master Format Categories**

UNIFORMAT II Categories			01 General conditions	02 Site work	03 Concrete	04 Masonry	05 Metals	06 Wood-plastic	07 Thermal & moisture protect	08 Doors & windows	09 Finishes	10 Specialties	11 Equipment
Level 1	Level 2	Level 3											
A SUBSTRUCTURE	A 10 Foundations	A 1010 Standard Foundations		■	■		■						
		A 1020 Other Foundations		■	■		■						
		A 1030 Slab on Grade		■	■		■						
	A 20 Basement Construction	A 2010 Basement Excavation		■									
		A 2020 Basement Walls		■	■		■		■				
B SHELL	B 10 Superstructure	B 1010 Floor Construction			■		■	■					
		B 1020 Roof Construction			■		■	■					
	B 20 Exterior Enclosure	B 2010 Exterior Walls			■	■	■	■	■		■		
		B 2020 Exterior Windows								■			
		B 2030 Exterior Doors								■			
	B 30 Roofing	B 3010 Roof Coverings					■		■				
		B 3020 Roof Openings					■		■	■			

Figure 6.4 Estimating format comparison

:: Filing drawing details;

:: Filing facility management and construction market information.[3]

In other words, the conceptual project cost estimate is based on unit prices for common functional systems or assemblies—not on individual parts and pieces of the system. The system will perform a given function regardless of the design details, construction methods, or specific materials used—therefore none of the detailed information is needed. For example, in conceptual estimating we would calculate the linear feet of a complete foundation system—not the cubic yards of excavation and backfill, square feet of forming, tons of reinforcing steel, cubic yards of concrete, and square feet of finishing. In the example shown in Figure 6.5, the design-build team made a decision that 30% of the exterior skin of one elevation would include some type of glass and glazing system; 30% of the entire elevation area equals 260 square feet. The quality standard selected by the team to meet the performance requirements resulted in a unit price of $30 per square foot. Therefore, the glazing system for that single elevation calculates at $30 times 260 square feet providing a budget of $7,800.

The design-build team does not need to know the design configuration, the size of each unit, or the exact placement in order to come up with the budget number. The team will have flexibility to fine-tune the glass and glazing design details after contract award. They will monitor the glazing system budget as they design it, using both the dollar amount and the quantity limits as benchmarks. In other words, they will design to budget. You will learn a lot more about this all-important design-build

Conceptual Estimating

- Very little design needed – Concept only
- Must meet performance requirements
- DB team to establish a design parameter for Glass & Glazing
    - gross quantity
    - quality standard
    - unit price & contingency
- Example: Glazing @ 30% of exterior skin

    Exterior skin = 864 SF

    30% of 864 SF = 260 SF

- Unit price for glass and glazing based on historical data = $30/SF including contingency
- Placement, manufacturer, quantity of units, and other details to be determined later

260 SF @ $30/SF = $7800

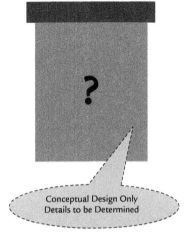

Figure 6.5
Conceptual estimating process

estimating concept later in this chapter. The team may have to adjust the selected glazing system elements up or down to meet all performance standards and stay on target, but the design parameters that determine cost and quantities remain stable, based upon the original agreed-upon conceptual design.

The Designers' Role

Although design-build estimators base an estimate on very few design drawings, a few elevations, floor plans, and a site plan, they are not merely guessing at what the design is going to look like. They must fully involve the design team so that the estimators have a complete understanding of the design intent before they put numbers to paper and create the conceptual estimate.

The designers must provide the estimator a full design narrative to properly assign a reasonable value to the various design elements. Their input is vital to the design-build estimating process. Traditional designers are often reluctant to provide these narratives because they are fearful that they will get locked into design decisions before they have fully vetted their options. But when estimators have to provide a guaranteed maximum or lump sum price so early in the process, locking as much of the design down as possible at the time they are locking down the price is not just advisable, but necessary. The designers and estimators must work side by side so they can communicate early in the process. That way, they can discuss available options from multiple perspectives—including quality, cost, schedule, aesthetics, constructability, productivity, and availability. Conceptual estimators should capture all design decisions and the logic behind them in writing. Of course, the team needs to build some design flexibility into the budget and design adjustments are inevitable. But confirm as many design details as possible as early as possible,to ensure that the project is financially feasible, both for the client and for the Design-build team.

Jim Zahn
Project Director
Ellerbe Becket
Minneapolis, Minnesota

The program is the single most important tool to infuse value into the project. A skilled design-builder recognizes this and invests the time necessary on this critical component of the process. It is imperative that your client understand their role in this process. It requires a major time commitment. Only after you have carefully validated the program can you efficiently explore placing the building on the site and the various design options available. In true design build, design proceeds from a detailed systems estimate developed at the programming stage. Armed with only a program, a good systems estimate, and a design concept, a good design-build estimator conceptualizes every single aspect of the project and develops quantities for each building element. This allows the project team to make the most informed design decisions possible.

Filling in All the Holes

The design-build estimator adjusts parametric formulas and unit pricing to meet the unique design challenges and performance requirements of each project. But trying to do traditional detailed quantity takeoffs from conceptual drawings has gotten more than one contractor or subcontractor in trouble. The design-build estimator must quantify and account for the entire scope of the work associated with the project whether she has drawings and detailed specifications for it or not. To go back to the owner or general contractor and announce that, "I didn't include the weather stripping in my bid because it was not shown on the drawings," is not an acceptable response in design-build. The owner should expect that the estimator is pricing a fully-functioning project that meets the performance requirements spelled out in the RFP, whether the details were shown on the conceptual drawings or not. That is why it is so important to have a consistent, standardized procedure for developing the conceptual estimate—so that important details are not lost. Alan Schorfheide, a highly-experienced estimator with the Korte Company, shares these fifteen steps as a basic guide for the design-build estimating process:

It is extremely important to bring your specialty and subcontractors on board early in the conceptual design phase, because they are an excellent resource for the kinds of parametric cost data that you need to come up with realistic budget numbers for your conceptual estimate. Trying to guess the cost of sophisticated specialty systems, without having the years of experience and intense knowledge needed to properly account for all of the variables, can get you in real trouble. Obtaining realistic parametric costs early will also mitigate your exposure.

1. Understand the owner's goals.
2. Understand where the design team wants to add emphasis to the project—finishes, features, etc.
3. Have a written or verbal understanding of the building footprint, floor-to-floor heights, desired structural framing system, mechanical system, and site area.
4. Develop a list of project priorities and desired betterments with the design team so that the entire team knows what features are of the highest priority.
5. Develop a list of risk variables the design must address in the estimate—geotechnical considerations, fireproofing requirements, undefined portions of the building, and governmental requirements.

6   Discuss the risk variables with the design team and develop a set of mutually agreeable assumptions associated with the various design components and systems on which you will base the estimate.

7   Begin the estimate with the building structure at the roof and work down to the foundation. In doing so, it is critical that the design assumptions you established earlier remain consistent throughout the estimating process. Even if your assumptions prove to be incorrect, keeping them consistent will make the estimate understandable and make adjustments easier. Keep a list of any assumptions that may need adjusting so you can discuss them with the design team before making any changes.

8   Develop the shell of the building by area—roof system and exterior enclosure.

9   Fill in the building from the core out—mechanical spaces, elevator shafts, restrooms, process areas, stairwells etc. As you think through each area, add the architectural, mechanical, structural, and equipment features of each system to the estimate. You need to think in terms of the entire system and all associated features, not in terms of the CSI MasterFormat. You also need to reconcile these features with previous structural and shell assumptions—if you uncover inconsistencies, step back, rethink your previous assumptions and reconcile as necessary.

10 Fill in the rest of the building by area or use by system, as you did in step 9.

11 Develop the site estimate, starting with parking or paved areas and moving onto landscaped areas. When you have the entire site filled up, add earthwork and utility systems that support the site features.

12 Always estimate physical quantities and a unit price for the quantities, not cost per square foot. This method tells the team what is in the estimate and makes it easier for the team to reconcile it as the design develops further. For example, if I have estimated the lineal feet of partitions I can compare that estimate with the next design iteration and the team knows if they are on budget. If I estimated a cost per square foot, nobody knows where we are relative to the budget.

13 When the entire estimate is done, think through the whole estimate in CSI format terms. By doing this you have looked at the cost in two ways—by system with building and site spaces "filled up" and then from a component perspective. This is an important cross-check.

14 Compare the estimate to other completed projects and look for inconsistencies. If found, review your assumptions and update. Perform another cross-check.

15 Report back to the team with the cost, included scope, and assumptions. As assumptions change, correct the estimate for team decisions.

The design-build estimating process entails more than merely generating the numbers; you must also clearly articulate what those numbers

> Conceptual estimating requires knowledge of the basic design elements associated with specific project types from a systems perspective and an excellent historical database of past projects that serve as models for establishing the systems pricing. This is what allows the design-builder to develop the project price first and the details of the design second.

represent—especially when there are few design details in the conceptual drawings. The more information you present clearly and concisely to the team, the more valuable and respected you will be as an integral part of that team, as the case study in this chapter illustrates.

COST AS AN ELEMENT OF DESIGN

Contractors, architects, and design engineers often have difficulty understanding that cost is an element of the design process in design-build. In the traditional project delivery approach, the design comes first and the cost estimate comes second. But as you learned in Chapter 3, in design-build the cost estimate comes first and the design follows—or at least the completion of the design. The notion that the project budget might actually lead the design frightens many architects and design engineers but for owners it only makes sense. Why, they reason, would an architect design something that the owner cannot afford to build? But in design-bid-build, this happens far too often. This is why owners like design-build—not only do they have early knowledge of their project's cost but they have a better chance of the design meeting their budgets as well.

Furthermore, design-builders frequently need to overcome architects' and engineers' belief that the budget is an unwelcome design constraint that should not interfere with the creative process. But in design-build, budget is always a reality and ignoring it is not an option. On the other hand, given that architects and engineers are professionally liable for the integrity of the design, they should speak up if they suspect that the budget is somehow truly undermining the adequacy of their designs.

> You will learn in Chapters 7 and 8 that the key to attaining this optimized interdependency between cost and design is a fully integrated, collaborative project team and some distinct design-build management skills.

In a properly implemented and managed design-build process, none of the critical elements is sacrificed for the sake of another—not the design, not the budget, and not the schedule. Rather, the design-builder optimizes and coordinates each one into an integrated, holistic, and comprehensive solution that serves all project requirements for quality, performance, function, cost, and time. Design-builders do not develop a design independent of the budget, but rather as an integral part of it. Therefore, they implement schematic design and conceptual estimating as simultaneous activities.

DESIGNING TO BUDGET: FIVE STAGES

The construction, engineering, architecture, and owner community all understand that design is an iterative and progressive process. But not all practitioners understand that the estimating process in design-build is also an iterative process—that is, many estimating cycles occur as the project moves from the programming stage to the construction stage. Although there is much debate about when one phase in the designing-to-budget process stops and another begins, the various stages can still help us define where we are in the process. Generally there are five identifiable design stages:

Programming Stage

In this first stage of the design process, the client defines the scope of the project, referred to as the program, usually as a narrative. In this project scope definition, the client identifies the overall purpose of the facility along with specific space requirements, room adjacencies, and functional needs. This summary includes known constraints as well as budget, schedule, size or height limits, regulatory restrictions or requirements, and other information that will help guide the design-build team. Some owners will define their projects before involving designers or builders, but most design-builders will tell you that this is the stage at which they would most prefer to get involved and where they feel they can add the most value. At this stage, the owner usually determines a "ballpark" estimate or initial budget to determine whether the project is feasible or not.

Conceptual Design Stage

The conceptual design includes a site plan with a rough building or facility layout and some sketches depicting a general concept of the project—size, mass, configuration, orientation, and the like. The terms "conceptual design" and "schematic design" are often used interchangeably, but in the context of the estimating process, it is best to distinguish between these two preliminary design steps. Sometimes the owner will include all or part of "the conceptual design" along with their program in the RFP. At this design stage, either the owner or the design-builder may first establish a budget for the project. If the owner establishes the budget at this stage and stipulates it in the RFP, then the responding design-build team needs to prepare an initial proposal design that meets that budget and adheres to it through to project completion, along with all other RFP requirements. When the design-build team helps to establish this initial budget, it serves as a cost model and becomes one of the critical controlling tools in the entire design-build process.

Schematic Design Stage

The design-build engineer or architect will review the owner's RFP and program to ascertain the requirements of the project. Based on this understanding, the designer prepares schematic design documents, which consist of preliminary drawings, outline specifications, and other documents illustrating the scale and relationship of the project components. At this stage, the designer often communicates from a systems perspective only and not the details. For example, she might indicate if the structural system in the proposal is concrete or steel, or that the mechanical system consists of rooftop units or hydronic water source heat pumps. In competitive design-build, these schematic drawings and outline specifications become part of the proposal deliverables and are the initial basis for the project estimate. Unless the owner elects to use direct design-build or QBS to secure their design-builder, this is most likely the earliest stage of design in which the design-build team participates.

Design Development Stage

During the design development phase, the design-build team develops the details of the design, consisting of drawings and other documents to fix and describe the size and character of the entire project as to civil, architectural, structural, mechanical, and electrical systems, materials and other such elements as appropriate. In competitive best value design-build, the design-development phase usually starts shortly after the contract award. Because the owner is on board as part of the team, it is an excellent opportunity to discuss, test, and evaluate alternative options that can work to optimize the overall project performance. It is also time to bring in major specialty and subcontractors to engage in constructability and best approach discussions. Value engineering takes place as a true "value adding" exercise and the team decides the final direction for the design. Estimating is a critical component of the design process during this stage.

Construction Document Stage

During the construction document phase, the designers prepare construction documents consisting of drawings and specifications setting forth in detail the requirements for the project construction. These drawings are sometimes referred to as working drawings, contract documents, or release for construction (RFC) drawings. In most cases, the number of construction drawings necessary to inform construction are far fewer in design-build than in design-bid-build, primarily due to the high levels of communication and coordination taking place among the team members.

In design-bid-build, the contractor is typically not involved in preparing any pricing during the design process. In fact, the contractor does not even see the design until it is complete and released for competitive bidding. In this traditional project delivery method, the owner will usually conduct a feasibility estimate early in the programming stage to confirm whether the project is economically viable or not. Once the project is deemed viable and the design process begins, the owner's architects and/or engineers are often asked to prepare an **engineer's estimate** that establishes the owner's initial budget for the project. Once this is done, the design continues until it is complete. Unfortunately, in design-bid-build, market-savvy construction cost estimators are not typically a part of the design team. By the time the design is completed, significant time has passed (months and even years) and little or no effort has been made to reconcile the final design with the original budget. The next cycle of pricing does not take place until the completed plans and specs are put out for bids and the project is ready for construction. Because of the time delay between the engineer's estimate and the completed design, the project bids frequently come in over the owner's budget and the project is usually sent back to the drawing board for redesign, lengthening the overall project schedule even more—or worse, it is released to the contractor to "value engineer" the design by cutting costs and finding ways to cheapen the project. Unfortunately, this attempt to bring the design

---

D-B DICTIONARY

Engineer's Estimate – Also called the architect's estimate, this estimate is usually prepared by someone on the owner's staff or by an independent consultant working for the owner or designer. The purpose of the estimate is to check for initial project feasibility and establish a budget. This estimate usually occurs many months, or even years, before the actual bidding.

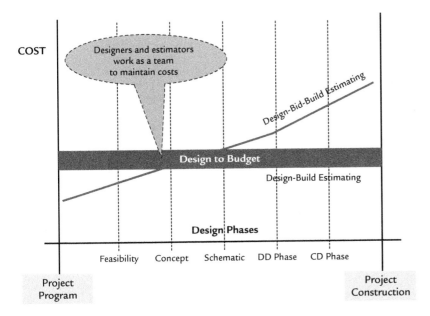

Figure 6.6
Designing to budget

in line with the budget has little to do with actually adding value to the project.

However, in design-build projects, the contractor is involved early in the design process, so the relationship between design and cost can be synchronized as a parallel activity. The estimating function happens simultaneously as the design progresses and provides for an ongoing opportunity to monitor and manage the design in relation to the budget and conduct true **value engineering** as it should be conducted—throughout the design process and not at the end of it.

This design-build estimating approach differs from the design-bid-build estimating approach, in which design and cost are treated as independent variables that seem to have no consequential connection. The diagram in Figure 6.6 illustrates the budget impacts resulting from these two different estimating approaches.

DEVELOPING THE BUDGET

Before you can design to budget, you need a budget to design to. As you have learned, either the design-build team or the owner must first establish the project budget. If the design-build team is hired at the very early stages of the project, before any design has begun, they can work with the owner's program and design concept to help create the budget and build a cost model for the project. The cost model defines and controls the money allocated into each of the building elements.

Cost models are project specific and developed from historical cost data associated with various project types. In essence, comparable projects of similar scopes and sizes create the basis of the model. The actual as-built costs associated with the comparable projects are broken down into square foot unit prices for categories and then into elements of

D-B DICTIONARY

Value Engineering –
A method for evaluating the cost effectiveness and value of the various methods, materials, and means by which we design and construct projects. The analysis considers the initial cost, maintenance cost, energy usage cost, replacement cost, and life expectancy.

BALANCING ACT

Courtesy of Hensel Phelps Construction Co.

THE FACTS

Project Name:
Hilton San Diego Bayfront Hotel

Project Type:
Hospitality Development

Location:
San Diego, CA

Description:
A thirty-story, 1.65 million square foot hotel complex with a 385-foot tower, 165,000 square feet of meeting space, 5,360 square feet of retail space, a 23,082 square-foot health club, 1,190 private rooms, a 14,000 square-foot restaurant, and a 4.3-acre public park, all situated on a 12.8 acre site on San Diego Bay

Cost:
$335 million

Duration:
30 months

Completion Date:
December 2008

THE TEAM

Ownership Group:
One Park Boulevard (Hilton and ING)

Lease Holder:
Port of San Diego

Development Team:
Phelps Portman San Diego LLC

General Contractor:
Hensel Phelps Construction

THE LEAD-IN

Hilton Hotels Corporation is a leading hospitality company that owns, manages, and franchises over 2,000 hotels across the country. The company's international arm, Conrad Hotels, has locations in Australia, England, Ireland, Egypt, Belgium, Turkey, Hong Kong, and Singapore. Though publicly traded, the chain was, for most of its history, led by members of the Hilton family from 1919, when founder Conrad Hilton bought his first hotel in Cisco, Texas. Cisco was experiencing an oil boom at the time and the hotel was booked solid from day one. Conrad and his partner, L.M. Drown, even rented their own beds, sleeping on chairs in the office. They also converted much of the hotel's public space into additional guest quarters. Making use of wasted space became a hallmark of the Hilton chain[4], and it still is today. When the Port of San Diego decided to build a major hospitality project on the old Campbell Shipyard along the waterfront near downtown San Diego, the Hilton Corporation stepped in to compete for the opportunity—and true to form, creating world-class accommodations while still maximizing space efficiencies was goal number one.

THE STORY

Although the San Diego Bayfront Hilton Hotel project did not use a standard design-build contract, the team was structured from the start to function and perform as an integrated design-build partnership, including a developer. After a lengthy RFP "development competition" in which there were five finalists (each with its own hotel operator partner), the Port of San Diego awarded the new hospitality project to the Hilton/Hensel Phelps (HP)/Portman Associates team, allowing them to exclusively negotiate the development of the property. The initial Hilton/HP/Portman Proposal submittal contained several renderings and concept floor plans of the project as well as an overall development budget that contained one line item for construction cost. That single line item cost was the first step in the intensive progressive estimating process that would follow the design from initial concept to completed drawings.

Immediately after the team was selected, they shifted into program verification studies and continued to develop the project's concept design. They gathered input from all ownership stakeholders, working through many options to finalize a workable concept and move on to the schematic stage of design. As the schematic design took shape, the Chief Estimator on the project, Mike Verrastro of Hensel Phelps, put together a preliminary cost estimate that the team could review in conjunction with the finalized program. A twelve-page set of drawings with no specifications was used as a baseline set of documents for this initial cost model. The estimator immediately divided the project into sectors (hotel tower, meeting spaces, restaurant, health club, etc.) so all parties could better understand the estimate. It also served as a living document that everyone could follow, provide input to, and comment on. Each sector was broken down into the various UniFormat systems. This initial estimate document was a one-page multicolumn spreadsheet with cost data from several other similar hotel projects which helped establish the "target numbers" for the project, even with a very limited Concept Design Package.

According to Verrastro, "One of the things that made establishing the targets so tough in the early stages was the ever-changing square footage of the project." Once all preliminary input from the Hotel Operations team and the Design team was incorporated and the floor plates were "frozen," it was time to begin developing the backup detail for the cost model. The conceptual design progressed to about thirty sheets of drawings. A two-day "discovery workshop" was scheduled to discuss all building systems, scope issues, and details. This is the point where the estimator must include as much detail as he can envision for the project because it is the only way to convey to the designers what you think it is going to take to build the project. Verrastro states, "The workshop session was extremely beneficial for me to help develop the scope of the project and to also build a working relationship with the individual design disciplines. I needed to solicit as much information as I could from them so I could anticipate what they were going to design even before any real design work had even begun." Procedurally, after these detailed scopes of work were developed, a comparison of systems costs was made to other projects to ensure that the Hilton project was within range of other comparable projects, and if not, why not. At one time there were as many as six estimators working on the project, trying to nail down as much cost information as possible. Once all of this work was complete, this detailed conceptual estimate and its associated design assumptions were presented to the entire design and development team. Each of these assumptions was fully vetted and

THE TEAM (continued)
John Portman and Associates (Atlanta)
Associate Architect:
Joseph Wong Design Associates (San Diego)
Civil Engineer:
Flores Lund Consultants
Structural:
Martin and Peltyn
MEP:
JBA Consulting Engineers
Interiors:
Nirsch Bedner
Environmental Consultant:
Kleinfelder, Inc.
Parking Consultant:
International Parking Design, Inc.
Traffic Consultant:
Linscott, Law & Greenspan Engineers

Courtesy of Hensel Phelps Construction Co.

decisions were made as a team so the designers had some guidelines as they progressed on to the schematic design phase.

The design team organized weekly workshop sessions as the schematic design took shape. It was critical for the estimator to attend these meetings and understand what the scope issues and design desires were. As the schematic design continued, the estimate was routinely updated and communicated to the design team. This cost input helped guide key decisions, such as structural system selection, major exterior facade material choice, as well as mechanical and electrical system options. Even with such close communication between the estimator and the design team, as the schematic design was completed, the estimate was approximately $15 million over the budgets set by the developers. Furthermore, the designers had even more "enhancements" that they wanted to work into the project—and the developers themselves wanted to believe that they could have these enhancements and still remain in budget. Verrastro acknowledges that, "As the estimator, you become the defender of the scope and cost you have been presenting to the team all along. The designers want to believe that there is conservatism in your numbers and that the design does not need to be modified, and typically the developers believe the same. To be successful in this setting, the estimator needs to allow the give-and-take of this process. The estimator must perform a balancing act, working with the designers to help achieve their goals and looking for ways to save money at the same time. The more information sharing and open communication you are able to foster with all parties allows you the opportunity for success in this dynamic and ever-changing, progressive estimating process."

In an effort to reinforce the reliability of the numbers, Hilton brought in their private cost consultant to develop a parallel baseline estimate.

Once completed, both parties compared scope, quantities, and pricing and agreed on a reconciled estimate. Working through this second opinion with the owner/developer stakeholders adds credibility to the plan and gives everyone confidence to make the right decisions regarding design adjustments. The earlier the detail is developed, accurately priced, and shared with the team, the better. Key decisions regarding scope reduction versus enhancement become very costly if they are attempted later in the design process. The project team did elect to simplify elements of the exterior closure of the hotel to save cost, but they managed to include and keep the interior design intact throughout the design process.

Once the project was in the design development stage, more details of the design were known and the contractor was expected to establish a guaranteed maximum price (GMP) for construction. The construction cost was over 65% of the total development cost. The financial market was interested in lending on the project, but only if the general contractor had a guaranteed price. At this stage there were about 100 sheets of drawings for the project and a deadline had been set. The estimator had three weeks to assemble a final GMP estimate package and present it to the development team. Everyone was counting on the progressive estimate to become the GMP. The project was still in the early design development phase and the most recent set of drawings was sent out to competent subcontractors able to perform detailed proposals. These sub quotes would provide backup to the assumptions the team had been designing under since the start. The estimator was being pressured to get the number down "as tight as possible" and to include everything necessary to create a cost proposal with "no holes." It was that simple and that risky at the same time.

Over the three-week period, every minute was used to define and detail every component of the conceptual estimate to ensure its reliability when both senior management and the development team reviewed it. The final GMP was set at $229 million for construction. This guaranteed cost was required to secure the loans, complete the design, and begin construction. No subcontracts were actually in place at this time because the design was still in progress.

Once the design was far enough along and the project buyout [the issuance of purchase orders for materials and equipment] began, commodity prices began to rise as the massive exporting of all major building materials to China dramatically increased during 2005 and 2006. Almost immediately after the GMP for the project had been set, the costs of aluminum, steel, cement, drywall and glass began to escalate and none of the subcontractors was on board yet. The on-site purchasing team began preparing bid packages for each component of the project, knowing they would soon find out how close the estimator had come to forecasting the final costs. With great anticipation, the first bid packages, including site utilities, earthwork, precast driven piles, and mechanical and electrical systems were released for bids. The subcontractor prices came in at a net over budget compared to what had been included in the GMP. It looked like the project might be headed for trouble as the next set of bid packages, including over $80 million dollars worth of steel, reinforcing, concrete, glass and aluminum were

Courtesy of Hensel Phelps Construction Co.

about to be released. These were pivotal packages which, if extensively over budget, would mean having to change the course of the design and the project itself. Obviously, nerves were a little on edge when the results of these trade estimates came in. Several of the bidding subcontractors were millions over the targets the team had set four months earlier. However, when the team analyzed all the numbers, it looked like the estimated GMP values were not going to lose much more ground. There was still over $90 million to complete the buyout and balance the design impacts. The team anticipated a long design-cost management effort that the estimators would handle one bid package at a time. The team initiated an intensive trend management program to make sure that the entire team was aware of exactly where the numbers were in relation to the targets at any point along the design process. The trend summary report tracked all changes, ideas, and value engineering options as the design progressed. The report kept the development and design teams current on choices, options, and impacts. The estimator prompted the team to finalize design decisions and commit to them so prices could be confirmed and locked in.

The estimator continually looked for savings and the designers and hotel stakeholders were looking to enhance items. There was constant feedback, both pro and con, as the teams balanced design interests with cost impacts. Throughout the process, the stakeholders agreed on various tradeoffs to balance design and budget realities. For example, the podium skin material was switched from precast to plaster saving $1.7 million. This was a large concession by the designer to help get the budget back in line. However, the team decided that the hotel needed a grand staircase. They directed the architects to proceed with the concept and the price was incorporated into the estimate. They also added another restaurant escalator because the hotel operator thought that it was important for traffic flow. Some of the modifications were the result of influences outside the team's control. For example, stringent new Title 24 of the California Code of Regulations requirements had to be incorporated into the design which had cost impacts on the mechanical, electrical, and plumbing packages. The teams agreed on several other design adjustments to help manage the GMP. Fortunately, as the design neared completion and purchasing began for the interior finishes, the market had softened slightly and the overages experienced at the beginning of the buyout were gained back on later purchasing. The on-site design management team working closely with the interiors design team in a huge effort to stay within the previously-established budgets as the buyout progressed.

THE CLOSE

The estimator plays a critical role on the design-build team and must remain fully engaged throughout the entire design process. He is the gatekeeper of all of the thoughts, ideas, changes, wants, and needs that might have an impact on the project's cost. The estimator must have the rare ability to work alongside the architects, engineers, and other stakeholders, as well as end users, to continually filter design considerations, anticipate their cost impacts, and judiciously communicate them to the team to inform their decision making. Not just any estimator has the skills, patience, or courage to perform this role. Fortunately, the San Diego Bayfront Hilton Hotel project had one of the best—someone who was not just an expert in developing costs, but also one who knew how to communicate effectively with designers and owners alike—giving meaning to that single line item cost for construction that was laid out in the original proposal. Everyone on the team could understand what it represented and how it could manifest into a magnificent 30-story, 1.65 million-square-foot hotel sitting on the water's edge, and featuring 1,190 guest rooms, and more than 100,000 square feet of flexible meeting space.[5] More than 1.5 million working hours were logged to complete the project along with 2,000 concrete piles and stone columns, 112,000 tons of rebar, 55,000 cubic yards of concrete, and 4,000 tons of steel.[6] It is clearly a project that the entire team can be proud of, not just because of its architectural beauty, but also for the superb balancing of that beauty with responsible development and estimate management.

From the Estimator

My role on the San Diego Bayfront Hilton project as a Senior Estimator was a very rewarding experience. It was energizing to work from the start with a world-class design and development team on a major hospitality property. I enjoyed seeing the design creativity put to paper early on in the process. I used every bit of my construction experience as I tried to piece together all the necessary detailed scopes of work for the project's cost estimates.

Mike Verrastro, Chief Estimator,

Hensel Phelps Construction Co.

Sample Project
Phoenix, Arizona
University Classroom Buildings

System Cost Model

CONSTRUCTION SYSTEM	249,610	S.F.	222,166	S.F.	250,000					S.F.
	PROJECT A		PROJECT B				YOUR PROJECT COST RANGE			
	TOTAL	CST/SF	TOTAL	CST/SF	Low	CST/SF	High	CST/SF		
DIRECT JOB EXPENSE	2,069,916	8.29	1,699,570	7.65	1,937,500	7.75	2,125,000	8.50		
SITE WORK	2,029,329	8.13	4,292,866	19.32	2,250,000	9.00	5,000,000	20.00		
SUBSTRUCTURE	1,647,815	6.60	1,805,884	8.13	1,687,500	6.75	2,125,000	8.50		
SUPERSTRUCTURE	11,396,713	45.66	7,881,295	35.47	8,875,000	35.50	12,000,000	48.00		
EXTERIOR WALL	4,570,050	18.31	4,690,303	21.11	4,625,000	18.50	5,500,000	22.00		
ROOFING	324,693	1.30	500,511	2.25	325,000	1.30	562,500	2.25		
INTERIOR CONSTRUCTION	6,269,275	25.12	3,983,884	17.93	4,500,000	18.00	6,250,000	25.00		
INTERIOR FINISHES	2,148,383	8.61	2,567,962	11.56	2,250,000	9.00	3,000,000	12.00		
BUILDING SPECIALTIES	257,048	1.03	191,789	0.86	250,000	1.00	275,000	1.10		
EQUIPMENT AND FURNISHINGS	91,996	0.37	45,938	0.21	62,500	0.25	125,000	0.50		
SPECIAL CONSTRUCTION	43,292	0.17	438,035	1.97	62,500	0.25	500,000	2.00		
CONVEYING SYSTEM	1,853,454	7.43	794,063	3.57	1,250,000	5.00	2,000,000	8.00		
FIRE PROTECTION SYSTEM	781,968	3.13	516,141	2.32	625,000	2.50	812,500	3.25		
PLUMBING	1,168,894	4.68	887,526	3.99	1,000,000	4.00	1,250,000	5.00		
H.V.A.C.	9,591,963	38.43	5,264,680	23.70	6,250,000	25.00	10,000,000	40.00		
ELECTRICAL	8,109,200	32.49	6,385,760	28.74	7,250,000	29.00	8,750,000	35.00		
SUBTOTAL	52,353,990	209.74	41,946,209	188.81	43,200,000	172.80	60,275,000	241.10		
DESIGN COSTS	7,900,000	31.65	0	0.00	6,000,000	24.00	8,500,000	34.00		
INDIRECT COSTS	15,249,733	61.09	12,342,047	55.55	12,960,000	51.84	18,082,500	72.33		
GRAND TOTAL	$75,503,723	302.49	$54,288,257	244.36	$62,160,000	248.64	$86,857,500	347.43		

Note - Indirect Cost Include:
General Conditions, Sales Tax,
Bonds, Insurance, Contingency
and Fee.

Figure 6.7
Sample cost model. Courtesy Sundt Construction

the construction. These cost model unit prices are used to establish the budgets for the new project. The sample shown in Figure 6.7 only uses two models to form the new price, but it is not uncommon for some models to use up to four or even five comparable projects.

> *Once we determine the product type, we query our historical cost database for similar projects. We then collect the "as bought" systems cost for those projects and cost index them to today's dollars. We put these onto a spreadsheet and develop a low range and a high range. By identifying the two extremes you can start a meaningful dialogue about the specifics of your project and develop a project-specific model for everyone to use. To get a better sense of the applicability of the reference models you can visit the projects and get a feel for what the budget did buy.*

Don Goodrich, DBIA
Vice President
Sundt Construction, Inc.
Tempe, Arizona

If the design-builder is brought on after the budget has been established, then the design-build team will create a cost model that reflects that budget and use it to help manage the design process. Either way, creating a cost model is the critical first step in the design-build estimating process. And it is critical that every member of the design-build team—contractors, architects, engineers, and owners, all understand how the cost model works and how important it is to design-build project success.

Purpose of the Cost Model

Failing to properly manage the design to budget is one of the biggest mistakes made by contractors new to design-build. Once they win the project, they send their design partners off to design the project, giving them the overall contract amount as a guide and the directive, "Just make sure the design does not exceed this contract price" and "We'll meet again when you have the design at the 30% stage for review." But architects are not any better at calculating construction costs and budgeting then contractors are at managing design. In all likelihood, the 30% design is (most likely) going to be over budget and the contractor is going to blame the designer and immediately send her back to redesign, and probably refuse to pay additional design fees to do so. This finger pointing begins to look a lot like what we see with the design-bid-build model—no wonder some architects are not exactly enamored with design-build, or contractors either. This wasted redesign time is nothing new in the traditional design process; but this time, the design-build team is paying for it and not the owner. If the design-build team is being led by a contractor, which most of them are, then the contractor is going to pay for this additional design time. Obviously, sending your design team off to design as usual, with nothing more than the contract price as a guide, does not provide the proper monitoring and feedback necessary to keep the project on target. If the project is contractor-led, then the

> One of the strongest arguments for choosing your design-builder through the QBS approach is to get the team on board early to make that initial tie-in between the project program and the cost model, before any design occurs. The design-build team can establish a realistic, fixed budget to match the program requirements based on past projects. The design-build project manager can then work directly with the entire project team, including the owner and all associated stakeholders and end users to help manage the design to the budget throughout the design process—eliminating costly redesign associated with budget busts.

Contractors are not accustomed to sharing project estimates and cost information with architects and design engineers. It is typically not a requirement in design-bid-build, especially under a lump-sum bid option. When first getting involved in design-build, many contractors guard their cost information closely, fearing that their design partners will leak this information to their competitors. But in design-build, it becomes vital that the design team has adequate cost and quantity information to help guide their designs to meet the budgets that have been set forth. Withholding this information can be disastrous.

TAKEAWAY TIPS

Hitting the Targets

Consider conducting a formal **Cost Model Briefing Session** with the entire design-build team so everyone understands the critical relationship between design and cost relative to the project. Explain the various elements of the model so everyone knows how to use it, what the boundaries are, how to measure them, what the tradeoffs might be, and how to test design progress against the model. The cost model is one of the primary design-build project control tools, and all variances from it should be agreed upon and properly documented. Continue to update and refine the model as the design progresses.

---

contractor is responsible for providing the cost expertise, information, and tools necessary to guide their architects' and engineers' designs to adhere to the budgets established.

The most reliable way to manage the design to budget is to create a cost model at the earliest possible design stage. As the design progresses, the team will manage it to fit the cost model.

The cost model sets the design parameters, the scope, cost, and size limitations for the project. The cost model establishes the initial benchmarks by which the project costs will be measured and managed. As the design progresses and the cost model evolves, quantity limits for each element of the design are often tied to the budget limits as a more useable guide for designers, but both should be consistently monitored as the design develops. In addition to helping guide the design-build team as they work through the various alternatives and options on their way to a final design, the cost model also serves as an excellent tool for directing the value engineering effort throughout the design process.

How the Cost Model Evolves

Once an appropriate historical cost model is chosen for the new project, the design-build team adjusts its various components to fit the unique circumstances of the project at hand. The adjustments could account for time, location, quality, and any other factor which might impact the accuracy of the model. Some design elements might need to be calculated from scratch if the historical information is not available or useful. Once the team makes these adjustments, they agree on the cost model and it will serve as a reference point and guide as the design continues to develop from the initial concept and/or schematic drawings submitted in the proposal through final committed design details needed for construction. The team updates and refines the cost model as the design is completed and eventually evolves through a series of estimate iterations transitioning from systems, to elements of the system, to assemblies, and eventually to components of the assembly as you might see in a detailed estimate. The first three stages are typically organized using the UniFormat structure. (See Figure 6.8.) Eventually the estimate will be converted into the MasterFormat organizational structure so work packages can be assembled for subcontractor and vendor pricing.

■■ PROGRESSIVE ESTIMATING PHASES

As you have learned, the cost model is only the first step of the design-build estimating process; there are several other conceptual estimating iterations necessary to take a design-build project from program to construction. A design-build project generally goes through five **progressive estimating** phases. Each of these phases corresponds to a complementary design stage and has a specific purpose, as shown in Figure 6.9.

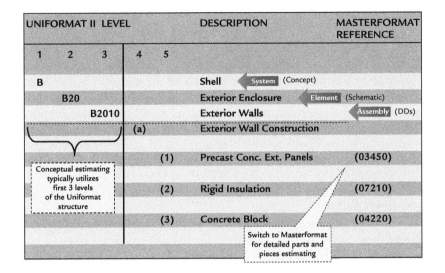

Figure 6.8
Cost model evolution using UniFormat

1. Feasibility Estimate

The purpose of the feasibility estimate is to establish whether the project and scope of work is economically and financially within the client's means. The feasibility estimate, as you learned earlier, is usually done by the owner's own staff or a hired consultant and usually employs very basic estimating methods that generate what many people refer to as a ballpark estimate, or the figure that might be in the very first conversation about the project cost. An estimating method commonly applied at this stage is called the Rough Order of Magnitude (ROM) estimate. This simple, fast method applies a dollar figure per typical end user unit associated with the project type. For example, for hospitals, the ROM estimate would allocate a dollar amount per patient bed. For schools, it would calculate a dollar amount per student. Parking structures would estimate a dollar amount per parking stall, and wastewater treatment facilities might apply a dollar amount per gallon of water treated. The cost data used for these

Type of Estimate	Design Stage	Purpose
Feasibility	Program	Owner makes project go-ahead decision
Concept	Concept	Establishes the cost model
Schematic	Schematic	Establishes design parameters in accordance with cost model
Design Development	Design Development	Confirms details of design and cost model alignment
Detailed	Construction Documents	Release for construction

D-B DICTIONARY

Cost Model Briefing Session – A formal meeting that takes place with all members of the design-build team shortly after the cost model is created to explain how it will be used to manage the design to budget throughout the entire design process.

Progressive Estimating – Term used to describe the design-build estimating process. The estimating task begins at the earliest stages of design with feasibility or concept estimating, and progresses as the design develops through the schematic estimating stage, the design-development estimating stage and then finally to the detailed estimating stage.

Figure 6.9
Estimating phases

types of estimates comes from past completed projects of similar scope. Contractors who specialize in specific project types such as schools, hospitals, parking structures, wastewater plants, highway projects, or office buildings are an excellent resource for ROM estimating information. This type of feasibility estimate will help the client make the initial go/no go decision for the project. However, after that, a more progressive type of feasibility estimate may take place, using various square foot unit prices to verify the initial "go" decision. If the owner elects to choose their design-builder by direct selection or QBS, then the design-build team might provide all of this information from the start.

2. Concept Estimate

The concept estimate (also called a program estimate) primarily uses square foot cost data that considers the overall project or components of a project, like the sample cost model in Figure 6.7 and may also help further determine project feasibility. Whereas the feasibility estimate can help the owner make the "go or no-go" decision for the project, the concept estimate is often accurate enough to establish the project budget even before any real design work has been done. In most cases, the program information and performance criteria will provide enough information for a reliable cost model that will accommodate enough design to meet all project and program requirements. When engaged early, experienced design-builders can use the concept estimate to manage the design to budget from start to finish, serving both the client and the design team in achieving the overall project goals.

3. Schematic Estimate

It is not unusual for the lines to become blurred between the concept estimate and the schematic estimate, depending on when the design-build team is engaged. Regardless of the timing, a cost model should be developed as early as possible to guide the rest of the process. In competitive design-build, the schematic estimate is usually the first step in the conceptual estimating cycle—especially if the owner has already established the project budget. The project program and scope has already been defined and presented in the RFP. Some owners opt to include some preliminary design in their RFP as well, although this is not recommended. However, even with limited information, and many yet-to-be designed details, the design-build team would start the estimating process by building a cost model to first confirm the owner's budget before proceeding with any design. After the cost model is completed and the budget is confirmed, the design-build team can proceed to the schematic design phase and develop their proposed design and corresponding estimate simultaneously. The schematic phase uses square foot costs at the systems and assemblies stage. This is when the value engineering process usually begins and a lot of the what-if estimating scenarios across alternative design options are tested in order to determine a direction to proceed with the design. Unfortunately, unless the design-build team is chosen as a result of direct selection, the competitive process does not allow the owner to be engaged during these early what-if discussions—just another reason to select your design-builder under a QBS scenario if you have that option.

4. Design Development Estimate

During the design development estimating period, the various design details are systematically checked and verified to meet both budget and performance requirements. It is not uncommon for the value engineering effort to continue into the design development stage and changes in material and even size and configuration of the project may occur. However, the goal during design development is to confirm and commit to as many design details as possible, as quickly as possible, mitigating as much risk and exposure as you can. Once the design elements are finalized and scopes of work are packaged and issued for pricing, they can be verified to meet the budget limits set forth in the cost model. Then each component can be secured with the prices locked in, either through purchase orders or subcontracts and the design can actually be released for construction.

> It is a good idea for the designers and estimators to co-locate during the schematic phase if they can. Not only does it increase the odds of them getting the job; it also helps solidify the goal of keeping the design-cost relationship in balance after contract award.

5. Detailed Estimate

At this stage of the estimating process in design-build, you will generally not need to conduct a full-blown, detailed estimate. Much of the design work is complete and most of the pricing has been confirmed during the design development estimating stage. At this point, the only reason to possibly conduct a detailed estimate is that you are self-performing some of the work and need a quantity survey to order materials. Otherwise, the detailed estimating has most likely already been done by your specialty and subcontractors. Once they have submitted their lump sum prices to you, you do not need detailed estimates unless you want to create estimate checks to verify bids.

Different estimators may have different names for each of these five estimating phases and define the transition from one to another differently, but the concept remains the same: the design-build estimating process is progressive. It is not a one-time task. But the notion that the project cost itself is a moving target through the various phases is false. The project price is established early and then you manage the design to align with that price.

The traditional design-bid-build approach assumes that the level of estimate accuracy depends on the level of design completeness. However, in design-build, the level of estimate accuracy does not depend on the level of design completeness. Instead, it depends on the effort put forth to manage the design to the budget. The accuracy of an estimate will remain constant throughout the various design phases if:

:: You start with a rock-solid cost model with sufficient input from all the right stakeholders and associated parties;

> Most design-builders have reported that they feel comfortable locking in a lump sum or guaranteed maximum contract price with an owner at approximately the 70% stage of the design development phase. However, in most competitive design-build scenarios, they are required to submit their price at the schematic stage. That is where contingencies come in. You will learn more about contingencies later in this chapter.

D-B DICTIONARY

Design-Cost Reconciliation – A design-build management process used to verify that the project design is progressing in compliance with the project budget displayed as a cost model.

- :: The estimate is based on reliable historical cost data, properly adjusted to meet the unique circumstances of the project;
- :: The estimate is managed by qualified management personnel trained in the verification procedure called **design-cost reconciliation**.

Such an estimate will remain stable even if the initial estimate was created at the very earliest stages of design, and even as the estimate becomes more detailed as the details of the design progress. This statement may turn what you know about traditional detailed estimating upside down, and because the concept is so revolutionary, the next chapter of the book focuses on design-build project management and design-cost reconciliation.

The Estimating Churn

Experienced detailed estimators often find it difficult to get used to the repetitive nature of the design-build estimating process. They are often initially disturbed when they have to estimate again, and again, and again. Their design-bid-build experience suggests that "something must be wrong" if you have to make an estimate more than once, or continually adjust it. They are accustomed to developing an estimate once, for bidding purposes only. The plans were complete and they could count all the parts and pieces. But in design-build, initially, there are no parts and pieces to count—they only become apparent as the design develops, and yet you are still required to estimate and predict what the project is going to cost, and stand behind your contract price. The details of the estimate develop as the details of the design develop. The estimating process is progressive and dynamic. Instead of a one-time exercise, it is an ongoing practice. Carter Vecera, an estimator with The Beck Group in Dallas, Texas, referred to this recurring process as "the churn," which is a fitting description of the design-build estimating process. Each stage of the estimating progression represents a design-cost reconciliation milestone opportunity and the constant updating is the best way to monitor and manage the design to budget. Figure 6.10 shows the progression from one milestone estimate to another as the design develops.

However, even though Figure 6.10 shows a smooth transition from one progressive estimating phase to another, in reality there are always minor adjustments necessary to keep the project on track and within budget. The critical key is to detect the variances as soon as possible so you avoid costly and unnecessary design iterations.

TAKEAWAY TIPS

Garbage In, Garbage Out

Most cost models and historical project costs are based on delivering projects under the less efficient design-bid-build method. As you complete more and more design-build projects, you might want to track historical costs completed under the faster, better, cheaper design-build approach separately so you can create a more accurate database that reflects this more efficient method.

Beyond Milestones and Phases

Even the intermittent milestone estimates are not enough to keep the tenuous design-cost relationship in balance. You need to introduce a

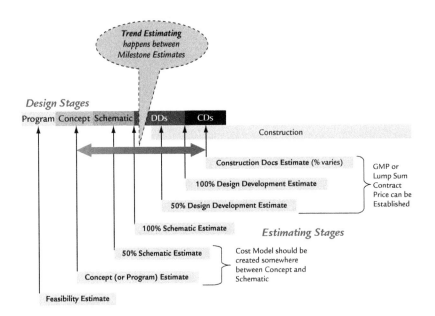

Figure 6.10
Estimating progression

process called **trend management** between the milestone estimates to keep the project on track, but you will learn more about that in the next chapter as well. For now, understand that performing these progressive milestone estimates alone is not sufficient to properly monitor and manage the design to budget as efficiently as possible.

Keeping the design on track with the budget is one of the most difficult aspects of design-build project management and often the root problem when design-build projects miss their profitability targets. Also, keep in mind that the various estimating phases may overlap one another just as the design stages overlap which adds even more complexity to the estimating process. It is not uncommon to have one element of the design in the schematic estimating phase and another part of the design in the design development estimating phase. For example, the site work and foundation design might be advanced to the design development

D-B DICTIONARY

Trend Management – A design-build management technique that tracks and documents cost fluctuations between milestone estimates as the project design progresses in an effort to keep the project on target with the budget.

> *In design-build you are not finished estimating just because you gave someone a price. Those first numbers developed at the schematic design stage may establish the overall budget, but there is still a whole lot of detail that must be fleshed out before materials can be ordered and subcontracts issued. There is this constant churn that takes place as the design advances—there are meetings and phone calls and adjustments made throughout the design process. It is very different from bidding a completed set of documents. Traditional estimators are often uncomfortable with the dynamic nature of the process. I have to constantly remind my staff and associates that this is the way it is supposed to go.*

Carter Vecera
Director of Preconstruction
The Beck Group
Dallas, Texas

stage with corresponding design development estimating going on while several of the architectural elements may still be in the schematic design and estimating stages. But as complex as the design-build estimating and design management process may seem, the good news is that there are tools and techniques you can use to achieve optimum project success for all parties involved.

QUALIFYING THE NUMBERS

In the design-build estimating process, determining the actual project cost is just part of the challenge. The numbers are important, of course, but not half as important as what the numbers represent. Not only do you have to quantify the project, but you also have to qualify your price. The words you put around that estimate—the clarifications and the qualifications—are just as important, if not more so, than the actual numbers. You have to document what the estimate is based on, so that everyone understands what the price represents.[7]

In design-bid-build, price qualification has already been done for you in the completed plans and prescriptive specifications. However, in design-build you develop your price based upon as little as the owner's program and some performance criteria, and perhaps some preliminary design. But none of those documents adequately describes and communicates what you have priced in your budget. And even though you have not worked out all the design details at the time of the estimate, you do have some sense of the quality range included in the cost model you used to establish your budget. In design-build, your estimate should always be accompanied by **outline specifications** that will communicate to your client the range of materials, products, options, and quality that you have priced and that they can expect to see as the design develops. The outline specs define the scope associated with each of the design elements in the estimate and are usually written as a narrative, as shown in this example:

> <u>Window System</u>: *Windows will be 1" thick gray-tinted thermal panes set in clear anodized aluminum frames with thermally-broken design. Sills will be metal flashing with drywall receiver. Spandrel glass will be provided at nonvision areas. Areas behind spandrel glass sections will be insulated. The window system will include integral aluminum doors including thresholds and appropriate panic hardware. Metal trim and flashing will match the window system on the exterior wall.*

If you do not include an outline spec or design narrative with the estimate, you open yourself up to a great deal of misunderstanding and unnecessary risk. Likewise the client is left to interpret your intent any way she likes and then you have to deal with the disappointment when her expectations and your deliverables do not match—even when your deliverables clearly meet the stipulated performance criteria. Communicating the design narrative can make design-build estimating a little tricky—and dangerous, if not managed correctly.

D-B DICTIONARY

Outline Specifications – A brief narrative outline corresponding to the design-build estimate describing the range of materials, products, options and quality that each cost category in the estimate represents. Also called a design narrative or project scope definition.

Crunching the Numbers

Design-build estimating requires a lot more documentation than hard-bid estimating. Because you do not have completed plans and specs, you have to make many assumptions and qualifications as you price the project. So you are always taking notes, jotting down your questions, your assumptions, conversations with other people, etc. It is very important because when you provide your first numbers the design may be very conceptual and it may be weeks or even months before you revisit that estimate. In between, you may have worked on ten different estimates. You have to quickly refresh your thinking regarding the logic and assumptions that you made when you last worked on the project, because many of those assumptions have now been clarified at the 20%–30% stage and you need to make sure they still jive. And you have to be careful not to make assumptions in isolation. Estimators have a tendency to do that for the sake of expediency, but it is a bad idea. You should always run your assumptions about the design by the engineers or architects. You have to do design-build estimating as a team. The best part about the process is when you are co-located in the same space and you can sit across the room from the engineers and bounce ideas back and forth and really get involved in the decision making about how the project will unfold. It really gets everybody's juices flowing and there is a lot more to it than just crunching numbers. It is my favorite part of the design-build estimating process.

MENTORING MOMENT

Michael Ng
Estimator
CDM Constructors Inc.
Rancho Cucamonga, California
Education: Cal Poly State University, San Luis Obispo
Graduated 2007
Major: Construction Management

ACCOUNTING FOR THE UNKNOWN

This chapter began by discussing the uncertain and risky nature of trying to put together a cost estimate when you have very little design information to go on. And you have learned about the various techniques that allow for the design-build team to anticipate what will be added during design. But how does the team account for all of those unknown design surprises that they are not able to anticipate early in the process? How do we account for the uncertainty associated with design impacts beyond our control, such as future decisions by seismic review boards or signoffs by permitting officials? How do we account for all of the unknown gaps in the design during the pricing stage, yet still come up with a reasonable project cost and budget for the project?

Estimates must always account for some level of uncertainty, whether the documents are complete or not. In fact, the less design that is complete at the time of pricing, the more uncertainty we have to account for and the more design that is complete, the less uncertainty that we have to contend with. In design-build, the mechanism used to account for this uncertainty is called **contingency**, and it is a part of every design-build estimate. Generally, the earlier the owner wants the design-builder to lock in a firm price, the more contingency will be included in the estimate. The contingency is an amount of money assigned to an estimate to cover the unknown. However, contingencies are not just fudge factors or random guesses. Estimators think them through and analyze them in a logical, and well informed manner.

D-B DICTIONARY

Contingency – An amount of money added to an estimate to account for various unknown factors at the time the estimate is initially developed, such as construction contingencies, design contingencies, and owner contingencies. As the design develops the contingencies generally decrease.

D-B DICTIONARY

Risk Assessment –
The determination of both quantitative and qualitative impacts associated with various known and unknown factors affecting the design and construction of a project.

Full Disclosure –
The act of revealing all relevant information about a design-build project and the process by which it is estimated, designed, and constructed.

Open Book Policy –
A philosophy that encourages transparency regarding all costs associated with a design-build estimate.

Contingency Development

Contingency development should always begin with a **risk assessment** for the entire project. To develop a risk assessment, the design-builder evaluates and analyzes the multiple factors that will influence the overall project performance to determine the quantitative and qualitative impacts to the project. Initially, most design-builders will do a risk assessment related to the RFP just to make a "go-no go" decision about whether to propose on the project or not. (You learned about this part of the process in Chapter 5.) However, once the decision is made to proceed with the proposal, a more intensive risk assessment should take place.

Contractors are accustomed to conducting risk assessments from a "construction only" perspective and taking various actions to mitigate those risks, either by assigning dollars to the estimate or by other means (insurance, subcontracts, etc.). But the design-builder must recognize and assess the integrated nature of a design-build risk environment so she can apply adequate contingencies, for both construction and design, to the estimate. The first step is to make sure that the right people are at the table when risks and contingencies are being discussed. However, this is sometimes easier said than done.

In a design-bid-build environment, contractors are notoriously guarded about letting any of their bid numbers or pricing strategies out to the public, and for good reason. Remember, in most cases they are required to compete based upon having the lowest price. However, in a design-build environment, this guarded approach can lead to gaps in the risk picture and can actually be dangerous to the success of the project. As you will learn in Chapter 8, the hallmark of design-build is trust, and nothing communicates trust more in a business relationship than **full disclosure** and an **open-book policy**—at least in terms of the project estimate and particularly with your own team members.

John Mc Garva
President and CEO
Western Water
Constructors, Inc.
Santa Rosa, California

We believe that full transparency is what builds trust, and trust is the key to any good design-build relationship. Voluntarily, we have initiated a web-based project management system that tracks all job costs on a daily basis. We set up the owner and engineer members of the project team with 100% access to the job costs as they develop, allowing them to see all costs on the job in real time. Nothing is hidden even though we are not required to disclose any of this information. All team members can independently drill down into the cost details on the project at any given moment and know exactly how every dollar is being spent. This allows for efficient and comprehensive communication on all things associated with the project budget. This is usually one of the more difficult issues on any project, but this system has really helped build trust among the team. Our experience with this system has resulted in a quality project that has been completed under budget and on time, with a satisfied owner and consulting design engineer. We exceeded all budget goals through the shared savings generated by cooperative value engineering. We hope this experience sets a precedent for future projects for our company.

Assessing Design Risks

Contractors are familiar with the notion of risk assessment. In fact, they are so comfortable with the process that sometimes they decide to assess the design risks on a design-build project without ever consulting their designers, which is a big mistake. Not being designers, contractors approach design risk from a purely construction orientation. They think in terms of what it is about the design that can interfere with their construction plans, such as:

:: Errors or omissions that the designer might make;
:: Redesign that will be necessary because the project is over budget;
:: Design delivery delays that may extend the construction schedule;
:: Having to pay more design fees to the architects and engineers because they have expended their design fees before the design is complete.

Contractors are not necessarily thinking about the unknown factors that could impact the design itself after the price has been committed or the types of issues that architects and engineers need to contend with as they move through the design process. In a design-build estimate, the design contingencies must reflect the designer's design issues as well as the contractor's design issues. Consider three distinct types of design risks when assessing estimate contingencies associated with design. They are design risks associated with:

1 **Design Services**: These are the type of services that most contractors think about when they are assessing design risks on a design-build job. For example, risks associated with design accuracy, timely design delivery, and completing design within the design budget.

2 **Design Impacts** Design impacts are the types of risks that architects and engineers think about. They include information that will influence the design direction and final design details—factors like loading requirements, environmental studies, seismic issues, regulatory restrictions, and others. In traditional design-bid-build, these impacts are typically dealt with before the plans and specs ever go out for bid, so their influence on the design has automatically been accounted for when the contractor estimates the job. But in a design-build environment, these impacts become part of the risk scenario and must be accounted for in the estimate through contingencies or other measures intended to mitigate the risks associated with them. Contractors have often been remiss in assessing these risks appropriately.

3 **Design Aesthetics** Risks that relate to design aesthetics often come into play on projects where "signature architecture" is part of the performance requirement. They deal with designer latitude and freedom in addressing the project aesthetics. Trying to assess the cost impacts of creative expression before the design is complete is next to impossible. In design-bid-build, the owner addresses this issue during the design contract and the ensuing bidding process after the design is complete. But in design-build, the estimate must reflect

If you are a contractor who has plugged in a "shot in the dark" design contingency on your last design-build project, you might be having an "aha moment." The risk exposure that you are taking on by not fully engaging your design team in the contingency development discussions is enormous.

D-B DICTIONARY

Design Impact Risks – Risks that influence the design direction and final design details, such as loading requirements, environmental studies, seismic issues, and regulatory restrictions.

Design Stability – The degree of a given design element's completeness at the time of pricing. It considers the likelihood that the design will change between the conceptual stage and the construction document design stage. The more likely that the design will change, the less stable the design.

a reasonable prediction of what this design impact is going to cost before it actually occurs. In some instances, this contingency can be built into the estimate and in other cases it may be identified as an independent add-on line item at the end of the estimate. But either way, it is clearly a contingency that you have to include if your client is expecting high-end architecture. Your estimate must account for both the added designer services and the construction impacts.

You assess design services and design aesthetics risks, and assign appropriate contingencies to mitigate them, in much the same way that you handle traditional construction risks. You consider two factors: the magnitude of the potential loss associated with the risk, and the probability that the loss will occur, and then calculate and assign a dollar amount to the estimate for contingency. Most contractors have a standard way of addressing risks in this manner. (Refer to Figure 5.2 in Chapter 5.)

However, the traditional probability approach is not sufficient for calculating **design impact risks**, risks such as loading restrictions, seismic issues, or life safety code requirements that influence the final design. This is because the accuracy and completeness of the design is dependent upon the availability of information at the time the estimate is prepared. Some of the information necessary to accurately design various elements of work might not be known at the time the project price is put together. So in addition to considering probability and impact, the reliability of the design should also be considered. The **design stability** approach, which considers the completeness, and therefore the reliability, of a design element at the time of pricing, should be used when dealing with design impact type risks. This approach considers the designer's confidence level that the design at the time of the estimate will remain stable throughout the design process—from the schematic design stage through construction documents. We all know that in design-build, the design at the time of the estimate is usually only at the conceptual or schematic phase and that the design will change as the various design impacts come into play. Some of those changes can be easily anticipated and accounted for even with limited information while others are not so easy to predict and the likelihood of an unanticipated change is higher.

In design-build, as you know, we must manage the design to the budget assigned by the cost model. But the cost model should account for design impacts and their influence on the final design. Some elements of the design will remain stable throughout the design process (such as finishes) and others will be unstable at the time of the estimate (such as the structural system). The design-build team must openly discuss the design stability of each major design element and its components, and then decide how best to account for that instability. The team might increase the contingency for the most unstable design elements. Or they might continue designing the high-dollar, high-impact design elements, until they are confident that a less conservative contingency will be sufficient. It is important to understand that the design contingency is not a mysterious, shot-in-the-dark percentage that everyone hopes will be enough. Instead, the entire team continually discusses and adjusts contingencies

Elements of Design – Structural System			
Slabs	**Columns**	**Beams**	**Shear Walls**
$ Value	$ Value	$ Value	$ Value
% of Project	% of Project	% of Project	% of Project
Design Stability Confidence Level	**Design Stability Confidence Level**	**Design Stability Confidence Level**	**Design Stability Confidence Level**
Adjustment Increase-decrease contingency OR Continue designing	Adjustment Increase-decrease contingency OR Continue designing	Adjustment Increase-decrease contingency OR Continue designing	Adjustment Increase-decrease contingency OR Continue designing

Figure 6.11
Design impact risk assessment

for all major elements, and then uses their risk assessments to make well-informed project decisions. See Figure 6.11.

One excellent strategy integrated design-build teams can use to capture risk assessment data is to convene a **contingency round table**. This meeting should include all of your of designers (architectural, mechanical, electrical, civil, etc.) as well as major subcontractors. You may even want to include your fabricators and major vendors, who might provide input that would significantly influence the design. Each of these team members will view the project risks from different perspectives and shed light on perils that the contractor might not have considered. This type of input is critical to making good contingency allocations in managing risks for both construction and design. This multidisciplined approach can help identify "holes and gaps" early in the process and may even enhance profit potential because it strengthens the overall risk mitigation strategy.

Avoiding Redundant Contingencies

Design-build considers many different types of contingencies. All parties involved in the project attempt to assess risk and assign appropriate contingencies to address them—not just the contractor or the designer. In addition to the construction contingencies and the primary design contingencies, consider also subcontractor, subconsultant, and supplier contingencies. But you must ensure that the parties are not "doubling up" contingency numbers, or creating **redundant contingencies**, in the estimate.

For example, if the subcontractors are already accounting for their associated risks when they give their prices to the general contractor, then the general contractor would not necessarily have to add any additional contingency. Or if the architect has already assigned figures to cover redesign efforts, then it would be redundant for the contractor to do the same. On the other hand, the architect might have only included contingency for architectural risks, but not for mechanical or electrical or other subconsultant design work. The potential for redundant contingencies

D-B DICTIONARY

Contingency Round Table –
A meeting of designers, subcontractors, fabricators, and vendors to discuss project risks from a variety of perspectives to assess risks and develop contingencies; representatives from all disciplines can voice their concerns for the project from their subjective perspectives.

Redundant Contingencies –
When duplicate contingencies are included in a design-build estimate; teams should work to prevent redundant contingencies.

Some contractors and subcontractors still create estimates from schematic drawings and treat those documents as if they are complete design, not properly accounting for the design adjustments that will inevitably occur as more design impact information becomes available. This is a classic example of the design-bid-build estimating mentality colliding with the integrated nature of the design-build process. We cannot approach design-build estimating from a "hard-bid" mentality and try to do traditional quantity takeoffs from schematic designs, and then after contract award, when our quantities are off, accuse the design team of changing the design after contract award—or worse yet, insist that they made errors in their original design because all of the details were not shown on the drawings from which the estimator calculated quantities. The schematic stage does not offer this level of detail. Design-build estimators cannot estimate in isolation. Assessing risks from a single-discipline perspective like this invites trouble. In design-build, estimating must be done by the integrated team, which must assess the design impact risks at the time of the estimate and account for those risks accordingly. Otherwise, we return to the finger-pointing scenario that so often characterizes design-bid-build.

makes it even more critical to develop contingencies as a team and not in isolation. In a contingency round table, all parties can sort through all of the various risks and discuss how best to account for and mitigate them. They can decide how much each contingency should be and how they should appear in the estimate. They will also decide who should take responsibility for each one and establish rules about how they can be used. This critical discussion must take place during the estimate phase to avoid misunderstandings and disagreements. Usually, the managing party will tell the other team members how to treat contingencies in their independent estimates to avoid redundancy. This communication should be abundantly clear and in writing. Any miscommunication about contingencies can cost the team the job. Or, worse yet, the team might get the job and then find out that they did not account for major risk, endangering both job success and profitability.

Owner's Contingency

As the design develops, owners, too, might find that they have overlooked important project features in their program or RFP. So a design-build estimate should include an owner's contingency as well. The design-build team should assign the owner's contingency to their estimate, unless the client specifically instructs the design-builder otherwise. For example, one design-build team might allot an owner's contingency of 4%, while another team might assign a 10% contingency on the same job. If the project is particularly innovative, a higher owner contingency might be in order, especially if very little design is complete at the time of pricing. On the other hand, if the design-build team has completed a great deal of design of those risky design elements, they can justify a smaller contingency on the same project. In a still-different scenario, if the client is undecided on certain design issues, or her program is incomplete, then the design-build team might increase the owner's contingency to account for possible scope creep. Whatever owner contingency number the team adds to the estimate, they will have to justify it.

Regardless of the level, the owner's contingency belongs to the owner. If the owner does not use it, it reverts back to her. For contractor and designer contingencies, the treatment of unused contingencies depends on the type of contract that has been executed. Under a lump-sum contract, any unused contingencies stay with the design-build team, because they assume the risk. But under a GMP contract, all saved contingency amounts flow back to the owner, unless the contract includes a split on savings. Then some portion of the savings would be used to incentivize the design-build team to be as efficient as possible with the GMP.

Displaying Contingencies in the Estimate

Although there are no hard and fast rules about how to incorporate contingencies into the estimate, often contingencies associated with the construction and design risks are built into the design element line items as part of the unit price, or as a percentage added to each design

Project Estimate Summary

Project Name: Sanford Office Complex
Location: Rome, New York
Owner: Sanford Insurance, Inc.
Project Description: Three-story office
Gross Area: 45,000 SF

Estimator: ALJ
Date: 25-Jun-09

Figure 6.12
Design-build estimate with owner's contingency

Division	Description	Subtotal ($)	Cost/SF	Percentage
A	Substructure	146,977	3.27	3
B10	Shell: Superstructure	704,321	15.65	16
B20	Shell: Exterior Closure	721,090	16.02	16
B30	Shell: Roofing	52,116	1.16	1
C	Interiors	796,987	17.71	18
D10	Services: Conveying	114,055	2.53	3
D20	Services: Plumbing	193,433	4.30	4
D30	Services: HVAC	689,745	15.33	16
D40	Services: Fire Protection	53,011	1.18	1
D50	Electrical	453,981	10.09	10
E	Equipment & Furnishings	111,000	2.47	3
F	Special Construction	8,900	0.20	0
G	Building Sitework	372,336	8.27	8
	Total Direct Costs	4,417,952	98.18	100
	General Conditions – 6%	265,077	5.89	
	Subtotal	4,683,029	104.07	
	Design Fees – 7%	327,812	7.28	
	Subtotal	5,010,841	111.35	
	Overhead – 4%	200,434	4.45	
	Subtotal	5,211,275	115.81	
	Profit – 3.5%	182,395	4.05	
	Subtotal	5,393,669	119.86	
	Owner Contingency – 3%	161,810	3.60	
	Project Total	5,555,480	123.46	

element line item total. Design services and design aesthetics might be handled as a percentage contingency and appear as an add-on at the end of the estimate. However, owner contingencies are almost always shown as a separate add-on percentage at the end of the estimate, as shown in Figure 6.12.

Obviously, the more transparent the contingencies are, the easier it is to see them and therefore manage them. If the owner has not already defined how they expect the contingencies to be handled in the RFP, then it is important that the design-build team clarify this as part of the proposal so that all stakeholders understand them.

ADDING VALUE AS YOU GO

The goal of design-build is to add value to a project as it moves through the design process. The integrated design-build approach allows for value engineering to happen as it should—during the design and not after. The United States Office of Management and Budget defines value engineering as:

> *An organized effort directed at analyzing the functions of systems, equipment, facilities, services, and supplies for the purpose of achieving the essential function at the lowest life cycle cost consistent with the required performance, reliability, quality, and safety.*[8]

In design-build, value refers to the amount of functional benefit the team can provide for the budget, as shown in Figure 6.13.

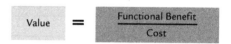

Figure 6.13
Value engineering concept

There are many ways to improve value on a project. In some efforts you might impact benefits and in other, you might impact costs. In still others, you might have the opportunity to impact both. For example you can:

:: Improve benefits and maintain the cost;
:: Maintain benefits and reduce the cost;
:: Improve benefits and reduce the cost;
:: Reduce benefits and reduce cost (if benefits remain within needs);
:: Increase benefits and increase cost (if cost remains within limitations).

As you learned in Chapter 5, you must know the owner and understand their challenges. Then you know how you can best help them meet those challenges, while still saving them money. These savings can then be used elsewhere on the project to add even further value. To do this, you might focus on a product or item itself and see if you can provide it for less money, as shown in Figure 6.14.

An example might be a stick-by-stick, field-erected light-gauge structural steel framing system versus a prefabricated light-gauge steel framing system. The prefab system might be exactly the same system but the cost to build it is less because of efficiencies in the manufacturing process. Or you might consider a function-oriented approach and focus on the needed function or performance. You look for alternative items that can deliver the same results at a lower cost, as shown in Figure 6.15.

> The more complicated and unique the design-build project, the more important it is that you properly assess risks and evaluate the assignment of contingencies to your estimate in an integrated fashion. More than one project has suffered because contractors have assigned design contingencies to a design-build project without ever consulting their own design team. In these instances, it is not unusual for the design risk assessment and the related contingency assignment to be grossly underestimated.

Product Oriented Focus

"How can the same item A be made at a lower cost?"

The result is item **A1**, a modified version of **A**

Figure 6.14
Project-oriented focus

An example might be to consider a prefabricated structural insulated panel (SIP) system that uses wood instead of steel, versus the light gauge structural steel framing system. If the SIP system provides the same function, at the same performance level, but at a lower cost, then you could include that option in your proposal.

When the owner presents her functional needs in performance terms, the design-build team can respond with multiple options and define the value that each brings to address the owner's specific needs. This is what makes design-build fun. The owner benefits because she gets to see an array of solutions that all deliver the same function and performance. A traditional design-bid-build price competition would never show these alternative solutions because every team bids on the single solution articulated in the completed plans and specs.

Of all the benefits the integrated, multidisciplined design-build process can provide, the greatest is adding value as you design. Allowing design-build teams to compete on the basis of their experience, expertise, innovation, creativity, and intellect results in solutions that otherwise could have never been envisioned as a group of isolated individuals—no matter how individually talented they are.

Figure 6.16 and Figure 6.17 are prime examples of what can happen when you get the right design-builder on the job. Figure 6.16 represents the design stipulated in the RFP bridging documents associated with the Fort Lauderdale Airport Interchange. Figure 6.17 illustrates the proposed solution presented by the design-build team of PCL Civil Constructors, Inc. (PCL) and Beiswenger, Hoch and Associates, Inc. (BHA), who ultimately won the project. This example demonstrates the added value benefit that the design-build process can deliver. You do not have to be a civil

> It is up to the design-build team to define value for the owner. Do not presume that the client knows value when they see it. You must point out how your solution provides a greater functional benefit for the dollars spent than the next team's. Some functional benefits are easier to describe than others, such as increased durability, flexibility, sustainability, convenience, or safety. But others are more difficult to describe and justify, like aesthetics, good will, and public perception.

Function Oriented Focus

"How can the basic function of item A be provided at a lower cost utilizing a different item?"

The result is item **B**, which is sometimes a completely different item

Figure 6.15
Function-oriented focus

Figure 6.16
Stipulated RFP design. Courtesy of PCL Construction Enterprises, Inc.

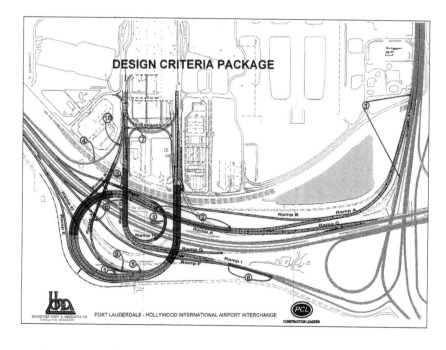

engineer to see that the proposed solution in Figure 6.17 is a more efficient and probably a more cost-effective solution than the RFP design. The proposed solution delivered by the PCL-BHA design-build team offered substantial advantages, including optimum interchange geometric features and cost savings for bridge structures. To provide uninterrupted traffic flow to the airport over the railroad tracks, a complex bridge arrangement was required to allow existing bridge deconstruction and adequate traffic maintenance, which saved the client $800,000. The collaborative effort of the integrated design-build team represents

Figure 6.17
Proposed alternative design. Courtesy of PCL Construction Enterprises, Inc.

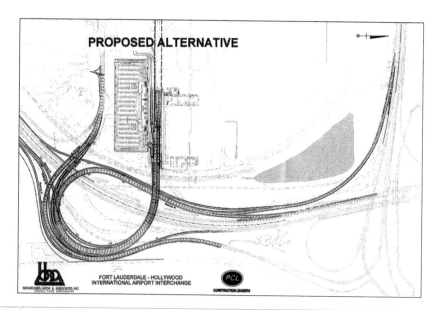

design-build value engineering at its best: it enhanced the project while shaving four months off the schedule.

However, remember that results like these are only possible if, and only if, the owner sets up the procurement as a value- and quality-based competition. You cannot create a low-bid contest, call it design-build because you have a single entity contract, and then expect the teams to spend their energy on finding you the absolute best solution. They will focus on what you reward—if you reward price over value, then that is how the teams will respond. If you want the teams to focus on value and quality, give them your budget, identify your project performance expectations, and ask the teams to bring you their best proposals.

ROLE OF THE ESTIMATOR IN DESIGN-BUILD

As you have learned, the design-build estimator must be a special individual. Not just any estimator can be a design-build estimator. The design-build estimator has a lot of influence and therefore must have considerable expertise and experience in all the major building systems and design elements. She must understand the factors that affect cost and the implications associated with various design options even before they have been determined.

> *The design-build estimator has to be an integral part of the design-build team. The estimator must be committed to the design-build process and understand that the goal is to provide the best possible facility within the owner's budget and schedule. Estimators must be included in design-build meetings with the owner and the design team so the estimator understands the owner's project goals.*
>
> *To build trust and rapport with the design team, the estimator should provide verbal input during the meetings purely from a cost perspective, not trying to tell the design team what to do or adding his personal preferences. A few positive, constructive comments generally establish a basis of trust within the team. To establish credibility and responsibility, the estimator's comments should always refer to the best interest of the project or the owner's project goals. I've always found that comments in the form of a question are more powerful than statements.*

Alan Schorfheide, DBIA
Senior Director of
Integrated Services
The Korte Company
Highland, Illinois

Conceptual estimating is as much an art as it is a science and requires special skills that go far beyond the technical aspects of putting together a price. It takes communication skills and the ability to see from multiple perspectives. As Jim Crockett has noted:

> *The conceptual estimator must assume the role of communicator and facilitator. Communication is a definite key. You've got to be able to get your point across and that starts with the owner. You've got to truly*

understand what it is he wants, even when he doesn't specifically say what he wants. The estimator must then disperse this knowledge through the rest of the team. It's this continuous communication string you have to keep going.[9]

Design-build estimating is not for the weak of heart. It is a complicated process that should only be undertaken by seasoned construction professionals with a lot of experience and who know what it takes to get something built. For this reason, most design-build conceptual estimators come from the ranks of experienced, senior level personnel. Only estimating veterans have enough experience to figure out the entire project scope without having all of the details worked out in advance, so they can develop reliable pricing models that can meet the project requirements. They must gain the design team's trust so they can help manage the design to meet the budget required.

Many companies separate their detailed estimating personnel from their conceptual estimating personnel. They often house the conceptual estimators in the Preconstruction Services division, while the detailed estimators are stationed on the Support Services or Construction Operations side of the firm. Wherever they are located, their roles and standard practices are very different. It is rare to find a detailed estimator and conceptual estimator in the same person. With such an important role to play, your estimator must have the proper tools, techniques, and mindset to accomplish the job and help the team stay on track from initial design concept and proposal budget through design completion and final buyout.

EMBRACING UNCERTAINTY

Having read an entire chapter about the uncertainty around a design-build conceptual estimate, you may wonder why anyone would want to use the design-build approach in the first place. Because even on complex projects, design-build, when implemented correctly, can still deliver faster, better, more efficient and cost-effective results than other project delivery methods.

Design-build can offer you even more control than other delivery methods, although it may not seem so at first. You have more control because design-build is an integrated, proactive, dynamic process in which you develop, monitor, and respond in real time. There is no lag between when the design development and design pricing; they happen simultaneously, and in an integrated fashion. Design-build develops and considers all critical factors, design, quality, performance, cost, and schedule, together. If any one of the project variables heads in a negative direction, the design-build team can make an immediate, integrated, and well-communicated correction. There should be no surprises.

Design-build requires constant coordination to deal with uncertainty. Those who follow the traditional project delivery method feel it has more "certainty" associated with it—but does it really? Here is how the traditional design-bid-build project starts:

- The client believes with some level of "certainty" that they have sufficiently captured their stakeholders' and end users' interests, and planned and programmed their project well;
- The owner hires a design firm to develop 100%-complete plans and specs, based upon her "certainty" that she has communicated the program needs sufficiently to the architect or engineer so the plans and specs contain no holes or gaps. The owner is also relatively "certain" that those plans and specs provide the most efficient and cost-effective solutions possible for the available budget and that these plans will provide the facility she needs;
- The designers complete the plans and specs, feeling "certain" that they have captured all the requirements their client communicated, that their plans and specs are complete and accurate and that they have created the best solution. They also are generally "certain" that the facility will perform for the owner as the architects and engineers had planned;
- When the owners and the designers put those plans and specs out to market for bidding, they are fairly "certain" that the project will come in on budget and they expect no change orders;
- The contractors proceed to review the plans and specs and develop their estimates, feeling somewhat "certain" that they will provide sufficient information and detail to accurately estimate the project and get it built on time. They are "certain" that the facility design represents what the end-users and stakeholders expect and will perform the way the owner had hoped;
- The stakeholders and end-users are fairly "certain" that the project being planned, designed, and built will meet their needs.

Now consider how operating from a sense of "certainty" in relation to planning, designing, and building projects has turned out. By some estimates, less than 45% of construction projects are delivered on budget and only 60% are delivered on time. Architects have reported that 60% to 80% of what they design never gets built because the project is over budget. Even if these numbers are only half right, it seems that the notion of certainty in relation to design-bid-build might be over-rated.

Yet people crave certainty, even about uncertain things. Why? Because it makes us feel better. We are uncomfortable working in uncertain environments, at least in the United States. But the traditional design-bid-build approach seems to create a false certainty, which might not be the best place to begin when planning, designing, and building major infrastructure and facilities. Perhaps the recognized and controlled uncertainty of design-build is a better place to start. Clampitt and DeKoch explain this notion of conscious uncertainty in their book *Embracing Uncertainty: The Essence of Leadership*:

> We are not embracing a kind of whimsical uncertainty that condones inactivity or a "whatever happens" kind of aimlessness. However,

demanding certainty where none exists is foolhardy, debilitating, and often dangerous.[10]

By first recognizing the uncertain nature of projects, we might actually be more inclined to embrace an integrated process like design-build. So the notion of a one-time estimating process, in an uncertain design-cost environment, would never make sense, and makes the progressive nature of design-build estimating seem more logical. At first this "churn" may seem uncomfortable—like something is wrong. In reality, it is the dynamic, progressive nature of the design-build estimating process that will eventually bring certainty to the project. But it does take a little getting used to. Margaret Wheatley writes about the nature of uncertainty relative to organizational change from her own experience:

> *But the real eye opener for me was to realize how control and order were two different things, and that you could have order without control. We're not comfortable in chaos, even in our thoughts, and we want to move out of confusion as quickly as possible. I now know that that's the place to be if you want to really be open to new thoughts, if you want to be totally open to a total reorganizing of your mental constructs. You can't get there without this period of letting go and confusion. For someone who's been taught to be a good analytical thinker, this is always a very painful moment.*[11]

Perhaps you have encountered your "painful moment" by now and are ready to move on. When you have the right planning and estimating processes in place and the right people on the team, the chaos and confusion that initially seemed so daunting in this uncertain environment of design-build actually begins to make perfect sense. Realize that in risk-oriented projects, we gradually transform this uncertain environment into a reliable, certain outcome—and that is where integrated, skilled design-build management turns chaos into order.

CHAPTER HIGHLIGHTS

1. Many owners prefer the design-build delivery method because they can determine the cost of the project early in the procurement process before spending a great deal of money on design services to complete plans and specifications. In design-build the price can be determined first and the design can follow.

2. Many owners are shifting away from trying to get the cheapest price to optimizing their project budgets. They begin by declaring how much money they have to spend and put the burden on the competing design-build teams to come up with the best solution for the specified dollars allotted.

3. Design-bid-build estimating performs a quantity take-off from 100%-complete plans and specifications. In design-build, the estimator must

perform conceptual estimating because the details of the design are not yet known.

4. Instead of using the traditional organizational framework of MasterFormat, design-builders use UniFormat, which is organized around broader functional elements of a building. Design-build determines project cost using unit prices for common functional systems or assemblies, rather than on individual parts and pieces of the system. The initial UniFormat estimate will eventually be converted to a MasterFormat structure as more details of the design become known.

5. The design team must be fully involved with the estimators so that the estimators have a complete understanding of the design intent before they create the conceptual estimate.

6. In design-build, none of the critical elements (design, budget, or schedule) is sacrificed for the sake of another. Each is optimized as it is coordinated into an integrated, holistic, and comprehensive solution that serves all project requirements for quality, performance, function, cost, and time.

7. Generally there are five identifiable stages of design: 1) programming, 2) conceptual design, 3) schematic design, 4) design development, and 5) construction documents.

8. The contractor is involved early in the design-build design process and the estimating function happens concurrently as the design progresses, leading to true value engineerings.

9. Creating a cost model is the critical first step in the design-build estimating process and it is critical that all members of the design-build team understand how it works. Cost models are project specific and developed from historical cost data associated with various project types. The cost model sets the design parameters for the project and should be consistently monitored as the design develops.

10. Starting with an appropriate historical cost model, the design-build team adjusts its components to fill the unique project circumstances, such as time, location, and quality. After the adjustments are made the resulting cost model will serve as a point of reference as design continues.

11. There are generally five progressive estimating phases associated with design-build: 1) feasibility estimate, 2) concept estimate, 3) schematic estimate, 4) design development estimate, and 5) detailed estimate. In design-build, the level of estimate accuracy is contingent upon the effort put forth to manage the design to the budget.

12. The design-build estimating process is progressive and dynamic, meaning that estimates must be monitored and adjusted. Each stage of the estimating progression represents a design-cost reconciliation milestone opportunity. The milestones alone are not sufficient to manage

design to budget, so trend management must be introduced between the milestone estimates.

13 In design-build, you must carefully define the project scope and qualify your prices. The estimate should be accompanied by an outline specification that will communicate to your client the range of materials, products, options, and quality that you have priced and they can expect to see as the design develops.

14 In design-build, the mechanism used to account for the uncertainty associated with every project is called contingency and it is part of every design-build estimate. Contingency development should begin with a risk assessment involving both construction and design personnel.

15 There are three distinct types of design risks: 1) design services, 2) design impacts, and 3) design aesthetics. To assess the design impact risks, use a design stability approach. The design-build team should discuss the design stability of each of the major design elements and their components.

16 In addition to construction and design contingences, there are also subcontractor, subconsultant, and supplier contingences to consider. Consider all the contingences together to avoid redundant contingencies.

17 Owner contingences are generally expressed as a single line item at the end of the estimate to account for anything the owner may have overlooked in the initial RFP. This contingency belongs to the owner, so the unused portions revert back to the owner, unless a portion is used to incentivize the design-build team.

18 In design-build, value means looking at how much functional benefit the team can provide for the budget. Some of the ways to improve value on a project are: improve benefits and maintain the cost; maintain benefits and reduce the cost; improve benefits and reduce the cost; reduce benefits and reduce cost (if benefits remain within needs); and increase benefits and increase cost (if cost remains within limitations).

19 A design-build estimator has a lot of influence and therefore must have considerable expertise and experience in all the major building systems and design elements. Conceptual estimating requires special skills that go beyond the "technical" aspects of putting together a price.

20 Design-builders need to become comfortable with the concept of uncertainty, especially early in a project. The concept of "certainty," while it gives us comfort, is often false certainty that does not represent reality. In design-build, early uncertainty is carefully managed so that, as a project proceeds, the outcome becomes certain and meets or exceeds the client's expectations.

RECOMMENDED READING

- ***Embracing Uncertainty: The Essence of Leadership,*** by Phillip G. Clampitt and Robert J. DeKoch, M.E. Sharpe, 2001.
- ***Leadership and the New Science: Learning about Organization from an Orderly Universe***, by Margaret J. Wheatley, Berrett-Koehler Publishers, 1994.

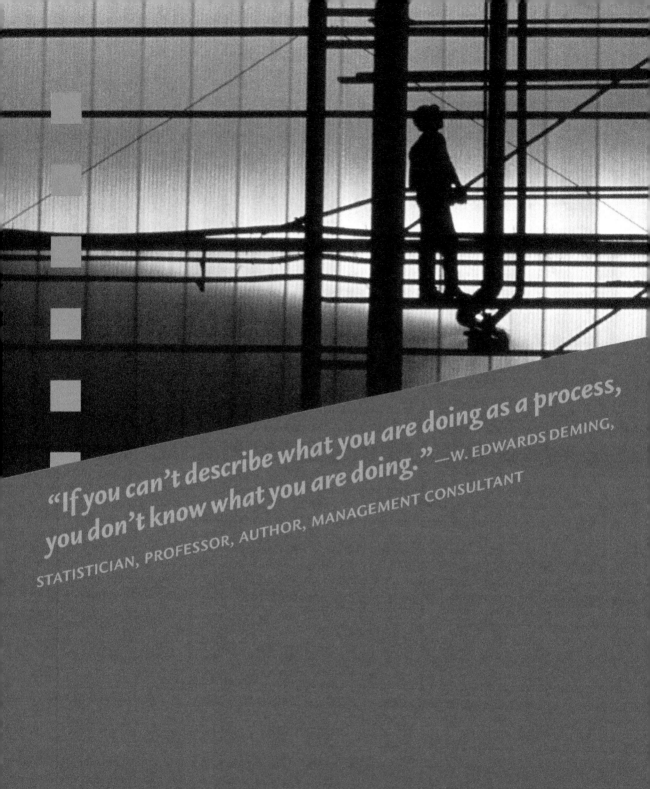

"If you can't describe what you are doing as a process, you don't know what you are doing." —W. EDWARDS DEMING, STATISTICIAN, PROFESSOR, AUTHOR, MANAGEMENT CONSULTANT

7
MANAGING THE DESIGN-BUILD PROCESS

In the first part of this book you studied the ways in which design-build project delivery is different from design-bid-build, including the distinctive procurement approaches associated with design-build. In Chapters 4 and 5 you learned how to prepare for and win a design-build project. In Chapter 6 you learned about the challenges inherent in the design-build estimating process, and, hopefully, you began to understand the practical implications of an integrated design and construction mindset—what it looks like and how it actually works. In this chapter you will learn about what happens after you are awarded the design-build contract. If you think pursuing and procuring design-build projects is different, wait until you learn how different managing them can be.

Design-build project management is first and foremost about managing people through a process, not just a project. Traditional construction management usually focuses on the project—the materials, methods, equipment, labor, quality, and safety. These elements are all very important, of course, but if your intention is to deliver a comprehensive solution addressing *all* aspects of the design-build contract—design, quality, performance, cost, and schedule—then you must expand your management focus. And remember, in design-build the design process does not begin until after all the team members come together, so we are starting from a very different point than we do with design-bid-build. This alone should hint at how very different the management task is. My objective in this chapter is to introduce you to some of the management challenges involved with integrated design-build as well as some of the unique techniques and tools associated with it.

I will also discuss why it takes some special skills, talents, and people to do it well. The best traditional construction project managers do not always turn out to be the top design-build project managers. Unless they can let go of the segregated services mentality, construction project managers may have a hard time adapting to the design-build process. The typical "control and command" skills needed to be successful in hard bid project management are not well suited to an integrated environment that depends on highly synergistic relationships based on communication, collaboration, and trust. Gary Hamel, in his book *The Future of Management* notes:

> *The current management model—centered on control and efficiency—no longer suffices in a world where adaptability and creativity drive business success.*[1]

Design-build requires adaptability and creativity. Learning how to manage projects from such a different perspective requires project managers to fully accept the philosophies at the core of design-build; for some, this will necessitate a significant mental shift. Chapter 8 focuses on how to make that mental shift, but for now I want to focus on the integrated nature of design-build project management and what it takes to deliver a successful project.

DESIGN-BUILD AS A PROCESS

The key to successfully managing design-build projects is to identify and implement a repeatable integrated design and construction process that consistently delivers quality projects on schedule and within budget. Design-build process management considers the entire project delivery package of activities from a performance-oriented perspective. In a way, design-build process management is about learning how to look at a project from a holistic point of view rather than piecemeal. Instead of just looking at the "dots" of management, in design-build we focus on the "connections" between the dots—how the various management tasks are related and impacted by one another. Design-build is an integrated process that connects all the dots. The management challenge then becomes quite different. All aspects of the project—design, procurement, and construction—must be managed against performance targets relating to design delivery, quality standards, financial budgets, schedule milestones, and other project deliverables. This concept of integrated process management is at the heart of successful design-build project management.

Shifting Paradigms

Traditional construction managers tend to view individual project performance criteria in isolation—even when it is obvious that one variable impacts another. Construction managers often look at cost or schedule or quality as if each was a stand-alone performance indicator, and they are, if your accountability perspective is limited to only one aspect of project performance at a time. In addition, the impacts of the developing design, which is a vital part of the design-build management task, do not even come into the equation with traditional construction management. In traditional construction management, design assessments and modifications are frequently made in isolation, which may help one

> The more you learn about design-build, the more you will come to realize that the design-build way of thinking can be applied to any type of project or activity. For instance, when planning a party with friends, you may find yourself saying, "Instead of trying to work with all the vendors separately, we should use a design-build approach. We can get the caterer, the facilities people, the entertainers, the wait staff, and other team members together to figure out how we are going to accomplish this task." It is a holistic, comprehensive approach to problem solving.

aspect of the project but may negatively impact the rest of the project. Trying to manage a project from such a fragmented perspective can lead to a multitude of gaps—as well as significant risk exposure. And what is worse, you may not even know about the gaps, so the risk goes unmanaged.

Design-build is intended to perform as an integrated process—a synergistic whole system, working in harmony to optimize value and deliver extraordinary results. To some, this goal might seem idealistic, especially if their expectations have been lowered because of their experiences with traditional design-bid-build. However, design-build, when properly performed and managed, can indeed deliver extraordinary results. Consider the results produced by the Skanska USA Civil Northeast and Parsons Brinckerhoff team highlighted in the case study on the following pages.

Many contractors, architects, engineers, and owners in the United States have not had the opportunity to manage a project from an integrated perspective. Design-build can give them that chance. However, achieving the exceptional results possible with design-build requires a paradigm shift for many. Thomas Kuhn, the author of *The Structure of Scientific Revolutions*, describes the power of a paradigm as follows:

> *A paradigm is more than a way of thinking—it's a world view, a broadly and deeply held belief about what types of problems are worth solving, or are even solvable.*[2]

You must believe that working and communicating as an integrated team is possible and that by doing so you can optimize performance and deliver superior results. In order to see the big picture, you must be looking at the project from a process perspective that requires multiple frames of reference. You can no longer do project management from a single discipline perspective or a single service perspective (design or construction or engineering). You have to do design-build project management from a design-build perspective and learn what it takes to engage the entire team. Only then can you effectively apply the management techniques and tools of design-build.

MANAGING DESIGN IN DESIGN-BUILD

The task of managing design within the design-build process is one of the unique aspects that can make design-build such a challenge. For example, in design-build projects, the entire team begins working together quite early in the development of the design so that many of the design decisions and major systems can be confirmed with input from all team members. As you can see in Figure 7.1 this is a significant change from the traditional process and requires a completely different approach to the management of the design.

In this instance the term "design management" does not refer to how designers actually perform design—they are professionals, and if you have selected individuals who share your company's values and have the right technical expertise to fit the project type and scope, then they

CASE STUDY

ALL ABOARD

THE FACTS

Project Name:
Corona Maintenance Shop and Car Washer Facility

Project Type:
Transportation

Location:
Corona, New York

Description:
Three-story 135,000 square foot train maintenance shop, a state-of-the-art 7,200 square foot car washer facility, a two-story 1,200 square foot signal relay building, and two circuit-breaker houses.

Cost:
$168,478,074

Duration:
48 months

Completion Date:
December 2, 2006

THE TEAM

Owner:
New York City Transit

Constructor:
Skanska USA Civil Northeast

Architect:
El Taller Colaborativo, PC

Engineer:
Parsons Brinckerhoff

Corona Maintenance Facility. Courtesy of Skanska USA Building.

THE LEAD-IN

On October 27th, 1904, New Yorkers celebrated an event that would change the city forever: the first cars of a new underground rapid transit railroad system left City Hall Station in lower Manhattan bound for 145th Street and Broadway. Dubbed "the subway," the underground transit system was a dream that had been in the making for more than thirty years. The nine-mile subway line that opened in 1904 took just four years to complete, quite a remarkable feat at the turn of the century considering the amount of maneuvering required around the underground network of sewers, gas lines, water mains, and steam pipes. By 2008, the system had expanded to 468 stations and 660 miles of track—delivering more than 1.6 billion rides per year.[3] The New York City subway system carries more passengers than all other rail mass transit systems in the United States combined.

THE STORY

Consider the challenges associated with designing and building new facilities for one of the busiest public transportation systems in the world—one that runs twenty-four hours a day, 365 days a year.[4] The design-build team of Skanska USA Civil Northeast and Parsons Brinckerhoff took on

Corona Maintenance Facility. Courtesy of Skanska USA Building.

THE TEAM (continued)
Specialty Contractors:
Dome-Tech, Inc., Kelly Masonry Corporation, Five Star Electric, Union Switch & Signal, Welsbach Electric Corp., Transit Technologies, Simmons Machine Tool Corporation, Permadur Industries Inc.

these challenges when it was awarded the design-build contract to design and construct the Maintenance Shop and Car Washer Facility at the 103rd Street – Corona Plaza station along the No. 7 Line running from Flushing, Queens to Times Square in Manhattan. Sometimes referred to as the "International Express" by locals because of the diversity of the population of the communities it serves, the line is also famous for

Corona Maintenance Facility. Courtesy of Skanska USA Building.

Corona Maintenance Facility. Courtesy of Skanska USA Building.

being the official train of the New York Mets and the US Open tennis tournament. The Corona yard is located at 126th Street and Roosevelt Avenue, directly across from the old Shea Stadium (now Citi Field) in Queens, New York.

The 103rd Street – Corona Plaza station is ranked as the 65th busiest station in the New York subway system, with a ridership of approximately 5.9 million annually.[5] Because of the project's scope, location, and activity level, the design-build team was faced with a number of challenges, not the least of which was that existing operations at the site were not to be disrupted during construction. The centerpiece of the project was a three-story 135,000-square-foot steel and concrete maintenance shop meant to service up to 400 rail cars; however, the project also included a 7,200-square-foot car washer facility, a two-story 1,200-square-foot signal relay building, and two circuit-breaker houses. The project scope encompassed civil, architectural, industrial, structural, mechanical, and electrical work.

Access and congestion were just two of the project's challenges. For example, the job site was located on a 100-year flood plain, which meant that the team had to design and construct the building foundations anywhere from five-to-nine feet above the existing grade. The maintenance facility and the other structures were installed on approximately 1,400 18-inch taber tube piles designed to support a thick concrete mat that would serve as a ground floor slab for the various structures and tracks. The slab for the large maintenance facility needed to be installed seven-to-nine feet above grade, which created some very difficult equipment access constraints. Further complicating construction was the fact that the building was situated as close to the wetlands and a park as legally allowable, due to constraints on available land in the area. The team had to build the structure with no access from the south or east sides of the job site.

The design-build team came up with a plan to build the structure in a "U" formation. Following the pile-driving operations, the concrete slab was poured, and the installation of the steel structure took place in a tightly coordinated operation. Building the structural frame in this manner allowed for an access road through the center that was large enough to accommodate the 365-ton Liebherr hydraulic crane used to set the 120-foot trusses spanning the shop facility. The team controlled access to the road by carefully scheduling concrete and equipment deliveries as well as the rebar and steel installation. Upon completion of the building's exterior, the team backed out of the access road and constructed the center of the structure from east to west. The new maintenance facility required staircases, an elevator, and building access ramps as well as 12,000 feet of rail track to allow trains to access the facilities.

The New York City Transit Authority's (NYCT) plan, which required the new car washer facility to be installed in the same location as the existing one, called for demolishing the existing car washer and constructing the new facility during the final year of the project. And although the RFP documents included a complex phasing plan to divert train traffic in the yard during each stage of construction, the design-build team put their heads together and came up with a better solution. The revised plan further reduced congestion, added efficiency to construction operations, and allowed the team to simultaneously construct the maintenance building and car washer facility, thereby avoiding costly delays associated with disrupted services.

To accomplish all of this, the design-build team proposed installing a temporary car washer on the lead tracks entering the yard. This allowed the team to remove the existing car washer facility from the construction zone earlier in the schedule, and it eliminated the complex phasing that was anticipated by the NYCT. In true design-build fashion, the Skanska USA-Parsons Brinckerhoff team took the opportunity to not only simplify the phasing of the car washer facility construction but to also upgrade the car washer operations and performance beyond what the owner had anticipated.

The existing car washer operations basically involved pouring water onto the trains. Detergents were not allowed because there was no collection system that would prevent the harmful chemicals from discharging onto park land below the facility; the existing facility also lacked a chemical treatment system to handle the effluent before it entered the sewer system. The temporary facility allowed for the use of detergents because it included a chemical treatment system that treated the runoff before it was discharged into the sewer system, providing a significant improvement over the existing car washer facility. Service ratings on train cleanliness improved dramatically after the temporary car washer was put into service.

The design and construction team faced several challenges relating to this aspect of the project, including: 1) installing the temporary car washer on elevated tracks 30 feet off the ground; 2) installing a watertight collection system to prevent runoff from reaching a park area below the car washer; and 3) installing a disposal system to monitor and treat the effluent to legally dispose of the discharge into the sewer system.

The team's approach provided a fully functioning temporary car washer facility outside the yard, which eased congestion in an already crowded and heavily phased construction site. It also allowed workers to build a new washer facility in the same location as the original without interruption of services. The same upgraded chemical treatment and collection system used in the temporary car washer was installed in the new facility. In addition to expediting the schedule, upgrading car washing operations, and improving train cleanliness ratings, the proposal resulted in an overall cost savings to the owner of over $3 million. Because the NYCT was so pleased with the performance provided by the temporary car washer, at the end of the Corona project the agency moved it to another site that was in need of a car washer facility.

THE CLOSE

The No. 7 line of the New York City subway system has been in use as a transportation route since the early 1900s. And in spite of some significant obstacles, that continuous run was not interrupted during the construction of the line's new maintenance shop and car washer facility—in large part due to a process that allowed for design and construction to overlap and that encouraged all team members to "get on board" to deliver some innovative solutions. In addition to conquering several major challenges and adding values above and beyond the requirements of the RFP, the Skanska USA-Parsons Brinckerhoff team built the Corona Maintenance Shop and Car Washer Facility as New York City Transit's first LEED-certified facility.

From the Owner

This project is a great example of how the implementation of the design-build method saves time and money. The design-build approach allowed the complex to be constructed on a fast-track schedule by overlapping the design and construction phases, and in doing so, resulted in significant cost savings. The stellar cooperation and dedication of the design-build team produced a high-quality, energy-efficient, and environmentally-friendly railcar maintenance shop and car washer facility. The relentless efforts of Skanska USA Civil Northwest and Parsons Brinckerhoff and their commitment to the design-build process in delivering this assignment on time and within the budget constraints were nothing short of excellent.

Robert L. Dondiego, P.E.

Lead Construction Manager

Line Equipment, Shops, Yards, and Facilities

New York City Transit

[Case study adapted from Skanska USA Building's submission for the Design-Build Institute of America 2008 Design-Build Award Competition, by permission.]

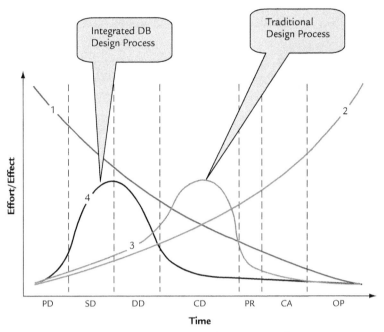

Figure 7.1
MacLeamy Curve. Courtesy of CURT.

The MacLeamy Curve was introduced in 2004 in a publication of The Construction Users Roundtable, titled "Collaboration, Integrated Information, and the Project Lifecycle in Building Design and Construction and Operation." The curve (shown in Figure 7.1) illustrates the advantages of making design decisions earlier in the project when opportunity to influence positive outcomes is maximized and the cost of changes is minimized.

will deliver the design that is needed. Rather, it refers to the following four aspects of design-build integration management:

:: design process
:: design-cost interface
:: design-construct interface
:: design-performance interface

We will cover each aspect of design-build project management in general terms here. Later in the chapter I will discuss how to apply specific design-build management techniques and tools to address the challenges associated with these tasks.

The Design Process

Unlike the construction industry, which has created a professional career path for construction management (over 100 universities now offer degrees in the field), designers have not really done a good job defining design management or developing standardized practices to go along with it. One thing that makes this difficult is that, while you can see the construction process, you cannot really see the **design process**. In construction, you can see what you have and what you lack—materials,

D-B DICTIONARY

Design Process –
A series of information inputs and outputs that result in a completed project design.

MENTORING MOMENT

Julie Halvorson, LEED AP
Project Manager II
Ryan Companies US, Inc.
Minneapolis, Minnesota
Education: Iowa State University – Graduated 2005
Major: Construction Engineering, Minor in Business

Respect Begets Respect

One of the most interesting aspects of working on a design-build project is that, as a design-build project manager, you get to work directly with the designers. There are more heads at the table and you come up with a lot more ideas that can work. The entire team has input and looks at the design from multiple aspects—constructability, value, cost, and schedule—not just the design aspect. Although I have only worked with our internal architects, I find that most designers are pretty receptive to the construction perspective as long as the contractors are respectful of the designer's perspective. I think the key is to always listen so you can understand where the other side is coming from. I work really hard to make sure I understand why the designers are pushing for an idea so that I can get to the root of the intention and then respond respectfully and openly. If I do have another approach that will get us to the same place but makes more sense from a construction perspective, then I make that contribution and it turns into a win-win. And even if they don't go with my idea 100%, they may take a portion of it in order to be responsive to my needs from a construction perspective. I find that if I respect their ideas that they are going to respect my ideas. At the end of the day we both know that we must have something that we can afford to build, otherwise we aren't being responsive to the client. I always try to stay focused on what will serve the client and the end user. With every decision I always ask, "Will this solution benefit the project or take away from it"—that's my bottom line as a design-build project manager.

equipment, or labor. If you do not have what you need then you do not proceed with the construction task. Design is different.

John Steele, an engineer by training, and Director of Adept Management Ltd. refers to the design process as a mysterious black box.

> *We have come to believe that design lives in this mysterious black box through which we input information and data and plans, and specs pop out the other end. We don't quite know or understand what happens in the box, and it just seems too complex to explain. Even architects and engineers aren't readily able to tell you what happens in there.*[6]

However, you cannot manage something that you cannot see. We must find a way to see what is happening in that black box. In design-bid-build, it is easy to think of design as nothing more than plans and specifications because the design is complete before construction starts. How the design process unfolds and is managed has no real impact on the construction process. But in design-build we must move beyond thinking of design as simply plans and specifications. If we try to manage the design by these two black box outputs only, then we will be very frustrated. By using a narrow concept of what constitutes design, we really never see what the problem is. We create an almost unworkable situation—our paradigm has limited what we can see as solvable.

Because design-build is such a fast-track process, some contractors may grumble about not having the drawings they need when they need them. Architects and engineers may complain that they cannot finish the drawings because they do not have the information they need.

Michael Schmieder, AIA
Project Design Principal
The Haskell Company
Jacksonville, Florida

I instigated an Integrated Design Review (IDR) process long before there was computer technology like Building Information Modeling (BIM) to do it for you. Being committed to an integrated design-build process and understanding the value of this interdisciplinary design and construction review, I incorporated the "RediCheck Interdisciplinary Coordination" manual (authored by William T. Nigro, AIA) early in my career, and I still use it today.

The Integrated Design Review is a deliberate process of coordinating the construction documents of all design disciplines towards the goal of eliminating conflicts and errors, which would otherwise result in a request for change orders involving dollars and days. Not only that, but in a design-build environment it helps to find the best approaches to serve the interests of both the design and construction entities, which ultimately serve the owner. The IDR includes participation from the team's construction representatives—to comment on issues of constructability and completeness, and to define building systems that are responsive to the owner's criteria but that are also simple, reliable, and maintainable. This elevates the IDR process to the highest level of coordination because their input "makes the design real."

For example, a project being discussed in an IDR session a few years ago had a long, mall-type space, with a certain center-to-center dimension between the two column lines on either side of the mall. It was going to be the iconic space that the facility would be remembered for. The construction superintendent asked if the column lines could be spread just one more foot apart. He explained that, given the structural footing schedule, spreading the columns and footings apart this small amount would allow him to rent only one crane for steel erection that could drive down the center of the mall space without rolling over the safe zone around the footings. His crew would be able to pick and set the steel for the entire building right off the flatbed trucks as soon as the steel arrived on site. Not having this extra distance between columns would necessitate him having to rent two cranes to work around the building perimeter. In addition, his crew would need to off-load the steel and "piece it" into place. Handling the steel twice is very inefficient and would add cost to the project—in addition to the cost of the extra crane rental. This simple suggestion returned over $100,000 in construction costs that could be used elsewhere on the project. These extra dollars helped offset a design feature the owner wanted at the main entry but at the time had been unable to afford.

Even though the design industry is moving towards BIM, nothing can take the place of good, rational thought to evaluate a design and confirm that it is responsive to the owner's expectations, is well-coordinated, and is constructible. In the end, it must be a simple, reliable, and maintainable design that is within the owner's ability to purchase.

If the design process is not well managed, designers may begin making assumptions in order to keep up with the demands of construction—adding unnecessary and often unrecognized risk to the project. No one can really see the true picture inside the black box unless the design process is managed throughout the project. When design, budgets, schedules,

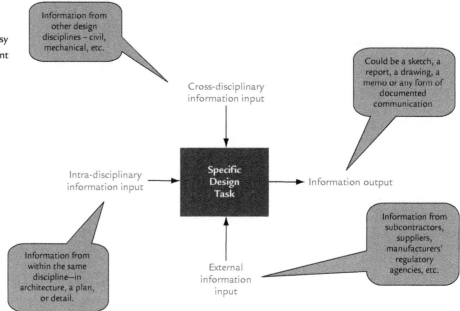

Figure 7.2
Information inputs and outputs. Courtesy of Adept Management Ltd.

D-B DICTIONARY

Information Outputs – The design deliverables associated with a project. In addition to plans and specifications, information outputs may include narratives, reports, calculations, memos, sketches, or any other form of documented communication that supports the design and construction processes. Design information may be released to the construction team before it is incorporated into a formal plan or spec.

Risk Register – A document that tracks the decisions made in relation to the identified risks of a design-build project. It is a dynamic tool designed to capture the logic associated with the decisions made.

performance, and quality are all combined into a single contractual obligation, it is crucial to find out what goes on inside that black box. The design impacts every other deliverable, and unless we can manage it, we cannot manage any of the other components of the project.

The mysterious activity that takes place within design's black box is simply the transfer of information. If we can identify the information needed to complete a design task then we can manage the flow of that information, and when we manage the flow of information, we manage the design process. Just as construction managers can verify if the necessary materials, equipment, and manpower are available before starting a construction task, a design manager can confirm that the information needed to produce a design element is available—thus making sure the architects and engineers are getting the information that they need, when they need it.

By defining the design process as a series of information inputs and outputs, as illustrated in Figure 7.2, we can make management decisions based on the information that we do or do not have. The information inputs and **information outputs** come from many different sources—both internal and external to the team. The design deliverables are not limited to drawings and specifications; rather, they may include memos, narrative descriptions, reports, calculations, meeting minutes, **risk registers,** sketches, and the like—whatever is needed to move the design and the construction forward.

The Design-Cost Interface

Once everyone is on the same page regarding the design process, managing the design-cost interface is the next big challenge for most design-build

One reason many owners decide to use design-build is the perception that there are no change orders in design-build. That perception is incorrect; however, in design-build there are certainly fewer change orders, which typically results in faster delivery and lower overall cost. Also, the disputes between the designer and the contractor are eliminated because they are on the same team. Whether a change order occurs or not is based on whether there is a change in the contract documents that the parties identify as forming the basis for the design. The DBIA contracts define these documents as the "Basis of Design Documents," and they are a snapshot in time that informs the parties as to what they intended when the price was set. As the project moves forward, the parties need to anticipate that there will be some changes in the parties' original assumptions. The changes may arise because the owner changed his mind about something or added requirements, or the change may arise through innovation and value engineering. Whatever the reason, the parties need to be able to anticipate change, particularly during design, and be prepared to manage that process.

Robynne Thaxton
Parkinson
Attorney
Robynne Thaxton
Parkinson PLLC
Mercer Island,
Washington

teams. The **design-cost interface** is the relationship between the design and the various cost components of a project. Design-build project managers must continually review the various elements of the design to ensure that the overall design stays within the original project scope and budget. Failing to properly monitor and oversee the design-cost interface can get a team and a project in real trouble. As a matter of fact, in my experience, design-cost interface issues lead to more team conflicts than any other type of issue.

You learned in the last chapter how important it is to create a good cost model early in the proposal process. A cost model is one tool for managing the design-cost interface. Later in this chapter you will learn about some useful techniques for managing this component of the design-build process.

The Design-Construct Interface

The **design-construct interface** is the relationship between the design and various aspects of construction, such as the constructability considerations, fabrication issues, availability of qualified labor, and the schedule. Most contractors, and designers for that matter, seem to best understand how to manage this facet of the design-build process.

One important part of managing the design-construct interface is reviewing the design from a constructability perspective. **Constructability reviews** must be conducted throughout the design process, to evaluate what construction-related challenges will be encountered because of the way the project is designed. These challenges could be related to the proposed materials, requested fabrication techniques, or the configuration of a design element. For example, there are almost always smarter, more economical ways to fabricate a specific design detail that will end up

Treating design as a series of information inputs and outputs allows designers to convey necessary information to the construction team long before the plans and specifications are complete. This is especially important at the proposal stage of design-build. If you are depending only on the drawings to communicate design then you do not understand the design-build process. The design discussions should be a collaborative team effort to flesh out the information required to successfully complete the project. When you collaborate in this fashion the excuse of "it wasn't on the drawings" is no longer legitimate.

D-B DICTIONARY

Design-Cost Interface – The relationship between the design and the various cost components of a project.

Design-Construct Interface – The relationship between the design and various aspects of construction, such as the availability of quality labor, fabrication efficiencies, schedule constraints, timely delivery of materials, and other constructability considerations.

achieving the same result. The easiest way is not always the best; however, there is no reason to design a detail that requires a complicated fabrication technique if another less difficult method of fabrication will accomplish the same overall design objective. In some instances, it is a simple case of standardizing the size of fabrication components. Bringing major suppliers and subcontractors into the discussions early can help head off many problems down the road.

Designers should also make it a priority, early in the project, to sit down with the construction superintendent to find out how he envisions the project unfolding—what his concerns are, how he intends to approach the project, and, logistically, what his primary challenges will likely be. If there is something that the designers can do to make the execution of the project more efficient, while maintaining the integrity and intent of the design, then the team should discuss that. This input should all take place before the design gets too far along—if not at the proposal stage, definitely during the initial design planning stages after contract award.

Managing the design-construct interface also involves overseeing the impacts of the design on the schedule and procurement needs so that the release of drawings is integrated with construction requirements. Early in the design process, the design-build team should discuss the need for equipment and materials that have long delivery times so that the items are on site when they are needed. For example, steel historically has a long delivery lead time. And although in design-bid-build the structural

Nat Killpatrick
Vice President, Sales
Texas Division
Burcamp Steel Company
Dallas, Texas

Quite often projects are called design-build, but that spirit is only generated between the architect and general contractor. Subcontractors and subconsultants are often selected and incorporated into a design-build project the same way as in a traditional design-bid-build project. This contradicts the whole idea of the design-build process and negates the contribution that qualified design-build subcontractors and subconsultants can bring to a project if they are allowed to participate as team members in the early stages. To reach the level of true design-build it is imperative to involve qualified subcontractors early in the process—they are essential to the success of the proposal process as well as the job as a whole. When the design-build firm and a sub are in sync and beholden to each other, the trust created allows both parties to collaborate on a level not seen in the typical bid-build scenario. This partnership allows both teams to look at the project as a whole, and make decisions based on the best outcome for the design-build team and owner. This takes a commitment of "We Live or Die Together."

This collaborative effort allows us to provide guaranteed pricing early in the design phase, better coordination throughout the project with fewer RFI's (Requests for Information), a faster submittal review process, and guaranteed material availability. A collaborative effort reduces the risk for everyone on the team. When subcontractors are allowed to work with design consultants directly there is less information loss between the design team and the subcontractor. This allows for a faster, more efficient project once the green light is given to start construction.

drawings are not usually started until after the architectural drawings are at least at the design development stage, in design-build the structural drawings must be completed much earlier. This allows the steel order to be placed with the vendor early enough to account for the long lead time so that the steel is at the site when the construction team needs it. Sequencing the steel design ahead of other design elements requires the designers to essentially design "out of order," at least from a traditional perspective. This is where the planning and management of the design process really pays off. Design planning is discussed later in the chapter.

It is not a matter of one element of the design-build process being more important than another. The point is that all elements of design and construction should be considered as the design is being developed. This is the beauty, and the challenge, of the design-build process. It takes a special mindset to embrace and incorporate so many different perspectives into a project solution but that is what gives design-build the advantage over traditional construction project management. I refer to this special mindset as the "genius of and," and you will learn more about it in the next chapter.

The Design-Performance Interface

Most design-build RFPs stipulate performance requirements. Managing the **design-performance interface** means confirming that all elements of the design comply with those performance requirements. In design-bid-build, this process might be as simple as confirming that the correct product was delivered to the site or that the specified dimension tolerances are met. In design-build projects, the RFP should include some **performance standards** against which the design and final product can be measured. This is vital information needed in order to properly manage the design-performance interface.

In some cases, the owner may even include some incentives around performance standards to increase the motivation for compliance. For example, in 2004 Arizona targeted an International Roughness Index (IRI) value of 41, with smoothness expressed as IRI in inches per mile for new highway construction projects. Incentives were earned for values below 38, and disincentives were assessed for values in excess of 48.[7] When a client ties performance incentives (or disincentives) to such standards, effective management of the design-performance interface becomes an even higher priority. And once again, you cannot afford to wait until the design is complete before this management effort kicks in. Failing to catch disconnects between the design and the expected performance early on can result in expensive redesign work that impacts all of the other management efforts. Someone on the design-build team must be charged with monitoring this interface as the design is being developed.

Later in the chapter, I will present suggestions for delegating these management responsibilities, as well as a recommended organizational model. For now, the key point to remember is that design-build process management is a series of ongoing tasks that require a team effort throughout the project—from the proposal stage all the way through to

---

D-B DICTIONARY

Constructability reviews – Reviews of the design, typically performed by contractors, conducted throughout the design process to evaluate what construction-related challenges will be encountered because of the way the project is designed.

Design-Performance Interface – The relationship between all of the elements of the design and the performance requirements for the project.

Performance Standard – A performance threshold or requirement against which a design-build project will be measured. In order to be useful, the performance standard should identify what methods will be used to measure performance against the standard.

If an RFP does not indicate performance standards or testing and measurement procedures by which the design and end product will be validated, then it is imperative that the design-builder stipulate the standards and measures they intend to use in designing and delivering the project. Do not risk leaving the substantiation component undefined. Every project must have a clearly defined way to confirm that the design and the end product meet the performance requirements spelled out in the RFP. If not, it will be very difficult to manage the design-performance aspect of the design-build process, and very difficult for the design-build team to prove that they fulfilled their obligations if there is a dispute.

project completion. Doing business as usual, trying to manage design and construction as independent activities, exposes the project and the team to needless risks. Knowing how to properly manage the design-build process from an integrated perspective will help the team avoid a multitude of mistakes and take advantage of opportunities to add value to the project.

I have not come across many companies, integrated or not, that have mastered all four of these design-build process management components. This is significant, because managing these four components is at the heart of design-build process management. However, as companies and teams grapple with these challenges through trial and error, many best practices are emerging. Design-build process management is an evolving discipline in which much progress has been made—reducing the risks for all involved.

ESTABLISHING STANDARD PRACTICE

According to Gary Hamel "management processes establish standard protocols for common management tasks and are the recipes and routines that determine how the work of management gets carried out on a day-to-day basis. They propagate best practice by translating successful techniques into tools and methods that can be broadly applied. Put simply, they are the "gears" that turn management principles into everyday practice."[8]

As noted earlier in the chapter, although there are many well-established tools and techniques associated with construction management that can be applied to the *build* part of design-build, there are really no industry-wide standard practices for managing a design-build project from an integrated process perspective. For example, there are standard procedures in construction management for processing RFIs and submittals and for tracking job costs against the construction budgets. Traditional construction managers also use work breakdown structures and critical path scheduling techniques for keeping the construction tasks organized and on track. These are all common approaches for managing the construction process, and many are used during the construction stage on design-build projects.

As an industry, we have made great strides over the past decade in standardizing the design-build procurement process (as discussed in Chapters 4 and 5), but I have not found a consistent set of practices used to manage the design-build process after contract award. Often, the individual design-build project manager is left to figure things out on his own. Typically, project managers develop and implement their own management techniques and tools for the design-build projects that they are in charge of. However, in most instances these techniques have not been formally adopted or institutionalized as standard practices for the whole company and therefore have not been incorporated into company training programs.

My objective in this chapter is to share some of the design-build process management practices that I have captured over the years. Even though these examples represent only a sampling of the many tools and techniques that may be used to manage the design-build process, I believe that they will aid you and your team in achieving consistent results on your design-build projects. If these applications are not already standard practice in your organization or firm then I encourage you to consider implementing them or others that accomplish the same goals. Furthermore, it is my hope that the techniques and tools presented in this chapter will soon become recognized as standard practices for all design-build projects.

Ideally, the concept of design-build process management, and the methodologies associated with it, will become part of university curriculums as required knowledge for the next generation of professional constructors, engineers, and architects. For now, you and your team can make use of the practices presented in this chapter to achieve consistent results for your design-build projects.

Most of the design-build project management techniques and tools that I am sharing in this chapter come from individual design-build project managers and industry leaders. Because of their generosity, young construction managers, architects and engineers may have the opportunity to learn these techniques and tools as part of their college education.

TEAM, TECHNIQUES, AND TOOLS

We manage the design-build process by planning, directing, and monitoring the "who," the "what," and the "how" of the design-build project. I refer to these components of the project as the team, the techniques, and the tools. As a whole, I term these three design-build management components the DB-T3. Abandoning any one of these three components can undermine the effectiveness of your design-build management methodology.

The Design-Build Team – Who

The design-build team is the foundation upon which the management techniques and tools rest. The design-build team represents the "who" of design-build process management. You must have the right people on your team and you must put the necessary mechanisms in place to properly manage the relationships and to encourage communication between team members in order to make the most of the design-build approach.

The Design-Build Techniques – What

Design-build management techniques are the standard practices used to direct the design-build process. There are a variety of applications associated with the techniques discussed in the next section but they are all intended to establish a standard practice for managing the design-build process; they represent the "what" of design-build management.

The Design-Build Tools – How

Design-build management tools are comprised of a whole host of checklists, guidelines, documentation forms, contracts, agreements, technologies, software and any number of instruments or devices used to manage the design-build process more efficiently.

DESIGN-BUILD MANAGEMENT STAGES

The design-build project management function can be broken into five basic stages. Although it is not uncommon for the edges of these stages to overlap, each one represents and requires some unique design-build

Every year, more construction management, architecture, and engineering students graduate from Cal Poly, having learned about the standardized design-build management techniques described in this chapter. Ideally, design-build project delivery and design-build process management will eventually become required courses at leading design and construction programs across the country. In 2008, the Charles Pankow Foundation, in cooperation with the Western Pacific DBIA Region, awarded education grants to three universities to initiate or significantly enhance their design-build curriculums. Strong industry support, as demonstrated by these organizations, is evidence that design-build education is becoming recognized as an asset in our industry. Universities are being encouraged to deliver graduates who have an understanding and knowledge of the design-build process.

teaming strategies, techniques, and tools. The following are the five stages of design-build process management:

1. proposal stage
2. post-award stage
3. design-development stage
4. design-construct stage
5. postconstruction stage

There are several management tasks that should be undertaken at each stage of the design-build process. In the following sections, I will identify a few of these management tasks in order to highlight the unique aspects of managing a design-build project. For each stage, I will present some recommended teaming strategies, management techniques, and tools.

PROPOSAL STAGE

Unlike traditional construction project management which begins after contract award, design-build management starts during the proposal stage when the team first comes together. The team develops a conceptual design in response to an RFP and builds a conceptual estimate around that design. The team will work together to develop the proposal documents and to prepare a presentation for the client. Several design-build management techniques relating to proposal development have already been introduced in previous chapters. For example, in Chapter 5, you learned about the design charette. In Chapter 6, joint risk assessments and contingency development roundtables were introduced.

However, in the proposal stage, there are also important management duties relating to team formation and unity. This is the time when the team leader needs to establish ground rules and set expectations for the team. Starting off on the right foot will help the team maintain positive relationships throughout the project. Typically, the newly developed team must very quickly develop a conceptual design, put a price on it, and deliver the proposal to the client. In such a high-pressure environment it takes discipline and commitment to diligently carry through with many of the team-oriented management strategies, but they play a vital role in the success of the project. If the team is not on solid ground at the very early stages of the process then the project may soon be headed for trouble. In the next section, we will discuss one team-oriented management task—establishing team focus and alignment. Then I will describe some tactics, techniques, and tools that will help you accomplish that task (see Figure 7.3).

Management Task: Establishing Team Focus and Alignment

Of all the management tasks associated with a design-build project, none is more important than creating team cohesion. Team members must be

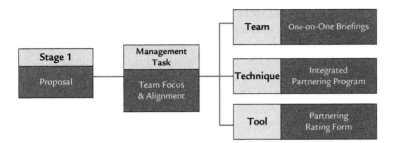

Figure 7.3
Proposal stage management task

committed to collaborating with other team members. Everyone should be focused on the same goals and aligned in terms of their approach to achieving those goals. Furthermore, everyone on the team should have a clear understanding of what is expected in terms of their contribution. It takes intentionality on the part of the design-build project manager or project director to achieve this focus and alignment. It does not happen by accident.

Teaming Strategy: One-on-One Briefings

The design-build project manager should meet individually with every member of the team to communicate to each person just how vital his role is in achieving success on the project. The purpose of these **one-on-one briefings** is to recognize the contributions of the individual team members and to clarify expectations. The most effective motivator for most people is an acknowledgement that their contributions make a difference, and these briefings can be used to help team members understand the significant role they each play in the project. According to Adrian Gostick and Chester Elton, the authors of The Invisible Employee, 79 % of employees in the United States leave their companies because of a lack of recognition.9 Gostick and Elton report in the same book that 90 % of senior executives claim that people are their companies' greatest assets.10

In design-build, the team members are the greatest asset, and spending time with them one-on-one can make a significant difference to the success of the individual, as well as the team. The thirty minutes that it takes to establish a **background of relatedness** with each team member will pay huge dividends as the project moves forward. You can start this process by having each person describe what a "win" on the project would look like for them personally. Then, show the team member how his personal goal relates to the overall project goals. Each team member can take ownership of the objectives when they are clear about why those objectives are meaningful to them. One-on-one briefings can also be used to clarify your expectations of each team member, and the team members can communicate their expectations of you—the design-build project manager.

Technique: Integrated Partnering Program

Some firms rely on a formal **integrated partnering program** conducted by outside consultants to establish their teaming and project

D-B DICTIONARY

One-on-One Briefing –
A meeting conducted by the team leader to discuss any concerns that an individual team member may have, and to clarify roles and expectations for the project. The briefing is also used to acknowledge the contributions made by the individual members of the team.

Background of Relatedness –
The array of common interests, aspirations, goals, and objectives found among a diverse group of individuals working together as a team.

Integrated Partnering Program –
A teaming process that focuses on building trust and encouraging collaboration through a formal set of practices to build productive working relationships.

> **D-B DICTIONARY**
>
> **Integration Manager –** The person primarily responsible for establishing and maintaining team cohesion throughout the duration of the project. The integration manager kicks off the project and implements many of the protocols and processes that an outside partnering facilitator would do.
>
> **Commitment Management System –** A formal process for tracking the promises and commitments made between members on a design-build team.

goals. Others organize and implement their own programs using internal facilitators. For example, some design-build companies have created an **integration manager** position. The person in this role primarily focuses on establishing and maintaining team cohesion throughout the duration of the project. The integration manager, along with the design-build project manager, kicks off the project and implements many of the protocols and processes that an outside partnering facilitator would do. The design-build integration manager is not just working with the construction team—general contractors, subcontractors, vendors, and owners representatives; they are also bringing all of the designers into the fold as well. Even when formal partnering is implemented, I still highly recommend assigning an in-house individual to serve as integration manager for the design-build team. As responsive and quick as you have to be in design-build, sometimes the formality of formal partnering becomes too cumbersome because the outside facilitators are not always present as the project unfolds. The integration manager, as a member of the team, is directly engaged in the project and can spontaneously facilitate interactions to help keep the team focused, efficient, and effective, as well as create, organize, and monitor the communications and **commitment management systems** that are so vital to the success of the team.

Tool: Partnering Rating Form

An effective partnering program includes evaluation tools. Team members should be able to easily report on progress and challenges as they arise. One of the tools often used to track and manage the team's performance is a partnering rating form (an example is shown in Figure 7.4). The evaluation criteria are established by the team as a whole—this activity itself helps create team alignment and cohesion. The example form reports on performance associated with many less-tangible elements such as communication, teamwork, cooperation, and respect; it also includes more technical measurement criteria like safety, job progress, and quality.

> If formal partnering is implemented on your design-build project, make sure that it continues through the life of the project and that it is facilitated by someone who is familiar with the integrated nature of design-build. Partnering that only takes place in a two- or three-day event is usually of little benefit to the team in the long run. It is very easy to slip back into old behaviors and attitudes unless the team has a long-term program in place to remind everyone of the shared goals and expectations.

The partnering rating form is a useful tool because it allows everyone on the team to voice their opinion regarding how they think things are going on the project. The time interval between reports should be short enough to be of benefit, yet not so short that the reporting effort becomes a burden. In addition to acknowledging successful performance relating to specific criteria, the team integration manager (or other team leader) must promptly respond to any emerging issues.

Some project leaders are very good at proactively creating a strong collaborative environment at this early stage, and some just let things happen as they may. However, there is a lot at stake when a team is trying to design and construct a complex project in an integrated fashion. A laissez-faire attitude is not the leadership approach likely to successfully manage the team or the process to the finish line. If you do not start off with an action plan that clearly establishes the project goals as well as

PARTNERING RATING FORM

Intercounty Connector - **Contract C**   Evaluation Period

Name: _____ (required)

Task Force: _____ (if applicable)

Circle Rating for Each Element. Comments are highly encouraged but not required.

(1) Communication — Open and honest communication among the group members is:	Non-Existent 1	Cautious/Guarded 2	Meeting Needs 3	Open/Free 4	I don't know N/A
	Comments:				

(2) Teamwork — The group encourages all of its members to participate:	Never 1	Infrequently 2	Often 3	Always 4	I don't know N/A
	Comments:				

(3) Cooperation and Respect — On this project, relationships among team members as a whole are characterized by:	Lack of Cooperation and Respect is the Norm 1	Cooperation and Respect Often Prevail 2	Cooperation and Respect Almost Always Prevail 3	Cooperation and Respect are Strong and are Being Nurtured 4	I don't know N/A
	Comments:				

(4) Issue Resolution — Team members and their counterparts identify issues and I find that the process of timely resolution or escalations is:	Not Functioning 1	Functioning, but Untimely 2	Established and Functioning 3	Exceeding Expectations 4	I don't know N/A
	Comments:				

Figure 7.4 Partnering rating form

PARTNERING RATING FORM

(5) Job Progress The process to monitor and assure the project's on time completion is:	Unresponsive 1	Marginally Successful 2	Meeting Expectations 3	Exceeding Expectations 4	I don't know N/A
	Comments: _____ _____ _____				

(6) Safety Safety for workers and the public is supported by all team members	Unsatisfactory support 1	Minimally supported 2	Supported 3	Safety is paramount 4	I don't know N/A
	Comments: _____ _____ _____				

(7) Environmental Commitments Are environmental commitments and requirements being met?	Not being met 1	Sometimes being met 2	Being met 3	Always met 4	I don't know N/A
	Comments: _____ _____ _____				

(8) Community Outreach The community outreach program proactively shares project information with the public.	Not proactive 1	Marginally proactive 2	Shares 3	Proactively shares 4	I don't know N/A
	Comments: _____ _____ _____				

Figure 7.4 (continued)

PARTNERING RATING FORM

(9) Quality Are quality expectations being met?	Major Deficiencies 1	Minor Deficiencies 2	Meets 3	Exceeds 4	I don't know N/A
	Comments:_____				

(10) DBE & OJT Training Programs Are the project's Disadvantaged Business Enterprise (DBE) and on-the-job training (OJT) programs effective?	Not Effective 1	Minimally Effective 2	Effective 3	Very Effective 4	I don't know N/A
	Comments:_____				

(11) Submittal Review Process Are time frames being met per agreed upon schedules?	Not Timely 1	Minimally Timely 2	Meets Schedule 3	Beats Schedule 4	I don't know N/A
	Comments:_____				

(12) Budget Are we assisting one another in meeting our budget expectations?	Never 1	Infrequently 2	Often 3	Always 4	I don't know N/A
	Comments:_____				

(13) Work Processes Are work processes effective?	Not Effective 1	Marginally Effective 2	Effective 3	Very Effective 4	I don't know N/A
	Comments:_____				

14. Material Clearance 15. Maintenance of Traffic 16. Erosion & Sediment Control

Your membership status on the Team: (CHECK ONE)
☐ ICC Team   ☐ Contractor   ☐ Contractor / Design Consultant

Figure 7.4 (continued)

individual accountabilities, then the project is at risk from the start. At a minimum, consider starting off the project by:

- Making sure that there is alignment between the core values and cultures of the separate entities that come together to serve on the project team and that behavior expectations that represent those core values have been clearly communicated.
- Executing a teaming agreement that defines the roles, responsibilities, and expectations of each of the team members during the proposal phase, and, if the proposal is successful, after the contract award.
- Writing a purpose statement that includes clear cut objectives for the project that go beyond the construction issues and get at what is important personally to the members of the team.
- Identifying specific measurable results that would indicate that the stated goals of the project have been met; continuously monitor progress toward those goals.
- Creating a culture of trust and appreciation by conducting one-on-one meetings with all team members to help them understand how important their individual contribution is to the success of the whole project
- Identifying the critical components of the project that must be addressed in order for the project to be successful, and creating a plan to resolve them.
- Creating a responsibility matrix identifying key accountabilities, team member assignments, and milestone dates. (You will see an example of a responsibility matrix in Chapter 8.)

POST-AWARD STAGE

All of your team's hard work during the proposal stage has paid off—you have won the job. The client has chosen your team to deliver their project. Now it is time to bring the client into the fold as a member of your team. The owner's interests should guide you as you maneuver the path from conceptual design to final design. The path is rarely a straight one, and, using the RFP and the proposal as guides, the design-build team leader and the client must lead the way. The team's rapport with the owner is critical, and the meetings that immediately follow the contract award can make or break the relationship. Effectively managing these engagements is critical to the success of the project. Although there are several key management tasks associated with this stage of the design-build process, two that I consider to be critically important are: 1) proposal and design validation and 2) design process planning (see Figure 7.5).

Management Task: Proposal and Design Validation

Shortly after contract award, you should meet with the client to review all facets of your proposal—administrative, technical, budgetary, schedule, and, of course, the design. It is critical to validate all aspects of the proposal relative to the RFP. Now is the time to discuss any discrepancies or

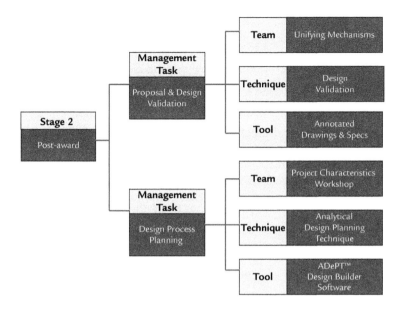

Figure 7.5
Post-award stage management tasks

misunderstandings and make the necessary adjustments so that everyone has the same expectations for the project deliverables. Open discussions regarding any concerns relating to the RFP, the proposal, or the design should be encouraged. The inflexible attitudes that often accompany the design-bid-build contract are not suitable for the dynamic, responsive engagement of an integrated design-build team.

> *The likelihood of success on a project is enhanced when all stakeholders collaborate early in the design-build process to ensure a balance of respective goals and objectives. When analyzing these goals and objectives the team needs to keep in mind key benchmarks such as the budget, schedule, quality, safety, and design integrity. Mutual understanding of these benchmarks, and others that may be specific to a given project, will allow an integrated team to make timely and objective decisions on the project. These decision points can cause some angst among team members—especially when trade-offs are being negotiated. But it is really a form of tough love—you have got to do the right thing for the project as a whole and let go of your individual interests. Prompt decisions that have team "buy-in" will allow a trusting partnership to develop amongst stakeholders and will prevent the "offense-defense" mentality that is common on many design-bid-build projects. This collaborative, trusting partnership will be the cornerstone for a successful project.*

Lou Palandrani,
LEED AP
Senior Vice President
Clark Construction Group
Bethesda, Maryland

Teaming Strategy - Establishing Unifying Mechanisms

Although the owner may initially kick off a project with a special event and pep talk of some sort, it is really up to the design-build project leader to clearly articulate the owner's objectives to the team and to identify the

> **D-B DICTIONARY**
>
> **Unifying Mechanism –** An activity, device, technique, symbol, slogan, or tool that serves to facilitate and encourage team cohesion.
>
> **Design Validation –** The process of reconciling the proposal design with the owner's expectations. Design validation meetings should occur shortly after contract award, before undertaking any further design efforts.

critical components of success. In addition the project leader must focus on creating a team culture and project identity that will unify the team around achieving those objectives. It is not that difficult to establish some **unifying mechanisms** that help create alignment and a common sense of purpose. For example, on one job site I visited, the design-build team was comprised of members from several different companies. The project team leader had banned any independent company attire on the project. Instead, the design-build contractor bought shirts, hats, and jackets for the entire team displaying only the team member's name, the project name, and project logo—along with an inspiring tagline that the team had created together. While hats and shirts do not make a team, this approach helped create a sense that everyone on the project was part of one cohesive group, which also came across in the attitude displayed by all of the team members. I felt that I could ask any one of them about any aspect of the project goals and they could respond to me even if it was not "their part." As far as I was concerned, they all worked for the same entity—which is exactly what the design-build project manager was trying to achieve. Furthermore, the owner was confident that all eyes were focused on his project and that it was receiving the attention from all team members that he felt it deserved.

Technique: Design Validation

Design validation refers to the process by which the entire design-build team, including the owner, goes through the proposal documents in detail to resolve any concerns before the team moves ahead with any further design efforts. The design presented in the proposal is rarely ever perfect. The design validation process is where the design-builder and the owner begin to reconcile expectations with deliverables. It is in everyone's best interest to make sure that the design-build team is going to deliver a design that meets the owner's expectations relating to performance, quality, cost, and schedule. This is especially true when the design-build procurement is competitive and the design-build team and the owner have not yet had the opportunity to sit down face-to-face to discuss each other's concerns about the proposal design. This can be a time-consuming process, and it calls for patience and effort on everyone's part. But it is well worth the investment of time and effort.

I took part in design validation sessions for one project in which twenty-five people were in a room and several others participated via teleconference. The meetings went on for several days. A representative from every design discipline took part (architectural, electrical, mechanical, civil, etc.) as well as the estimators and cost control people. The project's senior management team was present, as were the major subcontractors. On the owner's side, all of the primary decision makers were involved, along with major stakeholders and end users. Every drawing in the proposal set was reviewed, and all sides had a chance to voice concerns. Some adjustments were made on the spot. Others required investigation and time to resolve, but all discussion items, assignments, decisions, and pending issues were documented. As tedious as this may sound, being thorough in this process can prevent numerous problems down the road if everyone comes to the meetings in a spirit of cooperation, openness, and trust.

Tool: Annotated Drawings and Specifications

Effective design-build teams use **annotated drawings** as a tool to facilitate the design validation process. Annotated drawings reflect all changes to the proposal documents (plans, outline specifications, etc.) agreed upon during the validation meetings. Any necessary adjustments to the RFP requirements are documented, and the annotated drawings replace the proposal drawings as the project's reference documents. In many instances, these adjustments to the proposal design do not result in any added dollars to the contract. Instead, the owner and the design-build team participate in some give-and-take negotiations resulting in the execution of a **zero dollar change order** to document the transaction.

These initial validation meetings are also a great time to discuss any value engineering (or value enhancing) suggestions that you may have come up with while preparing the proposal. This is actually one of the greatest advantages of the design-build project delivery method. You actually get the chance to offer up improvements to the design that the owner may not have thought of. For instance, on a recent mixed-use dormitory project, the owner's RFP bridging documents called for aluminum window louvers around the building to provide shading. However, during the initial validation meetings the design-build contractor and designer suggested to the owner that, the louvers might not be the best solution for shading the windows and instead might create unexpected problems down the road, such as water penetration, which could lead to mold issues. In addition, the louvers would likely become an attractive nesting location for birds, which could become a nuisance to the dormitory residents. Instead of the louvers, the design-build team suggested using argon filled low-E glass to achieve the same window shading result. While low-E windows are more expensive than regular windows, their cost on this project was almost the same as the cost of the louvers, when the labor costs for the louver installation were added in. The low-E windows also improved the aesthetics of the overall design. This is an example of design-build at its finest. Everyone walked away from the table feeling good about adding value to the project while maintaining the project objectives for cost, time, and quality.

Management Task: Design Process Planning

Ideally, design process planning would begin at the proposal stage; however, it rarely starts that early—unless the design-build team is selected without competition or on a qualifications basis. Regardless of when it happens, the design process planning should always occur before the team gets into the design-development stage.

TAKEAWAY TIPS

Smile for the Camera

Consider videotaping all proposal and design validation sessions. Even when notes are taken, it can be difficult to capture every discussion, and even harder to discern the logic and rationale for the concern or adjustment after the session has ended, when memories of the conversations have faded. (Of course, all participants should be made aware that the meeting is being videotaped for archiving purposes.) Later, as you try to accommodate an adjustment to the design that was agreed upon at the session you can reference the tape to get a better perspective of what the actual goal is and not waste time addressing the wrong issues or missing the intention altogether.

D-B DICTIONARY

Annotated Drawings – Project drawings that reflect the adjustments and modifications made to the proposal design immediately after contract award at the design validation stage.

Zero Dollar Change Order – A document used to modify the project's contract based on design or scope changes made at the design validation stage. Zero dollar change orders do not result in an increase in the contract price.

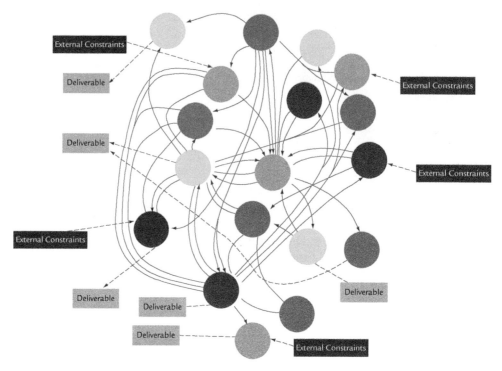

Figure 7.6
Typical design process.
Courtesy of Adept
Management Ltd.

For some, "design planning" is an oxymoron. After all, many of us think of design as a free-flowing activity that cannot be managed or controlled lest its creativity be lost. The diagram illustrated in Figure 7.6 depicts the complex and iterative nature of the design process. Is it any wonder that managing the design process is such a challenge?

However, as you learned earlier in the chapter, managing the design is all about managing the flow of information. If you can identify what information is needed to perform a design task then you can manage the flow of that information accordingly. Design planning will not eliminate all of the iteration that must take place in the design process, but it will make the process more efficient and effective—eliminating unnecessary loops. And remember, the information does not always have to result in a drawing or a spec. It may simply be written information from another design discipline, or a supplier or manufacturer. The key is to get what you need, when you need it.

D-B DICTIONARY

Project Characteristics Workshop –
A meeting attended by all primary design-build team members; participants discuss all aspects of a project including unique characteristics, risks, and challenges, and everyone gets to hear the different perspectives.

Teaming Strategy: Project Characteristics Workshop

One of the quickest ways to bring a design-build team together is to conduct a **project characteristics workshop**. This is the first step in design process planning. A project characteristics workshop provides an opportunity for key members of the construction team and the various design disciplines (civil, architectural, structural, etc.) to voice concerns and ask questions about the project. This is really where everyone on the design and construction team begins to understand the magnitude of the task at hand. Everyone involved in the workshop develops a broader understanding of the special challenges and risks associated with the project—before the design process commences. Through an open discussion, the team can begin to appreciate how each of the disciplines is related and how important it is to communicate in order to fully address the design from an integrated perspective.

Technique: Analytical Design Planning Technique

While there are many possible approaches to planning the design process, the Analytical Design Planning Technique (ADePT), developed by Adept Management Ltd., is the best I have encountered. ADePT makes use of a Dependency Structure Matrix (DSM) analysis, sometimes referred to as a DSM analysis. DSM analysis provides insights into how to manage complex processes by highlighting information flows, activity sequences, and iterations. DSM analysis reveals how various design tasks are connected, and it can help design-build teams streamline their processes based on the optimal flow of information between different interdependent activities. ADePT takes the mystery out of the design process. It is an approach that helps untangle the iterative knots of the design process so they can be assessed and evaluated in order to make decisions about the best way to proceed with the design.

The design-build team begins by developing a **design work breakdown structure (WBS)** for the design. Work breakdown structures are commonplace in construction, but not so in design. A WBS is an organizational framework which depicts a project's critical design tasks, as well as their relationships to each other and to the project as a whole. The team uses the WBS to identify the information that each design discipline will need to complete the tasks associated with any given design element. Once the information needs are known, the team identifies the dependencies among them. This enables designers and constructors to come up with the best way to move through the design process and most effectively sequence activities to support the needs of the project. The team can see the design information flows and better manage the process. The essence of the ADePT technique is shown in Figure 7.7. The design planning technique generally entails the following four steps:

1. Establish a work breakdown structure for the design and identify a clear division of design responsibilities.
2. Identify all information dependencies and prioritize them.
3. Schedule all deliverables to be produced by the design team and record key project stages, target dates, and durations.
4. Sequence the design process to exploit the availability of design information; identify key risks; and decisions required by the team and client/owner.[11]

The value of this technique usually becomes obvious even before the team fully understands how radically it can improve the team's ability

D-B DICTIONARY

Design Work Breakdown Structure (WBS) – An organizational framework which depicts a project's critical design tasks, as well as their relationships to each other and to the project as a whole.

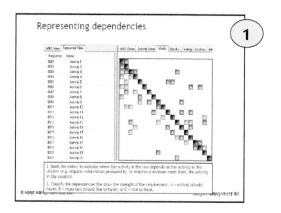
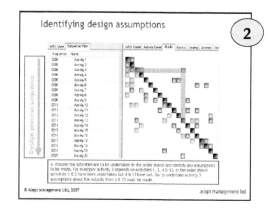
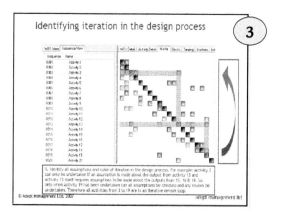
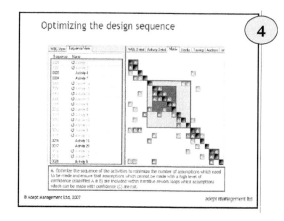

Figure 7.7
Analytical Design Planning Technique (ADePT™).
Courtesy of Adept Management Ltd.

to mitigate risks and deliver projects on time and within budget. The in-depth and detailed discussions that occur between design disciplines and contractors in an ADePT design planning session can greatly contribute to team alignment and cohesion. The level of appreciation and understanding that team members have for one another after one of these sessions is remarkable—a real eye opener for anyone who has never experienced the process.

Tool: ADePT Design Builder™ Software

The ADePT technique can be used with or without software. However, using a software tool can make the management task much easier. The Adept Management team developed the ADePT Design Builder software as a tool for developing and manipulating the WBS quickly and consistently. The tool allows project planners and design managers to effectively plan their projects, ensure the design is integrated with procurement and construction, and monitor projects in a meaningful way.

In the post-award stage, it is important to remain flexible while still pressing for resolution on as many open issues as possible. Once you have gone through the proposal, validated the design, and planned your design process, you are off to the races. Now the team can go to work to advance the design and start building.

DESIGN-DEVELOPMENT STAGE

The design-development stage is the most challenging stage of design-build process management. In most cases, the design-build contract is awarded at the conceptual or schematic design stage, and although there may be projects for which the conceptual design documents are fairly well developed, they are still a long way from completed construction drawings. During the design-development stage, the team must further develop the conceptual design and use that to generate the final documents needed for construction.

The proposal only establishes the basic parameters for the design. That means that all of the details of the design still need to be worked out. For example, as part of a conceptual design, a design-build team might assign a design parameter for structural steel of eleven pounds per square foot for the building structure. At the time the contract is signed the team would not have determined the shape or weight of the individual steel members. However, the project budget will be based upon these types of conceptual design parameters. Consider also the example shown in Figure 6.5 in Chapter 6, in which a design parameter for the glass and glazing was established at 30% of the exterior skin of the building. The number of windows, the style of windows, the type of glazing, and the exact placement of the windows were not yet determined.

After contract award, and as we move through the design-development stage, our challenge as design-builders is to design to the budget that we established at the proposal stage. This requires a phenomenal amount of team communication, coordination, collaboration, and finesse. As a team we must work through each one of the design elements, advancing the information from conceptual parameters to final construction details—confirming that our choices meet the performance and quality requirements spelled out in the RFP, while maintaining the project budget and schedule. This reconciliation process is at the heart of the design-development stage of the DB process (see Figure 7.8).

Management Task: Design-Cost Reconciliation

Managing the design to the budget is referred to as **design-cost reconciliation,** and it is a vital design-build management task. The diagram in Figure 7.9 depicts this design-development management challenge.

In Figure 7.9, each tick on the line from "RFP" and to the "100% Final Design" represents an element of the design development process. As the team moves toward the final design, they must work closely together to make sure everything stays on track—performance, quality, cost, and schedule. If the design becomes out of balance in terms of any

---

Once a design WBS model is created for a particular project type, it can be used for similar type projects with just minor modifications. However, it is best to start the ADePT process from scratch on a new project because thinking through the design process together as a team is what brings alignment and buy-in regarding the decisions that are made along the way. This initial engagement is usually held as a design process planning workshop.

One of the best ways to learn about the Analytical Design Planning Technique is by attending an ADePT Awareness Workshop. Information regarding these workshops and the ADePT suite of design manager software is available at http://www.adeptmanagement.com/amltechnologies.

D-B DICTIONARY

Design-Cost Reconciliation – The process of verifying that the design meets the requirements of the project budget throughout the various design stages.

Figure 7.8
Design-development stage management tasks

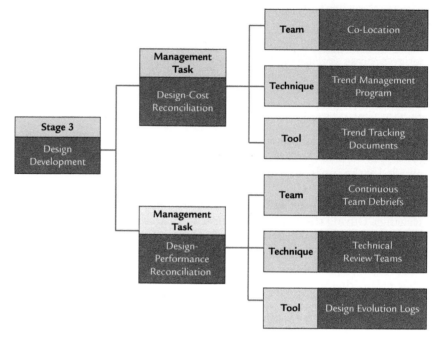

one of those elements, the whole project plan is in jeopardy. Both time and dollars will be wasted trying to bring the design back in line and on budget. This is why it is critical that the contractor's estimators and cost control personnel work directly with the designers.

In practical terms there will always be instances where the design details will take a project off budget. When that happens, the team will

Figure 7.9
Design validation and reconciliation

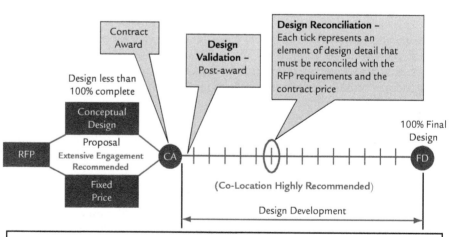

have to adjust other design elements to get the project back in line with the budget—without sacrificing quality or performance. It is a dynamic process that requires give and take. The design-build team must be flexible in order to manage this process—without it they will have a very hard time maintaining good relationships and doing their job.

Teaming Strategy: Co-Location

One of the most effective ways to facilitate the discussions that must take place during design-development is by co-locating the team. To co-locate means to bring all of the key players together in the same space. Designers and contractors share offices, even if they are from different firms. In many instances, co-location office space is rented near the project, and the team members move in furniture and equipment needed for the project. In other instances, multiple job trailers are joined to create a mini-office complex to accommodate the entire team on site.

When co-location is physically impossible (for example, when team members are from multiple regional offices or from firms outside the United States), there are ways to co-locate virtually using video conferencing and other electronic remote access tools. However, although these are useful tools, they never deliver the level of engagement that you get when everyone is in the same room. With proper planning, even the most highly dispersed teams can usually arrange to co-locate during the most critical parts of the design-development stage.

Technique: Trend Management Program

Even though there are no (non-owner initiated) change orders in design-build, plenty of changes occur as the team moves through the design development stage. With each design iteration the team must constantly check to see if the project is still on track to stay within the project budget. As the team works through the various details of the design they will have to trend the costs of these changes to see if it is on target budget or off target in relation to the cost model. While **trend management** is not necessarily new in construction, in design-build, the trending function must integrate design and cost management together into a single task.

The diagram in Figure 3.10 in Chapter 3 illustrates the concept of trend management. The fixed line represents the project budget. The squiggly line represents the trend. Some design iterations put the project over budget and some put the project under budget but ultimately, it must all balance out. It is critical that the design-build team know where the project is in terms of cost at any given point during the design-development process. Otherwise the team cannot properly vet design options and make necessary adjustments in a timely fashion. Letting the design progress move forward without a proper trend program in place is very risky.

TAKEAWAY TIPS

Tear Down the Walls

Consider removing all interior walls in your job trailers or rented office space when co-locating your team. Install shorter cubicle type partitions instead, creating a completely open space in which there is no hiding out. Subcontractors love it because they can walk in the door and go directly to the person who can answer their questions most quickly.

Many design-build firms are now creating special design-build centers where teams can co-locate during the proposal development stage and the design-development stage. This puts the people and the technology in same place. These "war room" environments encourage high levels of communication to facilitate true design-build integration.

D-B DICTIONARY

Trend Management – A process of measuring the cost and schedule impacts throughout the various design and construction stages of the project. Various trend tools are used to identify, evaluate, and manage changes that occur during design development, project buy-out, and construction that impact costs, quantities, and schedule.

Notice of Trend

Project No._____ Project Name:_____

Initiator Name:_____
Trend Title: _____
Description/Scope:_____
_____
_____
_____

Reason:_____
_____

Cause: A/E Design Development ☐ Regulatory/Tech. Code ☐
        Client Preference   ☐ Construction Site Requirement ☐
                                 ☐ Supplier / Supplier Requirement ☐

Other_____

Send to: _____
Trend Engineer_____ Date Rec'd:_____

Staff Hours			Estimated Cost Impact ($000)
Group	To Evaluate	To Execute	
A/E - Design	_____	_____	Direct: _____
Procure.	_____	_____	Engr./Indirect:_____
Const.	_____	_____	OH &P _____
Est./ Sched.	_____	_____	Total _____
PM	_____	_____	
Other	_____	_____	Estimated Sched. Impact
Total	_____	_____	_____ wks (+,-)
			Critical Path _____ (Y,N)

Comments:_____
_____
_____

Discipline / Group:_____ Location:_____
                                 Phone: _____
Effect on Other Disciplines _____
_____
_____

Effect on Cost _____
_____
_____

Effect on Schedule _____
_____
_____

Trend # _____ Client Issue # _____

Disposition	Project	Client	Reference & Date
Approved: Scope Change	☐	☐	_____
: Other Trend	☐	☐	_____
Disapproved:	☐	☐	_____

Recommended Action: _____
_____
_____
_____

Trend Engineer, date _____ Project Controls Specialist, date _____ Project Manager, date _____ Client, date (as required)

Figure 7.10
Notice of trend. Courtesy of ODC Synergy Inc.

Tool: Trend Tracking Documents

There are a number of trend management documents that can be used to help manage the design-cost interface. The trend management process typically starts with a team member recording the proposed change on a notice of trend form (see Figure 7.10), which identifies the change, its cause, the justification for it, who initiated it, and what the impacts are. The trend notice is often accompanied by a trend estimate report (shown in Figure 7.11). The estimate report details the cost impact data, which is summarized in the trend notice. And, finally, a trend log summary (shown in Figure 7.12) documents all trend activity and the overall budget impacts associated with them. (Trend log summaries are also referred to as trend register logs) Companies may use different terms to identify these tools, but the concept remains the same—there must be a logical and efficient way to track the cost impacts associated with the various adjustments that are being made to the design during the design-development phase.

Management Task: Design-Performance Reconciliation

Whereas the design-cost interface is usually managed by someone from the construction ranks of the team, oversight of the design-performance

TREND ESTIMATE REPORT

Project: I didnt do it Detention Facility

The following change of scope:
    Add Resilient Tile Flooring at:
1. All Dayrooms
2. All Main Corridors
3. All Medical Areas except Cells & Wards
4. Revise VCT at Admin Offices to Carpet

Trend No.: 9
Initiated by: AJP
Estimator: JDB
Approved: 03/01/09
Rev. No.:

Would result in the revised costs noted below should it be accepted. Revised costs are based on project costs as currently indicated in Estimate Revision No. 03, dated 21-JAN-09.

Sector	Parameter/Item of Work	Qty	Unit	Unit Cost	Total Cost
Original Budget:					
6.22	Resilient Tile Flooring at Administration Building	5,923	sf	0.95	5,627
	Carpet Allowance at Administration Building	379	sy	24.00	9,096
	Resilient Tile Flooring at Support Building	16,999	sf	0.95	16,149
	Resilient Tile Flooring at Industries Building	200	sf	0.95	190
	Resilient Tile Flooring at Special Housing	235	sf	0.95	240
	Resilient Tile Flooring at General Housing	6,292	sf	0.95	5,977
	Resilient Tile Flooring at Gym/Activities Building	6,067	sf	0.95	5,764
					0
					0
					0
					0
					0
					0
					0
					0
					0
				Subtotal	$43,043
Revised Budget:					
6.22	Resilient Tile Flooring at Administration Building	4,527	sf	0.95	4,301
	Carpet Allowance at Administration Building	614	sy	24.00	14,736
	Resilient Tile Flooring at Support Building	30,430	sf	0.95	28,909
	Resilient Tile Flooring at Industries Building	200	sf	0.95	190
	Resilient Tile Flooring at Special Housing	4,672	sf	0.95	4,438
	Resilient Tile Flooring at General Housing	27,416	sf	0.95	26,045
	Resilient Tile Flooring at Gym/Activities Building	7,318	sf	0.95	6,952
					0
					0
					0
					0
					0
					0
					0
					0
					0
				Subtotal	$85,571
Trend Estimate Report					
Hensel Phelps Construction Co.				NET CHANGE	$42,528

Figure 7.11 Trend estimate report. Courtesy of Hensel Phelps Construction.

Ididntdoit Detention Facility
TREND LOG SUMMARY

Trends Based On Estimate Revision No. 03, Dated 21-JAN-09

Printed: 15-Mar-09
Time: 8:57 PM

PROJECT SECTOR:
GROSS FLOOR AREA: 345,528 SQFT

TE#	DESCRIPTION	Action			PENDING	BUILDING	SITEWORK	TOTAL	REJECTED
	Estimate Totals --							52,626,111	
1	Delete Gyp Bd Ceilings at Housing Buildings	Accepted	WGF	1/22/1999		(311,000)		(311,000)	
2	Delete Painting of Cell Interiors	Rejected	WGF	1/22/1999				0	(338,000)
3	Adjust Scope of Detention Package					686,776		686,776	
4	Adjust Scope of Precast Cells					(887,771)		(887,771)	
5	Decrease Height of Industries Building to 16'	Accepted	CVF	3/2/1999		(30,000)		(30,000)	
6	Various Mechanical Revisions					318,135		318,135	
7	Various Electrical Revisions					101,565		101,565	
8	Delete Topping on Tee's at Housing	Accepted	HPCC	3/2/1999		(312,827)		(312,827)	
9	Add VCT Flooring at Housing & Corridors					42,528		42,528	
10	Adjust Sitework for Reivsed Grading Plan	Accepted	HPCC	3/2/1999		(168,863)		(168,863)	
11	PVC Water Mains I.L.O. Ductile Iron	Accepted	HPCC	3/2/1999		(56,008)		(56,008)	
12								0	
13								0	
14								0	
15								0	
16								0	
17								0	
18								0	
19								0	
20								0	
21								0	
22								0	
23								0	
24								0	
25								0	
26								0	
27								0	
28								0	
29								0	
30								0	
31								0	
32								0	
33								0	
	TRENDED TOTAL CONSTRUCTION COST					(617,465)	0	52,008,646	Target Value:
							Current Delta:	-0.59%	$52,315,111
		Pending Trend Estimates Total:			0	Rejected Trend Estimates Total:			(338,000)

Figure 7.12
Trend log summary.
Courtesy of Hensel Phelps Construction.

interface is usually handled by the design manager. However, the reconciliation concept is the same. As noted earlier in the chapter, the primary purpose of this management task is to see to it that all elements of the design comply with the performance requirements spelled out in the RFP. This effort must be systematically implemented throughout the design-development process, and usually requires the expertise of both internal team members and external team members, such as specialty contractors, subconsultants, and vendors.

Teaming Strategy: Continuous Team Debriefs

Team integration goes beyond simply sharing project information; it requires team members to share expertise, knowledge, and experience in order to come up with a solution that addresses the project challenges from multiple perspectives. Continuous team debriefs are one of the best ways to find the gaps in the project design from a performance standpoint—in other words, what works and what does not. The team must engage in these conversations and "think out loud," together as the design

Trend Management

Typical Internal Trends

- Design development changes
- Work quantity adjustments
- Changes associated with constructability or fabrication issues
- Escalation associated with materials, labor, or equipment costs
- Labor and/or equipment productivity adjustments
- Items that may have been included in the scope, but left out of the proposal

Typical External Trends

- Client driven scope changes
- Unanticipated changes in site conditions
- Industry standard changes
- Permitting process changes
- Unanticipated regulatory constraints
- Owner or agency specified changes
- Review board impositions
- Changes associated with financial, tax, or legal issues
- Force majeure events

Figure 7.13
Internal and external trend sources

Trend management is one of the most effective methodologies used to manage the design-cost interface. However, the changes that affect the trend are not limited to those initiated by the design process. Trends can be attributed to anything causing a change in cost—an owner-directed scope change, an error in the original estimate, price escalation, delayed deliveries, etc. All of these changes, and the cost impacts that they generate, should be tracked as they occur—whether they originate from internal sources or external sources; examples of each are listed in Figure 7.13.

is developing—not after it is complete. The continuous team debriefs occur throughout the design-development stage at regular intervals—sometimes even daily, depending on proximity of the individual team members and the complexity of the problem. The intent is for the problem solving to be done communally, not in isolation. The debriefing sessions are not intended to take much time at any given sitting, maybe 45 minutes to an hour. Often these sessions only involve 3 or 4 team members who serve as a sounding board for an individual team member working on a specific individual design task. Sometimes a longer session involving the entire team might be in order. The key is not to sit in a cubicle and try to come up with the right answer on your own—involve other members of the team regularly in order to access the different perspectives.

D-B DICTIONARY

Technical Review Team – A group of senior level personnel who are uniquely qualified, based on their background and experience, to provide a review of one or more elements of the project at various stages of design development. The members of this review committee are typically not directly associated with the project.

Technique: Technical Review Team

Designing and delivering projects fast does not mean that you should shortcut the review process. As a matter of fact, you should do just the opposite. One way to do that is to convene a **technical review team** to offer a second set of expert eyes to evaluate the project design. A technical review team is usually comprised of senior level personnel who are uniquely qualified, based on their background and experience, to provide a review of one or more elements of the project at various stages of design development. The members of this review committee are typically not associated directly with the project and this outside perspective often allows them to spot design flaws that project team members have overlooked.

The review team evaluates the design and documents their comments. The project team then meets with the review committee to explain their design rationale as it relates to the project's unique characteristics

Philip J. Sheridan, PE
DBIA
Project Executive
Clark Civil, LLC
Bethesda, Maryland

The design and construction coordinator serves as the bridge between the design members of the team and the construction operations members of the team and ensures that the design is produced in a timely manner to support planned construction operations. The design and construction coordinator should also make sure that all necessary aspects of safety, quality, environmental awareness, and durability are incorporated into the design. The coordinator must fully understanding the means and methods to be employed in the field so that work is done correctly the first time, in the most efficient manner possible—assuring maximum return for the design-build team and the client. It is vital that coordination and communication between the design members and construction members of the team, including input from key subcontractors, is supported at the highest level. This requires an engineer or construction manager who speaks the language of both the designers and the builders. This person should thoroughly understand the design process, and he should know the job sequence and schedule; the contract requirements; the key team leaders, managers, and staff; and all the construction means and methods available and considered. It's like having one foot in the design tent and one in the construction tent. You really must understand the language of both, and respond accordingly in the interest of the project as a whole.

> One of the great advantages associated with co-location is that it allows for team members to engage in continuous debriefs very easily and as needed throughout the design process.

as well as to the client's requirements. After discussions, the review team's comments and the design-build team's responses are reconciled and recorded. These reviews occur several times throughout the design process. A review team is a great way to link the project team to senior management early on so that there are not any surprises down the road. It is important to have a technical review team involved with the project whether you are using in-house designers or designers from other firms. It is a smart way to enhance team cohesion while mitigating the potential for design flaws and the risks associated with them.

Tool: Design Evolution Log

Throughout a design-build project, many decisions are made that influence the direction of the design. The decisions start in the design planning stages and continue through design development, reconciliation, technical reviews, and into the construction phase. Effective design-build teams use a **design evolution log** to record all design-related decisions, along with the rationale for those decisions and the people involved in making them. Different companies refer to this tracking tool by different names (e.g., risk register) but the goal is the same—to record the decisions associated with all design modifications, from start to finish. Some companies try to keep up with all of the design decisions by way of meeting minutes, emails, and memos but that method is extremely ineffective and unreliable. Because design-build is such a dynamic process there

> **D-B DICTIONARY**
>
> **Design Evolution Log –** A document that tracks the changes in the design throughout the design-build process and the logic associated with them.

must be a standardized method for capturing all of the discussions and recording the final decisions made.

Failure to manage cost, schedule, and performance through the design-development stage presents the greatest liability associated with the design-build process. In the design validation and reconciliation diagram in Figure 7.9, each of the ticks on the line represents an element of the design development process but it also represents a risk exposure until that element is reconciled with the RFP requirements and the contract price. **Design commitment** occurs when the team confirms that a design feature or element complies with the performance requirements of the RFP, meets the project budget, and fits within the project schedule. At that point, the risk exposure is removed. The concept of design commitment is a very important one to understand. Once design commitment is achieved for a portion of a project, that piece can be released for construction, purchase orders for materials can be issued, and subcontracts for labor can be let. The team must work through the entire design development phase together until all elements of the design have been committed.

DESIGN-CONSTRUCT STAGE

In design-build it is common practice for construction to begin before the design is complete. In some instances, the foundations of a project may be completely designed, submitted for a building permit, and constructed before the full architectural plans have been completed. Obviously, this cannot happen without cooperation from the local regulatory authorities, but it is not unusual. The pace of a design-build project is usually very fast, and the overlap between design and construction adds a new twist to construction operations in design-build projects. The project team must become accustomed to making decisions and solving problems on the fly.

When a problem arises in the field during a design-bid-build project, the contractor usually sends off five or ten RFIs to the architect and then waits two weeks for a response. This is not how it works in design-build projects. In design-build, if we have a problem in the field we get the architect, engineers, subcontractors, owner's representative, and anyone else necessary, to immediately address the issue on site. Ideas and potential solutions are sketched out on legal pads (or a handy 2 × 4); options are discussed, decisions are made on the spot, and the work continues. The whole procedure is then documented and paperwork is filed for record. Kind of like an RFI in reverse. Instead of submitting a question in writing first and then waiting for the decision in writing, the decision is documented in writing with background regarding the question associated with it. It is a lively and dynamic environment.

One of the factors that will directly impact how smoothly things go in the field is how well the design is coordinated with the procurement schedule. Although contractors build from the ground up, architects and engineers traditionally design from the top down. This upside-down relationship can cause real problems when the contract requires a delivery

---

Some design-build teams engage the services of a third party peer review team for their technical design reviews. This may be in addition to the technical reviews performed by their own partners. The key is to make sure that more than one set of expert eyes scrutinizes the details of the design as it develops.

D-B DICTIONARY

Design Commitment – Design commitment is made when a design element is verified to comply with the requirements of the RFP by the designer-of-record and has been priced and confirmed within budget. Only then can purchasing decisions be made and construction planned.

Design-build teams often try to rely on meeting minutes to track design decisions, but minutes are a poor substitute for a design evolution log. Design specific decisions should be captured and documented throughout the entire design-build process utilizing one tool. Dates, discussions, decisions, and rationale associated with the design should all be recorded along with the people who were involved in those decisions.

Figure 7.14

Design-construct stage management task

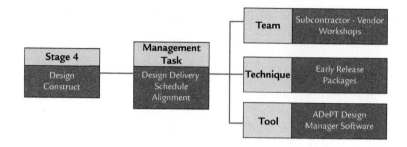

date that necessitates starting construction before the design is complete—which is typical with design-build. Therefore, the design planning process must address the need to coordinate and sequence the design schedule to complement the procurement and construction strategy. This task becomes one of the most challenging management efforts associated with the design-construct stage of the process (see Figure 7.14).

Management Task: Design Delivery Schedule Alignment

A common complaint heard from contractors regarding the integration of design with construction is that the design is often lagging behind the required procurement schedule. On the other hand, architects and engineers often feel like they are being rushed to deliver design before it is ready to be released. However, the reality is that clients are demanding shorter and shorter construction schedules, and construction cannot begin unless materials are delivered and labor contracts have been issued. That means that a serious effort must be made to coordinate and align procurement needs with design sequencing. This is no easy task.

Teaming Strategy: Subcontractor-Vendor Workshops

When attempting to coordinate the design release schedule with the procurement schedule, be sure to get feedback from key subcontractors and vendors. One of the most efficient methods for achieving this goal is to conduct **subcontractor-vendor workshops.** These sessions, often lasting several hours to a full day, help the team develop an overall procurement strategy that will allow the design-build team to deliver the project as expediently as possible. Several related trades may meet together, along with the appropriate design and construction team members, to hash out the various procurement options. Using the overall design plan as reference, the team can begin to determine what information is needed to align the design sequencing and release schedule with the procurement strategy. For example, the designer may initially specify a product that meets the required performance for the project but has a delivery lead time of several months. A knowledgeable vendor or subcontractor may be able to suggest another product that can meet the same performance but with a much shorter lead time. The goal is to find a way to fix (or commit) as quickly as possible those design elements that are critical to getting the project started. In order to do this, the design-build team

> **D-B DICTIONARY**
>
> **Subcontractor-Vendor Workshop –**
> A planning session involving the design-build team along with key subcontractor and vendor partners, specifically designed to evaluate and assess various procurement options as they relate to the construction needs and design constraints associated with a project.

must make some assumptions and establish some early design parameters so the design can progress faster than it normally might.

Technique: Early Release Packages

Working together, the architects, engineers, and constructors on a design-build project can plan elements of the design to accommodate an accelerated procurement schedule for critical items. The team identifies those material and equipment items that most severely constrain the construction schedule, and then uses **early release packages** to move forward with procurement and construction for those critical items. With design-build projects, the team has an opportunity to get a jump start on the project because they can identify the critical items before the design is complete. The use of early release packages is one of the reasons why design-build is known for delivering projects faster than any other project delivery method (see Figure 7.15).

Tool: ADePT Design Manager™ Software

When attempting to align the design sequence with the procurement and construction schedule, the goal should be to develop an optimized schedule. The ADePT design planning tool discussed earlier in the chapter helps to do just that by allowing the design-build team to evaluate the critical design elements as they relate to the procurement and construction needs. The team can then make decisions about what they can and cannot do to adjust the information flows and to reduce the amount of design iteration necessary. Once the plan has been optimized it can be exported to scheduling software, and a **target design schedule** can

> **D-B DICTIONARY**
>
> **Early Release Package** – A bidding package or scope of work for critical design elements that includes all necessary design and specification criteria needed to secure pricing and contract commitment earlier than would be normally achievable under the design-bid-build model.
>
> **Target Design Schedule** – A schedule for design delivery that has been systematically coordinated and aligned with the procurement and construction needs of a project through the collaboration of various members of the design-build team.

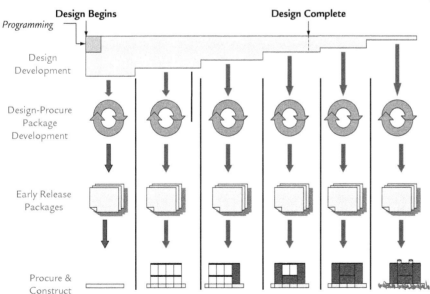

Figure 7.15
Early release package

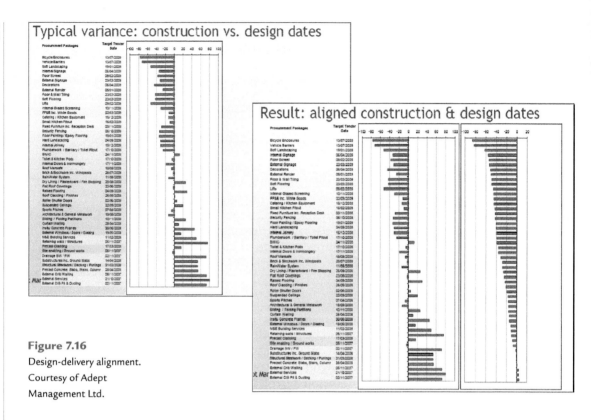

Figure 7.16
Design-delivery alignment.
Courtesy of Adept Management Ltd.

be produced. The impact is dramatic, as can be seen in Figure 7.16. The ADePT Design Manager helps keep the design work on target with the aligned design schedule. The ADePT Design Builder and Design Manager tools help the team see where the information outputs are at any given time and manage them to the agreed upon timeline.

POSTCONSTRUCTION STAGE

The postconstruction stage of the design-build process has a number of management tasks associated with it, especially when additional services are part of the contract (e.g., design-build-operate) or if it is a LEED Certified project. In these instances the postconstruction management responsibilities actually start at the beginning of the project and carry through to the end. For example, the people who will be responsible for the commissioning, maintenance, or operation of a facility should be brought on board early in the process—ideally at the proposal stage. Commissioning and operations personnel, in particular, often have knowledge and insight that no other member of the team will have. Their input throughout the design process, and even during construction in some instances, is vital. Do not make the mistake of leaving them out of the loop until after the design and construction is complete. If you fail to engage them during design you may risk making mistakes that could end up costing a great deal to correct. In addition to management tasks associated with integrating back-end personnel into the project, there

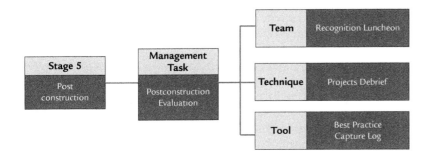

Figure 7.17
Postconstruction stage management task

are a few design-build close-out tasks that I think are worth mentioning, as well. I am not referring to the tasks typically associated with a project—such as startup coordination and the creation of as-built drawings—these are typical and must be handled whether the project is design-build or design-bid-build. Instead, I am talking about managing two of the most neglected aspects of project close-out—acknowledging the team for a job well done and capturing the lessons learned (see Figure 7.17).

Management Task: Postconstruction Evaluation

Anyone who has ever completed a construction project knows just how hard it is to stay focused once all of the major work starts to wind down. It can be difficult to concentrate on the current project when you are ready to move on to the next exciting job. However, because experience is indeed the best teacher, it is critical to capture the lessons learned from every design-build project you complete; this is especially important when you consider that the whole art and science of design-build process management is just beginning to be appreciated.

Another aspect of design-build that should always be evaluated at the end of a project is the performance of the team and the effectiveness of the relationships developed on the project. Were there certain designers, subcontractors, or vendors that you worked particularly well with, and, if so, why? Were there others that you would not engage again? Do you have a method for assessing these relationships and passing that evaluation on to other teams within your company?

In design-bid-build these close-out management tasks might not be a high priority, but in design-build, the teaming aspects of the process can make or break the project. In many cases, you will be working with the same partners on other projects, and the stronger your relationships become the more effective you will be as a team.

Teaming Strategy: Recognition Luncheon

As I mentioned earlier in the chapter, the most effective motivator for most people is an acknowledgement that their contributions make a difference. In our haste to move on to the next project we often forget what it really was that made the project a success—the people. In design-build we have an opportunity to celebrate the contributions of the entire team—all disciplines and all participants. A recognition luncheon to honor the

> **D-B DICTIONARY**
>
> **Project Debriefs –**
> A process in which the design-build team reviews project performance and makes note of what worked well and what did not. Lessons learned and best practices are noted and captured for future use.

accomplishments of the team is just one way to recognize the contributions made by individual team members. There are many other ways to show appreciation, as well. I know of one company that hosts a weekend retreat for all team members and their spouses at the end of all of their design-build projects. These team members enjoy the recognition and are anxious to be selected for the next project—especially because they know they might be working with others who were also part of a successful project. These experiences help build a background of relatedness and can make a huge difference in terms of the team's effectiveness and cohesion on the next project. The bottom line is that the team is the most important thing in design-build. So whether it is a lunch, a dinner, a picnic, or a special outing—be intentional about this, and do something at the end of the project that acknowledges your team for a job well done.

Technique: Project Debrief

Project debriefs take place shortly after the completion of a project—ideally within a 30 day period. This process is nothing new to most contractors; however, we are no longer talking about simply debriefing the construction portion of the project. In design-build, we need to debrief the design and the construction as an integrated process, along with the team performance. These sessions should be facilitated by someone who is good at soliciting honest feedback. The idea is to walk the team through the project from the proposal stage through the postconstruction stage, discussing what worked and what did not work throughout the entire project.

Although it is often difficult to regather the team and dedicate the time necessary to fully vet the process and experience, it is one of the best things that you can do to improve your performance, and the performance of your company, as a design-builder.

Reinventing how to manage a design-build project with each new contract is risky business. The only way that you can get better and begin to perform consistently as a team on project after project is to figure out what works, develop it as a regular practice, and then make it a standard operating procedure. Conducting project debriefs at the end of each job will help you do that.

> Several years ago I invited team members from a very successful design-build job to my classroom to share with my students how they worked together throughout their project. I asked a representative from the construction team, one of the principle architects, an engineering member of the team, and the owner to take part. As I queried them about what worked and what did not work, I realized that this was the first time that the team had ever really thought about the questions that I was asking. With prompting, the team discussed some specific project challenges and how they worked together to resolve them. But, up to that moment, they had never discussed how they made it all work. In other words, their project debrief happened right there in front of my class.

Tool: Best Practice Capture Log

Many companies conduct some type of project debrief session at the end of each project, and most admit that they do a poor job of doing anything with the information that is gathered. Collecting best practices and implementing them as standard operating procedures really fits in with efforts associated with quality management and continuous improvement. However, in many organizations, no one person is charged with organizing and setting up a reliable capturing process. And getting the project personnel together long enough to capture the information is a difficult challenge—it takes real discipline and commitment.

By using a **best practice capture log,** companies can make efficient use of this valuable information to train their staff and ensure consistent performance on future projects. A best practice capture log is used to record the lessons learned on each design-build project and to make that information accessible to others within the organization. Some firms use electronic chat rooms that allow team members to share lessons learned on an individual basis. Although this can be helpful during a project, a better long-term approach is to include that information in the company's ongoing training program so that effective teaming strategies, techniques, and tools become institutionalized across the company.

In the preceding sections, I have highlighted a few of the distinctive management tasks associated with the design-build process, and I have identified some of the most effective teaming strategies, techniques, and tools used by effective design-builders across the United States. Most companies have established some best practices for managing design-build projects, but many are trying to use techniques intended for design-bid-build projects—with mixed results. The majority of companies have not yet institutionalized a standardized approach to managing the design-build process, and all of the unique risks and challenges associated with it. I believe that the examples in this chapter can guide you as you develop your own design-build management protocol. In the end, a project will be successful when the team delivers the design and performance, hits the key milestones, controls the budget, and ensures that all risks are recognized and mitigated or removed.

THE DESIGN-BUILD MANAGEMENT TEAM

As I have tried to emphasize throughout this chapter, design-build is all about the team. (You will learn more about the importance of the team in Chapter 8.) No one person can know everything there is to know about how to best manage an integrated process. Individuals must rely on other team members to contribute their expertise in order for the process to work. In Chapter 6, I suggested that design-build teams use a team estimating approach. I also recommend a team management approach for design-build projects. One of the most critical components to be managed is the **team integration.** As a matter of fact, this vital role should be clearly identified as a distinct design-build management accountability. Relationship management is a significant component of success, and it requires setting up and monitoring protocols and systems that ensure that the team continues to operate as a whole and does not slip back into the segregated services mentality that can lead to risk gaps and missed opportunities.

The traditional control and command approach will not work. The job of the design-build project manager is to empower the team to collectively solve project problems and deliver superior results time and time again. An effective design-build project manager gets the most out of individuals and empowers the design-build team to achieve together what they could never achieve as individuals. Nowadays, the integrated design-build team seems more an imperative than an option. In a competitive environment of limited resources, increasing demands, and complex risk scenarios, all players on the team must be at the top of their game and be fully committed to the project. There is no better way to make that happen than to ensure that you have the right leadership in place.

> **D-B DICTIONARY**
>
> **Best Practice Capture Log –**
> A tool used to document the most effective methods, techniques, and means associated with managing a design-build project.
>
> **Team Integration –**
> The process through which a design-build team moves beyond the information sharing and cooperation associated with a typical project and into a synergistic, highly engaged collaboration where integrated solutions are discovered through the collective sharing of expertise, and experience.

PIONEER PERSPECTIVES

Michael C. Loulakis
President and CEO
Capital Project Strategies, LLC
Reston, Virginia

Doing Something With It

I've often been asked about how I came to be involved in construction law and so active in design-build. I can honestly say neither decision was part of a grand plan. In reality, they came from a series of fortunate events that somehow worked together and put me in the right place at the right time. Here's how it all happened.

While I can't remember exactly when, I knew that I wanted to be a lawyer. When I was admitted into the College of Liberal Arts at Tufts University in 1972, I felt that I was well on my way. That was until my father had a "sit-down" with me and expressed his concerns about my going through four years of liberal arts at this incredibly expensive school without a back-up plan in the event that I didn't get into, or could not afford, law school. Dad had done some digging and learned that Tufts had a small, but well-respected, engineering school. He suggested that engineering was a great way to train for being a lawyer, as both professions required problem-solving skills and analytical thinking. I decided to take his advice and entered Tufts as an incoming member of the civil engineering class, which was the single most important decision in my professional life. I was able to use my engineering background as a stepping-stone to a legal career. After I finished my coursework at Tufts I got a job as an engineer at GEI (a Boston-based geotechnical firm). It was while I was at GEI that I was first exposed to the world of construction claims and litigation, and I found it an interesting application of engineering and legal skills. I applied to law school while at GEI, decided to go to Boston University School of Law, and told everyone that I was going to be a construction lawyer.

There is no question that my engineering background gave me an advantage in applying for jobs. In 1979, Jon Wickwire and Don Gavin hired me to be an associate with their small construction law firm in Tysons Corner, Virginia. That was the second most important decision in my professional life. The firm was on the cutting edge of construction law and wanted its attorneys to not just bill hours, but to find ways to better the construction industry. Jon Wickwire in particular pushed me to find something that I was genuinely interested in and "do something with it." It took me quite a bit of time before I found something that truly interested me. That area turned out to be design-build project delivery.

My first real exposure to the world of design-build occurred in the power generation sector. I was fortunate to form friendships with some of my partners who did work in the energy regulatory area. They had clients who were developing hydroelectric and cogeneration plants, and needed help in putting the Engineer-Procure-Construct (EPC) agreements together. I thought it was interesting work and, by the mid-1980s, had drafted and negotiated several of these EPC contracts.

As I worked on those projects, I started wondering why this integrated form of project delivery was not used in other sectors of the construction industry. I also became quite interested in some of the liability and contractual issues surrounding EPC and design-build. It was such a new area that there were only a handful of reported cases that discussed the topic.

I saw a major change in the dynamics of design-build in the early 1990s. Organizations like the National Society for Professional Engineers started putting on design-build programs, and I was fortunate enough to be asked to participate and give my views. It was at that time that some public and private projects started experimenting with design-build in lieu of at-risk construction management or design-bid-build. Interest was definitely building in the integrated project delivery industry, and I was becoming more interested in it as well. I was aware of a few others who were trying to think through the issues of design-build. In particular, I was impressed with Jeff Beard, who had written an in-depth report for American Society of Civil Engineers (ASCE) on the subject.

Jeff invited me to lunch at some point in 1993, and he brought with him Preston Haskell, whom I had never met before. Jeff and Preston told me about their vision for a new organization that would be focused on advancing the design-build process, and asked me if I would be interested in participating. Shortly after that lunch I became actively involved in what became known as the DBIA.

Saying "yes" to Jeff and Preston was the third most important decision in my professional life. Working with DBIA gave me a forum to contribute to a major industry initiative. I met a "dream team" of industry players who were pioneers in the design-build arena and ultimately helped make DBIA a vital and relevant industry organization. In addition to Jeff and Preston, I think of Rik Kunnath, Kraig Kreikemeier, Don Warren, Grant McCullough, Ed Wundram, and Stan Smith as some of those who made major early contributions to the design-build body of knowledge, and committed incredible amounts of their time to the cause. As one looks beyond the early years of DBIA, it is evident that a host of others stepped up to continue these traditions.

One of the most enjoyable things I worked on for DBIA was the development of its Manual of Practice and standard form contracts. From almost the moment I met him, Preston felt that DBIA should develop its own set of contracts for design-build, and he wanted them to be done quickly. My view was that DBIA was too young an organization to jump out and create its own forms, and that we would be better off with a deliberate process, evaluating the forms sponsored by other organizations (AIA, AGC and EJCDC), letting the industry know what we thought regarding how those forms squared with best practices, and then creating our forms after we received industry feedback. Preston ultimately agreed, and under his leadership as chairman of DBIA's Manual of Practice Task Force, I drafted what became known as the DBIA Design-Build Contracting Guide. Content for the guide came from many sources, including Bennett Greenberg and the DBIA Board of Directors.

After DBIA published the guide, Bennett and I turned our attention to drafting DBIA's standard form contracts, which we published in 1998 and 1999. I remember how great I felt when these forms released and were generally received by the industry as balanced, respectable forms of contract. In late 2008 and early 2009, Bennett and I worked together again, along with several other attorneys, to revise the forms and consider some of the more important issues in today's design-build world—including BIM and sustainability. The latest versions were released in spring 2009.

It is clear that design-build has become a mainstream project delivery system, and best practices are well established in many organizations. However, design-build can be far more effective than the way we currently see it practiced. More needs to be done to integrate the entire project team. The first generation of design-build leaders were focused on the integration of design and construction—this was a hard step to accomplish given the historical attitudes of architects and engineers toward design-build. With Integrated Project Delivery and Alliance Contracting being used more frequently, the construction industry is signaling that integration of owner and major subcontractors is vital to project success. Best practices need to be developed to support these new approaches to integration. There is also a need for procurement reform that will enable public agencies to use qualifications-based selection for design and construction on an integrated basis—as full integration is hard to do with traditional competitive-based selection processes.

I really don't know how it will all play out. But what I do know is that design-build will look very different ten years from now than it does today. And I expect it will be far better than we can imagine.

■

Roles and Responsibilities

Companies may use a variety of job titles for the team members responsible for the various design-build process management tasks—design-build project manager, integration manager, integrated project leader, integrated team leader, or design-build process manager. That is really for each company to sort out as this new paradigm for integrated project management evolves. The more important thing is that as an industry we recognize that the integrated design-build process requires something different in terms of personnel and accountabilities.

I recommend that you consider the following design-build management personnel when structuring your team:

:: **Integrated Project Leader (IPL)** This relatively new job title has started to replace the more traditional title of project manager, which is now more commonly used on "construction only" projects. The IPL title more accurately conveys the additional accountabilities associated with the position. The integrated project leader is charged with supporting the team in its efforts to achieve an integrated, comprehensive solution that meets or exceeds the owner's expectations. The person in this role also often performs all of the tasks associated with the role of a design-build project manager as noted below.

:: **Design-Build Project Manager (DB-PM)** In addition to monitoring job performance associated with construction (cost, schedule, and quality), the DB-PM also monitors design performance. The DB-PM may also be directly involved in coordinating owner and end-user input.

:: **Integration Manager** As noted earlier, this individual is responsible for managing the relationships among the design-build team members to ensure high levels of communication, collaboration, and integration. An integration manager is also typically involved in the initial partnering sessions and is charged with monitoring performance associated with criteria noted on the partnering rating form, such as, teamwork, respect, cooperation, etc. The integration manager may be responsible for more than one project team at a time.

:: **Design Manager** Also called a design coordinator, this person is responsible for managing the design work flow, monitoring the project scope, and seeing to it that the target design schedule is met. The design manager is also often responsible for coordinating subcontractor design with the designer-of-record, making sure there is good communication between the design team members and the construction team members, and ensuring that the owner is kept in the loop throughout the entire design process.

:: **Designer-of-Record (DOR)** In addition to their traditional responsibilities (e.g., approving the final design, stamping the drawings, etc.), in design-build the DOR has responsibility for managing the design-performance interface and confirming that all design details comply with the requirements spelled out in the RFP and proposal. The DOR is also responsible for deconflicting all drawings to ensure that designs provided by the various design disciplines are coordinated and work together without clashing.

:: **Design-Construct Coordinator** This person is responsible for managing the design-cost interface tasks and the design-construct interface tasks. The design-construct coordinator is also often involved in coordinating subcontractor and subconsultant input. The design-construct coordinator may participate in or manage the procurement process as he has direct influence on when design commitment is achieved.

:: **Construction Manager** In design-build, the construction manager role most resembles the traditional project manager role. In addition to overseeing the traditional project control items (such as the construction schedule, budgets, and quality), and supporting the superintendent and field operations personnel, the construction manager is also now responsible for implementing the design in the field.

Figure 7.18 illustrates a model design-build project team organizational structure. Keep in mind that the requirements for design-build management personnel correlate to the size of the project, just as is the case with construction project management. But the needed oversight is the same regardless of how many people you actually have managing the process. On smaller jobs, the integrated project leader might perform the role of the design-build process manager, and the design-manager might also be the designer-of-record. Although the actual team configuration as well as the specific titles and job duties may vary by project, the design-build management tasks remain the same, and someone must be assigned these responsibilities.

Figure 7.18

Design-build project team organizational structure

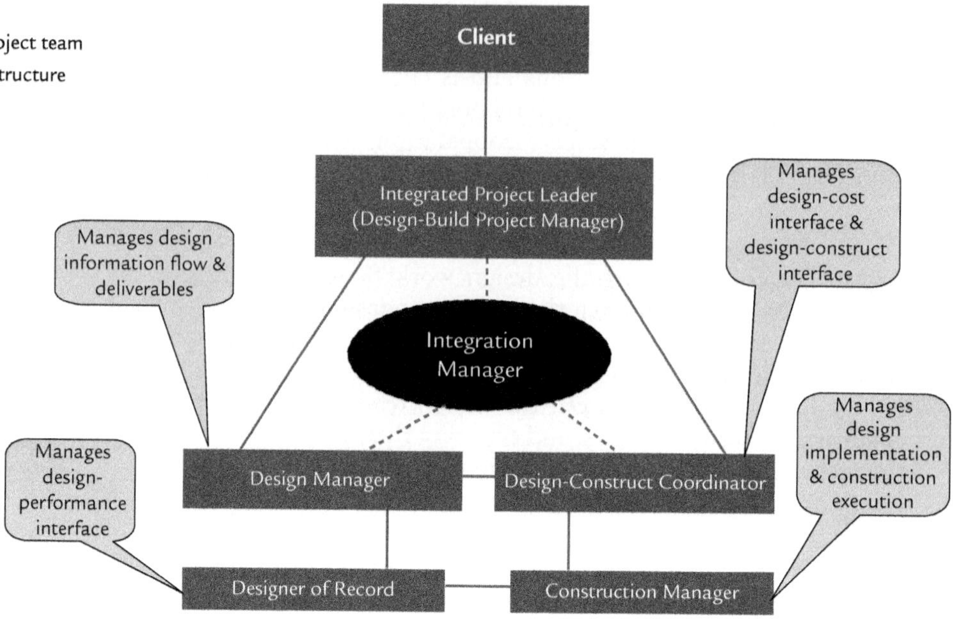

```
                              Client
                                |
                    Integrated Project Leader
                   (Design-Build Project Manager)
                                |
                       Integration Manager
                                |
          Design Manager — Design-Construct Coordinator
                |                            |
        Designer of Record          Construction Manager
```

- Manages design information flow & deliverables
- Manages design-cost interface & design-construct interface
- Manages design-performance interface
- Manages design implementation & construction execution

The role of the design manager (or design coordinator) in design-build is unique. In my experience this role is often filled by an architect or engineer with a construction management background. These individuals make excellent design managers on design-build projects because they understand both disciplines and have a unique interdisciplinary perspective. In essence they are capable of thinking with both sides of their brains—the creative side and the analytical side.

The Design-Build Project Manager

The role of the design-build project manager involves managing both construction and design. As I have tried to make clear, the design-build project manager must really be a process manager—and he must manage the design-build process in an integrated fashion. Not just anyone can be a design-build process manager—it takes a unique character. You will learn a little more about the characteristics of a true design-build project manager in the next chapter, but for now, it is enough to understand that a traditional project manager cannot be expected to perform as an integrated design-build project manager without training.

Design-build project managers can come from either the design or the construction ranks as long as they are able to think independently, despite their discipline inclinations. Although this is a real challenge, several very good design-build project managers have told me that it is possible to achieve—and a must if you want to have any credibility with the team. Joe Flores, Director of Human Resources and former Senior Project Manager with the Beck Group of Dallas, Texas explained it this way:

> *When leading a design-build team, if you come from the construction side then every time there's a decision to be made, the team—both architects and contractors are going to expect you to lean toward the contractor's interest. And if you come from the design group, the team will expect you to side with the architects or engineers. But you can't do that. You have to find a way of processing and thinking through decisions that serve the entire project and not just your particular discipline's viewpoint. It is tough, especially when you first start out. But eventually you learn how*

I think an effective design-build leader must be able to communicate a vision to the team that will speak to each member, and call them to their best work as a part of the whole. The vision and the message must be strong enough to capture the imagination and the dedication of each team member. And the vision must be able to withstand obstacles and overcome the resistance of time and history. The leader must be willing to invest the time and energy to put a stop to quick, automatic responses in favor of seeking out new solutions specifically targeted for the delivery of the vision. For teams or team members new to design-build they must know that their ideas and input are valued and will be considered throughout the development of the project. The leader must create, within the team, a sense of empathy and consideration for all effected by both the process and product of the project.

The design-build leader must understand that true success in design-build lies well outside of the contract requirements. The "wow" factor must be identified early and revisited often throughout the process.

John A. Martin, DBIA
Division President
Hintco Inc.
Austin, Texas

to think from a broader perspective and that perspective ends up being the owner's perspective. Sometimes it may appear that you are leaning toward the architects viewpoint or leaning toward the contractors viewpoint—but your job (as the integrated project leader) is to always consider the issue from the owner's perspective first and then sort through all of the options the team presents to you, making the best decision you can to serve your client's expectations and needs within the constraints of the contract[12]

I think Joe hit the nail square on the head when he described what it takes to be an effective design-build project manager (see Figure 7.19). It also helps to be a **systems thinker**—to have the ability to see how all of the dots connect to create the whole solution. But most importantly, the design-build project manager must have the ability to build trust on the team and create a vision for the project so that the individual participants can see how their input contributes to the overall solution.

The Design-Build Superintendent

Just as the administrative management team needs to adjust their thinking and behaviors to accommodate the integrated nature of design-build, so do the field management personnel—particularly the superintendent. I have been privileged to work with several individuals that I consider to be true design-build superintendents. Like the design-build project manager, the design-build superintendent is a unique character. True design-build superintendents are always trying to add value to the project. They embrace a proactive approach to problem-solving that serves not only the company that they work for but also the client—from

D-B DICTIONARY

Systems Thinker – Someone who can think through solutions from a broad perspective while considering many viewpoints at the same time. A systems thinker has the ability to leverage information and options to come up with a resolution that serves multiple interests.

Figure 7.19
Design-build project manager

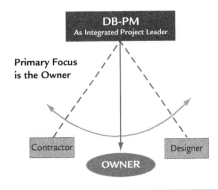

the first stages of design, right through to the final field installations. This is quite a departure from the traditional hard bid superintendent who typically focuses only on the requirements of the completed plans and specifications. But given that the design is not 100% complete when construction begins, the superintendent has an opportunity to provide input in advance of the next stages of design completion. A good design-build superintendent and design-build architect working as a team and staying in communication can find ways to improve a project in a number of ways—tweaking details to improve productivity, aesthetics, quality, the budget, the schedule, or safety. The key is to get the superintendent involved early—very early. Even though they are not designers, they can bring insightful opinions to the design process.

Michael Cebulla,
LEED AP
Superintendent
Ryan Companies USA, Inc.
Minneapolis, Minnesota

Being a superintendent on a design-build project makes you feel like you are really a part of the entire project solution and not just the guy who gets stuck trying to make things work. I actively engage with the architects from the very beginning—during the schematic phase—and then on a daily basis during construction. I first make sure that I understand exactly what they are trying to accomplish with the design. Once I understand that I can communicate it to the trades in the field and get their input to feed back to the designers so we can optimize not only the design but do so in the most efficient way. Then we can reduce cost, save time, and add real value for the client. I go for the win-win across the board. I can't imagine doing it any other way.

You can see that it takes some special skills and talents to perform well in design-build. If you do not have the proper skills and talents, then all of the techniques and tools in the world will not make much of a difference. Techniques and tools can help keep you out of trouble, but the team, and your ability to integrate the people on the team, is what will make the biggest difference. That is the ultimate job of the integrated project leader.

BUT ARE YOU READY?

To optimize your potential success with design-build, your organization and individual team members (particularly the team leaders) must fully embrace and understand the integrated nature of the design-build process. It takes a certain maturity to do design-build the right way, and it does not happen overnight. Some firms are more prepared than others to make the transition from a segregated services culture to an integrated services culture. Without a cultural shift it will be difficult to implement and use many of the design-build teaming strategies, techniques, and tools that I have described in this chapter—potentially keeping you from optimizing your project results.

In order to get better at something you must first know where you are. And even though I have just scratched the surface in describing the types of talents, skills, behaviors, techniques, and tools needed to perform well in design-build, it is a good place to start to evaluate your own design-build readiness as a company.

Design-build project management is an evolving practice and there is still much to learn and discover. And no matter where you are in the evolutionary process, you can improve, but the first place to start is not with techniques and tools—it is with the team. Start with the team, teach the techniques, and then use the proper tools to add efficiency.

CHAPTER HIGHLIGHTS

1. Design-build project management is not just about the product, it is first and foremost about managing the people through the process associated with delivering the product. This requires distinctive management applications, unique techniques and tools, and a project manager who can work in a highly synergistic environment.

2. The primary objective of design-build process management is to identify and implement a repeatable integrated design and construction process that behaves predictably and consistently and delivers a quality product on schedule and within budget.

3. Instead of managing a project from a single discipline perspective or a single service perspective, design-build project managers take a more holistic view. They manage the following four unique design-build process functions: 1) the design work plan, 2) the design-cost interface, 3) the design-construction interface, and 4) the design-performance interface.

4. Managing the design process means managing the flow of information inputs and outputs. Information inputs come from many different sources, both internal and external to the team. Information outputs may include items such as memos, narrative descriptions, reports, calculations, meeting minutes, risk registers, and sketches.

5. Managing the design-cost interface is a big challenge that is primarily applied to the critical design development stage of the process and entails reviewing the various elements of the design as it progresses and confirming that it stays within budget and scope.

6. The design-construct interface involves a number of tasks including the review of the design from a constructability perspective which can entail considerations associated with production, fabrication, labor availability, schedule and procurement.

7. The primary design-build management task associated with the design-performance interface is to confirm that the design, as it develops, complies with the performance requirements as spelled out in the RFP.

8. Managing the design-build process is done by planning, directing, and monitoring the "who" (the design-build team), the "what"

(design-build techniques), and the "how" (design-build tools); this process can be referred to as the DB-T3.

9. The five stages of design-build process management are: 1) proposal, 2) post-award, 3) design-development, 4) design-construction, and 5) postconstruction.

10. During the proposal stage it is critical to solidify the team. One vital management task in this stage is establishing team focus and alignment. Some techniques to achieve this are: conducting one-on-one briefings, forming an integrated partnering program, and utilizing partnering rating forms.

11. Two management tasks that are vitally important during the post-award stage are: 1) proposal and design validation and 2) design process planning. The goal of the design validation process is to confirm a meeting of the minds among all team members, including the owner, regarding the acceptance of the proposal design in response to the RFP. Design process planning should happen prior to the design-development stage so that information needed for design can be identified, thus making the design process more efficient and effective. At this stage the team may chose to conduct a project characteristics workshop and use techniques such as the Analytical Design Planning Technique (ADePT) developed by Adept Management Ltd.

12. The most challenging aspect of design-build process management is moving from the conceptual stage to the design-development stage. The team must design to the budget established at the proposal stage while also continually making sure that all performance and quality requirements in the RFP are being met.

13. One important management task in the design-development stage is design-cost reconciliation. In order to be successful in this endeavor the design-build team may choose to co-locate and use trend management techniques.

14. Design-performance reconciliation is another critical management task performed during the design-development stage. Conducting continuous team debriefs, convening a technical team review, and keeping a design evolution log aid in this process.

15. During the design-construct stage there is usually an overlap between design completion and construction startup. Creating a design delivery schedule that aligns with construction needs is a critical management task in this stage. Activities that support this endeavor are subcontractor-vendor workshops, utilizing early release packages, and using ADePT design manager software.

16. The evaluation and debrief that takes place during the postconstruction stage helps the design-build team improve by looking at both the project and the team itself. Hosting a recognition luncheon, conducting a project debrief, and creating a best practice capture log are worthwhile endeavors during this stage.

17. The role of the design-build team leader is a critical one. The design-build project leader needs to empower the team to collectively solve problems and deliver superior results.

18. Ideally the design-build team should contain the following design-build management personnel: an integrated project leader or design-build project manager, a design manager, a designer-of-record, a design-construction coordinator, and a construction manager. It is also recommended that the team incorporate the services of a unique position known as an integration manager whose job is to help manage the team relationships to ensure high levels of communication, collaboration, and integration

19. Design-build project managers can come from either the design or the construction ranks as long as they are able to think independently, despite their discipline inclinations. The design-build project manager should also be a systems thinker, be able to build trust on the team, and be able to create and communicate a vision for the project.

20. The design-build superintendent looks to add value to the project using a proactive approach to problem-solving that serves both the company and the client. He should be involved early in the project and maintain communication with the design-build architect.

RECOMMENDED READING

:: ***The Future of Management,*** Gary Hamel, Harvard Business School Press, 2007

:: ***The Structure of Scientific Revolutions,*** 3rd Edition, Thomas S. Kuhn, University of Chicago Press, 1996

:: ***The Invisible Employee,*** Adrian Gostick and Chester Elton, John Wiley and Sons, Inc., 2006

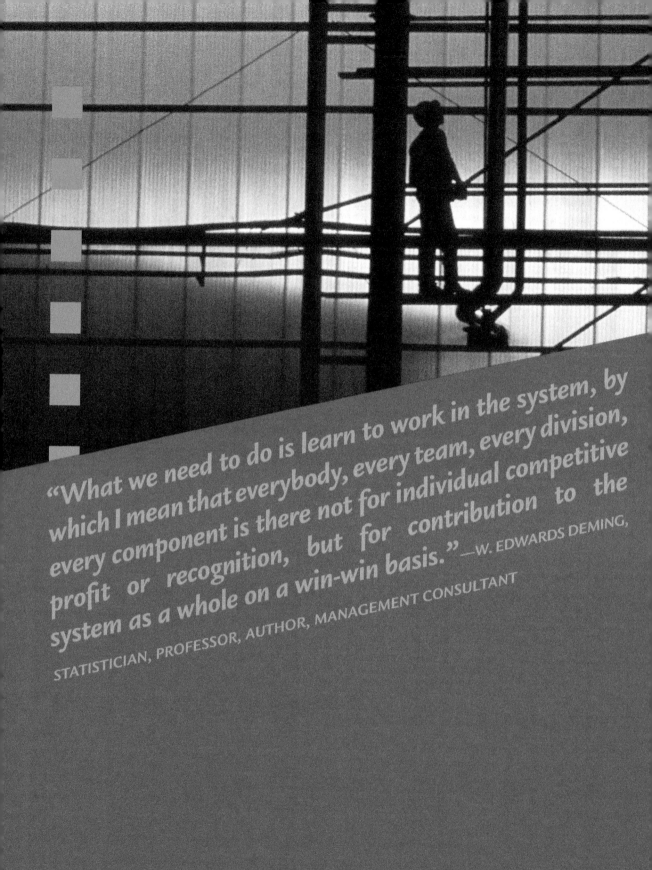

"What we need to do is learn to work in the system, by which I mean that everybody, every team, every division, every component is there not for individual competitive profit or recognition, but for contribution to the system as a whole on a win-win basis." —W. EDWARDS DEMING, STATISTICIAN, PROFESSOR, AUTHOR, MANAGEMENT CONSULTANT

8
THE DESIGN-BUILD TEAM

In the last chapter you learned that you need more than tools alone to become a design-builder; you need to understand techniques as well. Tools enable you to apply techniques more efficiently. But without the right team, you are missing the critical component for successful design-build project delivery. In design-build, everything starts with the team.

Many of us participate on successful teams in both our personal and professional lives—in sports, on a project, or in academic competitions. Good teams make a difference regardless of the venue. Teams work because the individual members share common values, beliefs, goals, and objectives. They can work closely together to plan, organize, and strategize to achieve their mission. Conversely, it is difficult to operate as a team when the members have different objectives and agendas and no common mission. Unfortunately in the traditional design-bid-build world, this is exactly the scenario that plays out. No matter how committed the contractual parties are to working well together and doing their very best jobs, the separate contracts, in which the owner deliberately partitions project deliverables and accountabilities into disconnected service packages, lacks coherence. The power of design-build lies in the single-contract accountability with the owner; this puts the designers and contractors on the same team with a solitary, integrated mission—serving the needs of the client. No other project delivery method is structured to accommodate highly collaborative teams and provide optimized, integrated project solutions like design-build.

In this chapter, you will learn how to put together and prepare a design-build team, what an integrated design-build team is, what it looks like, and how it behaves. You will also learn what it takes to turn a good design-build team into a great design-build team. Effective design-build teams don't just happen by accident—you must work with intention and take deliberate actions to create the environment of trust to deliver projects at the highest level. In the end, the payoff for you, your team, and your client will be well worth the effort.

Figure 8.1
No gaps game

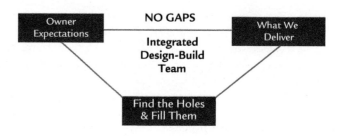

THE NO GAPS GAME

You can think of design-build as a "No Gaps Game," in which the owner has a set of expectations and it is the design-build team's job to deliver on those expectations in a seamless way. The design-builder must find the "gaps" between what the owner expects and what the design-build team delivers and fill them. The best way to provide this complete solution is with an *integrated* design-build team. See Figure 8.1.

To achieve an integrated design-build team, regardless of the individual team member's discipline (contractor, architect, or engineer), members must develop an **interdisciplinary fluency**, the ability to think across multiple disciplines. In addition to bringing their own discipline expertise to solve problems, team members readily understand, appreciate, and respect what members from the other disciplines bring to the table as well. But to have true interdisciplinary fluency, members must also be able to *think* through problems and address them from a multidimensional point of view. They must consider the various disciplines and see the connectedness between all of the inputs and influences affecting the outcome. And most importantly, members do not feel threatened by those inputs or influences. They are focused on solving the problem and finding the best value—not on protecting their individual positions or agendas. They are constantly listening to learn from one another so they can develop a comprehensive perspective on the problem and the solution, finding the holes and gaps between expectations and deliverables as they go.

But before a team can get on with the design-build work necessary to deliver extraordinary project solutions, members must adjust the traditional design-bid-build mindset before the team can reach the pinnacle of efficiency and collaboration.

MAKING THE MENTAL SHIFT

To successfully transition from design-bid-build to design-build project delivery, it is critical that team members make a significant mental transition. Transforming a group of individuals who are accustomed to working in a fragmented, often adversarial, low-bid environment with conflicting agendas, into an integrated, highly collaborative team operating in an environment of trust and transparency, focused on delivering the highest and best value, is no easy task—and it certainly is not going to happen by accident. Traditional design-bid-build is rooted in

---

D-B DICTIONARY

Interdisciplinary Fluency –
The ability to think across multiple disciplines when addressing problems and not just from one's own single discipline perspective.

a firmly-established **mental model** that influences how we behave and respond to one another as team members. This model persists even when we switch to a project delivery method designed to accentuate cooperation, collaboration, and cohesion. The first step in breaking this mental stronghold is to recognize that it exists.

According to Peter Senge, author of *The Fifth Discipline Fieldbook*, (1994), "A mental model is one's way of looking at the world. It is a framework for our cognitive processes of our mind. In other words, it determines how we think and act.... Mental models are usually tacit, existing below the level of awareness, and they are often untested and unexamined."[1] These mental models can keep us from fully embracing new ways of doing business (like integrated design-build) even when it appears that they are in everyone's best interests. "New insights fail to get put into practice because they conflict with deeply held internal images of how the world works, images that limit us to familiar ways of thinking and acting."[2] But at the same time, once we understand their power we can begin to deconstruct the old framework and build a new one.

A simple example of a mental model comes from an exercise described in Peter Senge's book *The Fifth Discipline Fieldbook*. In this exercise, pairs of seminar participants are asked to arm wrestle. They are told that "winning" in arm wrestling means bringing their opponent's arm to the table and they are each challenged to "win" as many times as they can in fifteen seconds. Most people struggle against their partner to win. Their mental model is that there can be only one winner in arm wrestling and that this is done by lowering their partner's arm more times than their partner can do the same thing to them. However, an alternative mental model would present a framework where both parties could win. If they stop resisting each other, they can work together flipping their arms back and forth. The end result is that they can both win and they can win many more times than if they were working against each other.[3]

As human beings we develop mental models as a way to maintain consistency and predictability in our lives. Thinking the same way we always have helps eliminate the anxiety and discomfort associated with change, but unfortunately, according to Senge and others, these mental models also keep us from learning from one another, growing, and achieving our highest potential. This is certainly true when teaming in design-build. Senge further explains this phenomenon:

> With a bit of thought most people can appreciate that none of us really sees reality as it is. Most people can also see how, over time, our personal perceptions can become self-reinforcing and gradually build up mental models, assumptions, and habitual ways of seeing—such as when we develop stereotypes of certain types of people. The problem is not these habitual mental models in the abstract but their pervasive influence in our daily lives. As the old saying goes, "The eye cannot see the eye." To do you need a mirror; you need to reflect. This is what we refer to as "suspending assumptions"—in effect, holding them out in front of you. Rather than asserting that what you believe is the truth, people who

---

D-B DICTIONARY

Mental Model – An established way of thinking and looking at the world that directs our behavior and actions.

practice suspending their assumptions recognize that their views of situations are, in fact their views. This subtle shift brings with it greater humility and ability to listen to others. In working teams, suspending assumptions breaks down the rigidities in thinking that otherwise thwart dialogue; as a result people become more willing to inquire into their own and one another's views.[4]

Although the design-build model puts the design and construction professionals on the same side contractually, the contract itself cannot change the negative opinions and judgments that have developed over the years and that can imperceptibly pit these two critical team players against one another, even when they are on the same team. Only when we consciously begin to challenge and examine the adversarial mental models linked with the traditional design-bid-build relationship will we be able to influence the nature of the design-build teaming from one of skepticism and mistrust to one of collaboration and full disclosure.

AUTHOR EDITORIAL

LABEL LIABILITY

Generally, labels are a good thing. They help direct us, and warn us of lurking danger in a medicine bottle or a dangerous curve in the road. However, sometimes labeling can influence our thinking in a negative way—especially when those labels create or reinforce stereotypes. We certainly know the negative effects of ethnic, religious, or gender stereotypes. But have you ever thought about the impact that a "discipline stereotype" might have on your ability to fully engage and collaborate in an open, honest relationship on a design-build project team?

To begin with, professional labels can conjure up negative stereotypes. On meeting a contractor, architect, engineer, or even an owner, feelings, perceptions, and images from past project experiences get conjured up, and they are not always pretty. Stereotypes that arise from labels can become a liability for team integration and team trust. They influence how we speak to one another, what information we disclose, and how we interact. They can get in the way of productive dialogue and positive relationships. If you cannot get past the label and get to know your team members as individual design-build professionals, regardless of their discipline, then you may never achieve the optimum results possible in design-build.

What is most unfortunate is that these stereotypes often get introduced to young college students early in their academic experience. They may begin to hear negative comments associated with the "other side" from faculty who unintentionally reinforce an "us against them" attitude that doesn't work in an integrated design-build environment. Young construction management students are still exposed to indiscreet faculty comments such as, "Architects don't know what they are doing," and "They design stuff that can't even be built and then contractors are left to fix it in the field." Young architecture students are still told by some faculty that, "Contractors only care about the money," and "They don't care what anything looks like, or quality, or the environment, or anything else." Although we often laugh a bit about all of this and don't really mean to cause any harm, these stereotypes get carried right into the young person's professional life before he or she even have a chance to experience the relationships firsthand. So before a young graduate gets to know the project players as individual human beings, his or her perceptions may already be tainted. And then sure enough, the contractor's actions, or the architect's actions, or the owner's actions will be interpreted to precisely fit the stereotype that the young person brought to the job site with them.

Under a segregated services model like design-bid-build, these negative perceptions may have little impact because you are only accountable for your little piece of the pie—you are not accountable for the whole. But in an integrated process like design-build, each member of the team must be accountable for the entire project solution—design, performance, quality, cost, and schedule. That is why it is critical that architects, engineers, and contractors who engage in design-build get past the stereotypes associated with their professional labels and start thinking from an integrated mindset that has everyone standing on a level playing field as equally valued contributors to the project.

Here's a thought: To mitigate the potential of label liability to contaminate your project, consider giving every member of the design-build team, including the owner representatives, project business cards, or project jackets or shirts that identify individuals as "Design-Builders" and leave off the discipline labels altogether. Then be sure that the label "design-builder" represents someone who is a highly collaborative team player committed to delivering exceptional project performance and results in an integrated fashion that serves every member of the team. Now that's a stereotype I wouldn't mind being tagged with!

All of this transformation takes time, of course. No matter how much sense design-build might make, it can be difficult to let go of old stereotypes. In their book *Sway: The Irresistible Pull of Irrational Behavior*, authors Ori Brafman and Rom Brafman use the term **diagnosis bias** to refer to the human propensity to label people, ideas, or things based on our initial opinions of them, and our inability to reconsider those judgments once we have made them. According to these authors, a diagnosis bias is an example of a sway mechanism.[5] In other words, once our minds are made up about how someone is, or how something should be, our perceptions may be forever influenced, or swayed, by these first encounters. Our opinions of architects, engineers, owners, and contractors have been molded around the low-bid mentality associated with designing and constructing projects using the traditional design-bid-build method, and most of those opinions are not very flattering when they are leveled from "the other side."

While it may be difficult to change people's attitudes and beliefs, it is not impossible. The first step is to first recognize that these natural human tendencies exist and are at play as we attempt to work together to deliver an integrated design and construction solution for our clients.

D-B DICTIONARY

Diagnosis Bias – The human tendency to label people, ideas, and things based upon our first impressions. These biases can influence how we behave and interact with one another, even when there is evidence that contradicts our initial opinions or opportunities to see a person, situation, or thing differently.

Once you begin to understand the human forces at work, you can consciously begin to see your design-build partners in a new light and behave and respond differently. It takes intentional work to convert a group of independent designers and constructors, accustomed to serving their own agendas under a segregated services model, into an integrated, cohesive team focused on delivering a comprehensive, optimized solution that serves all parties involved in the contract.

⁙ TEAMING FOR RESULTS

No matter how good you are at architecture, engineering or construction, you cannot deliver a design-build project without a team. And by executing a single-source design-build agreement, regardless of who holds the prime contract and serves as the controlling party, your designated group of designers, contractors, subcontractors, vendors, and consultants must function as a team. But ultimately, the best way to assess a team is to measure its results against its goals and objectives. Only then will you know whether it was a good team or a bad team. While it is nice if everyone also gets along well and looks good—it takes more than matching shirts and hats to make a team successful.

MENTORING MOMENT

Ryan DelaRiva, LEED AP
Field Engineer
Pankow Builders, Ltd.
Honolulu, Hawaii
Education: California Polytechnic State University, San Luis Obispo – Graduated 2008
Major: Construction Management

Passionate for People

I've always loved the technical side of construction. The depth of knowledge that my co-workers and mentors have amazes me. I get excited knowing that I'll continuously be introduced to more and more knowledge as my career advances. I also believe it will be extremely important to keep up with all of the developing technologies, strategies, and standards, such as BIM and LEED, as these advancements will be common practice in the industry in just a short period of time and they are important in terms of my overall professional development. However, this knowledge and technology will be useless to me if I lose sight of my true passion, which is interacting personally with all of the different people involved in a project. I can't think of any other field that puts you in an environment with so many diverse types of people and professions than the design and construction business. You can't follow set calculations or standard operating procedures in order to balance the priorities and expectations of people on a team, people who at times may have conflicting expectations, opposing agendas, and very likely, strong variations in personalities. New tools and technical knowledge will never be able to manage this. I embrace the challenge. While still in college, I participated in an interdisciplinary class called the "Collaborative Process" and it was the perfect class to end my college career. It helped me take a step back and look at what I've always brought to the table in group settings (my temperament, experiences, and interests), how it can affect an entire group's makeup, how I can improve it, and finally, how I can excel in a leadership role with this understanding. This is something I'll strive to improve as I continue my career, because from what I've observed, this is the most valuable asset that I can bring to the industry—and one that can really set you apart.

Not Just Any Team

A design-build team is not just any team. To build the right team, you should intentionally blend diverse backgrounds, experience, and viewpoints along with the discipline knowledge required to address a client's project needs at the highest level. Give serious consideration to the task of creating the design-build team from at least two perspectives—cultural fit and technical expertise.

Cultural Fit

Even before you start searching for technical expertise, it is imperative that you find firms that share your cultural values, business ethics, and corporate philosophies. If this compatibility is not there, then it will be quite difficult to optimize the team's technical expertise and attain the strategic advantage that you may be looking for.

When trying to identify your design-build teaming partners, start by looking at your past working relationships. However, in some areas you will have to develop new teaming relationships to respond to specific business interests and project needs. Consider developing a **Strategic DB-Teaming Initiative** to build up your design-build partner list. Focus your attention on the market sectors and project types that you intend to pursue. Look for companies that have a reputation for *best-in-class* technical knowledge and then perform the due diligence necessary to check for cultural fit. Making the effort to nurture these relationships before pursuing a project will save time and reduce risks in the long run. Apply this same strategy when looking for the right specialty contractors, subconsultants, and vendors to partner with on your design-build projects.

> **D-B DICTIONARY**
>
> **Strategic DB-Teaming Initiative** –
> A plan to identify teaming partners before pursuing a project.

Technical Expertise

Obviously, it is critical that a design-build team match the proper technical expertise with the technical needs of the project—and the more technical the project, the more important this is. For some projects, this expertise can come from the ranks of your own employees if you are a full-service design-build entity. On the other hand, if you are a single-discipline firm, than you will have to partner with an outside company to provide this expertise. On some projects the best arrangement is to combine internal knowledge with external expertise to deliver the best technical solution.

When looking to outside firms to find the right technical skills, it is important to consider both experience and performance. Just because a company has the experience in a particular project type does not mean that their performance record will meet your expectations. Furthermore, qualifying firms is one thing, but picking the right individual team members is another. It is always best to check with past clients, subcontractors, consultants, and vendors who can give you some insight into both the firm and the individuals' expertise and performance.

> Even after you have secured expert services to deliver the technical components of the design, you should still conduct independent technical reviews to supply an added layer of oversight, whether the design team is in house or under separate contract. This is essential, especially if you are working with a firm for the first time. As you learned in the last chapter, this important risk mitigation measure should become a part of your standard design-build project management protocol.

TAKEAWAY TIPS

Share and Share Alike

Consider inviting your design partners, major subcontractors, and vendors to your next design-build training session. On many occasions I have been asked to provide in-house company training to joint sessions of partnering design and construction firms. This helps achieve an equality of understanding regarding the design-build process with the companies that you work with most often. Having this common understanding is a critical component of successful design-build teaming—not to mention the reduction in training costs to any one firm. You may want to go even one step further and invite past and potential new clients to one of your design-build training sessions—especially if conducted by an outside, independent training group. In addition, you could organize a design-build orientation session once you win the job, just to make sure everybody is on the same page regarding the design-build process before the project gets underway. The key is to share your company's design-build training endeavors with your teaming partners whenever possible. It will save time, help match goals and objectives, and strengthen the relationship.

As you build your design-build team, remember that diversity is critical. You want to make sure that you have a rich pool of knowledge and perspectives to draw from—taking full advantage of the collective intelligence across the team. It is important to include some partners who can serve as provocateurs to stimulate ideas and innovation, to make sure that the team stretches beyond simply achieving a good solution and presses on to develop the best solution—ultimately increasing their odds of winning the job.

WHO SHOULD BE ON THE TEAM?

Initially, the design-build team consists of at least three parties—the owner, the contractor, and the primary designer (architect or engineer). It might also include representatives of these three principal players. For example, the owner might be represented on the team by a senior executive, upper management, or an operations person. The contractor is often represented by a project director, project manager, and senior estimator. And the designer is usually represented by a principal in the firm along with a lead architect or engineer. However, as the project develops, additional team members will be woven into the mix. In addition to major vendors, critical subconsultants and specialty contractors will be brought on board as needed. Their participation and involvement is a vital part of the design-build teaming arrangement and the sooner they are involved, the more likely it is that the project pursuit and eventual delivery will be successful.

It is extremely important to engage these team members at the proposal stage. Although every project is different, there are some major divisions of work that require the early involvement of subcontractor and consultant technical experts. In addition to the primary players, the most common early teaming partners are:

:: Structural consultants and contractors
:: Mechanical consultants and contractors
:: Electrical consultants and contractors
:: Fire sprinkler consultants and contractors
:: Curtain wall consultants and contractors

Not only will these technical experts provide input and guidance to the team but they will also be responsible for delivering various components of the actual design by way of shop drawings and other supportive

For those of us in the construction industry, the construction project is a wonderful thing, a unique combination of brains and brawn, of art and science, of form and function—each project an individual tribute to the numerous architects, engineers, and construction companies whose contributions led to its very existence. The role of the specialty contractor is enhanced in design-build because it focuses on the strengths of each member of the team, using their knowledge, their experience, and their skills to ensure the project goals are met. Having a working knowledge of constructability, costing, product application, and life cycle is important on any job but is more pronounced in the integrated model. This knowledge allows us as specialty contractors to collaborate more effectively with each and every trade to ensure the best interests of the client are served. In an atmosphere of transparency, team members must be able to trust and be trustworthy. This requires design and constructability competence in the type of project being built but it also requires the ability to communicate effectively— openly sharing information and ideas in order to not only meet the owner's expectations but exceed them. While studies have shown the early involvement of the specialty contractor has resulted in better quality, better cost controls, and better schedules, we have found that design-build is just a much better way to build. Our results are better, our productivity is better, our percentage of repeat business is higher and our employees enjoy the experience more.

Tom Sorley
Chairman and CEO
Rosendin Electric, Inc.
San Jose, California

documentation. As the project moves forward, other specialty team members may also be brought in and out of the process as necessary.

Strategic Partners

In addition to the primary players and major specialty contractors, sub-consultants, and vendors, it is also important to think about other partners who can support the team in achieving its goals. Many of these **strategic partners** are third party associates whose participation and input is often crucial to the project's success. They include fire marshals, code and permitting officials, regulatory agencies, community organizers, and any other individuals who can have a considerable influence on the project. Getting these parties involved early in the process and informing them of your goals and how you intend to achieve them can have a profound impact on the project outcome. Having an understanding of their concerns and getting their important early feedback on your ideas and plans will assist you in moving forward as allies instead of adversaries. Having these individuals and their representatives, as well as end users, on the team will help build trust, enhance cooperation and collaboration, and add efficiency to the entire process.

The beauty of design-build is that it allows you to customize your team by pulling together the best people at the right time to deliver a project. However, once you have identified the team players, you should

---

D-B DICTIONARY

Strategic Partners – Individuals with a potentially significant influence on a project's outcome. These team members often represent third party entities such as end user organizations, special permitting agencies, code enforcement offices, political associations, community groups, and others. Engage these strategic partners as early in the design-build process as possible.

Do not be afraid to remove people from the design-build team who just are not "getting it" or delivering as promised. Team members who display constant negativity or an unwillingness to cooperate and be forthcoming should be removed as soon as possible. This is true regardless of which group they represent—construction, design, or even the interests of the owner. Bad attitudes and obstructionism can often sabotage the entire effort and put the project at risk. The senior leadership team, consisting of representatives from the three primary parties, should agree in advance to move quickly if team dynamics start to turn sour because of one or two individuals on the team.

D-B DICTIONARY

Teaming Agreement – A vehicle for structuring the relationship of the teaming parties that identifies the roles, responsibilities, and expectations of the design-build team members relative to cultural issues, corporate and legal considerations, risk management, financial matters, and any other items of interest to the parties.

ask them one more question before going forward: "Is there anyone not on this team already whom you believe needs to be here?" This question should be repeated at different stages of the process to make sure that the right people are there when you need them.

ESTABLISHING THE GROUND RULES

Once you select your teaming partners, it is important to make sure that each member has an equal understanding of the various assumptions and expectations regarding their accountabilities and the nature of the association.

According to the Design-Build Institute of America:

The success of any design-build project is largely dependent on the ability of the constructor, designer, and subcontractors to establish a working relationship that will permit the design-build team to satisfy the owner's requirements and at the same time permit each party to realize the benefits of its contract. To reach these goals, designers, constructors, and key subcontractors must establish in advance of any decision to pursue the project, the necessary ground rules for working as a team and develop the trust and confidence in each other that is the cornerstone of any successful relationship. The vehicle for structuring the relationship between the parties is known as the **Teaming Agreement**.[6]

The teaming agreement should be negotiated when the design-build team is first formed and before the start of any competition. As you learned in Chapter 3, the design-build entity itself can be structured in several different ways and the structural configuration will influence the team members' viewpoints and how they participate. The entity may be led by a construction firm or a design firm, in which case the team members will come from different companies, and therefore several, if not all, of the team members may be working together for the first time. On the other hand, the design and construction professionals might all be with the same firm, as with an integrated design-build entity. But even then, the team should lay out how it will move forward on the project and manage their commitments. You might consider implementing a teaming agreement with both your internal and external team members, including your strategic specialty contractor and subconsultant partners. A well-thought-out teaming agreement will establish the framework for defining the roles, responsibilities, and risks during both the proposal development phase and the post-award phase of the engagement.

Proposal Development Phase

Putting together a design-build proposal takes a great deal of effort and time and can be expensive. It is important to have a comprehensive proposal preparation plan to increase the odds of winning the project and to mitigate as many risks as you can anticipate. Let's consider just a few of the important questions that a good teaming agreement will answer.

Who Will Do What?

You must dedicate personnel and resources to create a good design-build proposal. There are many tasks that must be performed to respond to the typical RFP. Once the team identifies these tasks, they must assign them to the appropriate team members to develop them and to create a schedule to ensure an on-time proposal delivery. These assignments must be properly coordinated to make sure that nothing falls through the cracks.

Who Will Be Responsible for Reviewing the Proposal Prior to Submission?

All proposal components (design, estimates, schedules, etc.) must be reviewed and approved before they become part of the released proposal package. Define clear accountabilities regarding which team members will review and approve each component and who will have the final authority to release the package for submission to the client. These responsibilities are typically assigned to senior level management personnel as needed.

Who Will Provide the Technical Oversight for the Design?

The teaming agreement must address who will perform technical oversight. Whether the design is prepared by internal architects and engineers or external architects and engineers, you need to ensure that a third party reviews the design before it is released for submission, especially when the design-build team is led by a construction firm without in-house design expertise. In Chapter 7, you learned about technical review committees comprised of senior design staff who were not directly involved in the project and who provide an unbiased assessment of the design before submission. A technical review committee is particularly important if the design uses a new technology or you are working with a design firm for the first time.

How Will the Costs Associated with the Proposal Development be Shared?

It is not cheap to put together a good design-build proposal. Different team members will invest varying amounts of time and dollars in the typical design-build pursuit. It is important to discuss the compensation and reimbursement terms, whether the team wins the project or loses the project. If the owner does offer an honorarium to the non-winning teams, it is important to discuss and agree how the honorarium will be split. It is not unusual for the design-build contractor to give all or almost the entire honorarium to the architects and/or engineers involved in the preparation of the conceptual design. And the more design work that is required at the proposal stage, the more likely it is that the designers would receive the entire amount of the honorarium.

How Will Confidentiality be Maintained Regarding Proprietary Information and Strategies?

Every design and construction firm has confidential business information and proprietary items that they need to protect. When two or more companies decide to partner for a design-build pursuit, they must often reveal confidential information. Every teaming agreement should clearly define under what circumstances this information can be shared and with whom and how it may be used in terms of pursuing the project.

Post-Award Phase

If all goes well, the project's proposal preparation phase will result in your team winning the job. Therefore, the teaming agreement must also address how the team will engage after the contract award. Numerous issues come into play once the contract is in hand, and some important questions will require attention.

Who is Responsible for Specific Tasks During the Contract Performance Phase?

The team must fulfill multiple obligations to deliver the project design and construction. The agreement should spell out as many of these obligations as possible, and should specify which individual or team of individuals has the specific responsibility for fulfilling them. It would be almost impossible to identify every item of work necessary to complete the project, but the agreement should attempt to identify the essential post-award items before the proposal is submitted. Because the design-build team is interdisciplinary, some activities will seem straightforward: architects and engineers will be responsible for preparing the design and the constructors will be responsible for developing the schedules and estimates. However, some items might be performed jointly, such as contingency development and integrated design reviews. In design-build such responsibilities are not as clear-cut as they are when the disciplines work under separate contracts with the owner. The use of a **responsibility matrix** to identify tasks and assignments, along with due dates, is an important tool that will clarify these relationships and responsibilities. Such a matrix can list everything from schematic design development, to permitting, to constructability reviews, to final approvals. It will also keep these tasks and accountabilities out in front of the team so everyone knows who is doing what, and when each task will be delivered. There are many different ways to organize and structure this document, which can be detailed or broad. Figure 8.2 shows one example.

How Will the Design and Construction Work be Coordinated?

Project management and coordination is at the crux of the design-build process. Much of what you learned in Chapter 7 addresses the techniques and tools available to manage and coordinate the design to ensure that it meets the owner's requirements for cost, schedule, quality,

D-B DICTIONARY

Responsibility Matrix – A tool used to outline the various design and construction tasks necessary to prepare the proposal and perform the contract obligations post-award, along with assigning accountability for each task and a due date for completion.

RESPONSIBILITY MATRIX - KEY DATE SCHEDULE
(Sample)

Project Director: Hayslip
Project Design Principal: Schmieder
Project Manager: Kernan
Project Architect: Allen

Schedule Date: 6/20/07
Revision Date: 12/3/07
Revision No: 11

Project: University Housing
Location: Jacksonville, Florida
Project No.: 93749271

Page 1

Comp. Date	Date Due	Item	Task/Event	Info. From	Info. To	Lead Discip.
3-2-07	3-2-07	1	Selection Committee Selects Haskell			
3-16-07	3-16-07	2	Project Kick-Off Meeting with Client	Team	Team	
3-27-07	3-27-07	3	Program Verification Meeting @ Haskell	Team	Team	
3-29-07	3-29-07	4	MEPF Systems (Program) Verification Meeting	Team	Team	
4-5-07	4-5-07	5	Site / Floor Plan Up-Date Meeting w/ Client	Team	Team	
4-5-07	4-5-07	6	MEPF Systems (Program) Verification Meeting w/Client	Team	Team	
4-13-07	4-13-07	7	Site / Floor Plan Up-Date Meeting w/ Client	Team	Team	
4-13-07	4-13-07	8	Preliminary "overall" Project Schedule to Client	Balz	Client	
4-20-07	4-20-07	9	Site / Floor Plan Up-Date Meeting w/ Client	Team	Team	
4-27-07	4-27-07	10	Site / Floor Plan Up-Date Meeting w/ Client	Team	Team	
5-10-07	5-10-07	11	Site / Floor Plan "Sign-Off" Meeting w/ Client	Team	Team	
5-10-07	5-10-07	12	Haskell D-B Contract Comments to Client	Balz	Client Rep	
5-14-07	5-15-07	13	Complete Code Search (Consultant)	Scott	Schmieder	A
5-15-07	5-9-07	14	Receive Soils Report from WPC	WPC	Schmieder	C, S
5-17-07	5-17-07	15	In-House Structural Meeting (Charette Room)	Haskell	Haskell	S
6-11-07	5-21-07	16	Receive Survey	Clary	Schmieder	C
5-23-07	5-18-07	17	Received Client "Sign-Off / Approval" of Floor Plans	Client	Schmieder	A
5-15-07	5-15-07	18	Prelim. Design - Courtyard Workshop w/ Client	Lycke	Client	LA
5-16-07	5-15-07	19	Request OF / OI Furniture Catalog Cuts From Client	Schmieder	Client	A
7-6-07	5-18-07	20	Received Client "Sign-Off / Approval of Site Plan	Client	Haskell	C

Figure 8.2 Responsibility matrix. Courtesy of the Haskell Company

6-18-07	6-18-07	21	Architectural Templates to A/E Design Disciplines	Dahlberg	Schmieder	A
6-11-07	6-11-07	22	Begin 60% CD's for Sitework	Jones / Skiles	Schmieder	C
6-22-07	6-22-07	23	Weekly Meeting With Client	Team	Team	
8-24-07	6-22-07	24	60% Design – Freeze Arch. Plan with Dimensions	Dahlberg	A/E Team	A
6-26-07	6-26-07	25	Issue Discipline Drawing List & "draft" mini-set	A/E Team	Schmieder	All
9-17-07	9-14-07	26	Long Lead Item Packages			
			a. Site Clearing	Jones	Schmieder	C
			b. Boardwalk	JB Construction	Schmieder	C
			c. Pool & Pool Systems	Lycke	Schmieder	LA
			d. Chiller Equipment / Modular Chiller Plant	Elkins	Schmieder	M, P, E
			e. Foundations (include underground MEP)	Patel	Hayslip	S
8-8-07	6-28-07	27	60% Design – Confirm Mech. Chase Sizes & Loc.	Elkins	Patel	M, A
6-28-07	6-28-07	28	Haskell Review SJRWMD Permit Package	Haskell	Jones / Skile	C
6-29-07	6-29-07	29	Submit for SJRWMD Permit	Jones / Skiles	SJRWMD	C
9-21-07	6-26-07	30	60% Design – Window Glass Sizes / Loc.	Allen	Elkins	A
8-13-07	6-29-07	31	60% Design – Prelim. Foundation Layout	Patel	Bertino	S
8-8-07	6-29-07	32	60% Design – Confirm Weights for Heaters & Tanks	Bertino	Patel	P
7-15-07	6-29-07	33	60% Design – Confirm Mech. FCU Closet Sizes	Elkins	Allen	M
7-15-07	6-29-07	34	60% Design – Confirm Mech. Exh. Chase Size & Loc.	Elkins	Allen	M
7-6-07	7-6-07	35	Preliminary Design – Complete A/E Design	A/E Team	Schmieder	All
8-22-07	7-6-07	36	60% Design – Building Piping Connection Points	Bertino	Jones	P
7-6-07	7-6-07	37	Preliminary Design – Building Elevations Completed	Allen / Moffett	A/E Team	A
6-29-07	7-9-07	38	Final Site Layout (SJRWMD) includes grading plan	Jones / Skiles	A/E Team	C

Primary Responsibility Code (Right-hand column)
C = Civil   LA = Landscape Architecture   A = Architectural   S = Structural
FP = Fire Protection   E = Electrical   M = Mechanical   P = Plumbing

Figure 8.2 (continued)

and performance. These strategies should be spelled out in the teaming agreement and should include assigned accountabilities. Taking the time to think through this coordination effort will give the entire team confidence as they move forward throughout the contract execution.

How Will Design and Construction Errors be Handled?

No matter how good a team you have, mistakes will inevitably occur during both the design and the construction phases. It is very important to acknowledge this inevitability up front and discuss how mistakes will play out. For example, at what point will such mistakes be charged against agreed contingencies and at what point will the individual team members be expected to bear the cost of the remedy themselves? In some cases the risks might be covered by various insurance and other third party protections, such as manufacturer warranties and the like. Do not limit this discussion to the architects, engineers, and contractors; include subcontractors and consultants, major vendors, and fabricators as well.

What is the Protocol for Engaging the Client in Design Reviews and Approvals?

Believe it or not, client engagement can be one of the trickiest aspects of the design-build process. Everyone needs to agree on how the team will coordinate design reviews with the owner. Communicate clearly and with one voice to the client; the last thing you want is for different members of the team to give the client mixed messages. In Chapter 7 you learned about the concept of design commitment and how important it is to coordinate any design release with the owner (even preliminary design), before it has been priced and confirmed to be within the established budgets. Establishing this extremely important ground rule at the very onset of the teaming discussions will save the team and the owner a lot of headaches down the road. Of course, team members must honor and commit to the ground rules they have established, which leads us to another important topic that the teaming agreement should handle:

What Will Happen if Team Members do not Honor the Commitments Spelled out in the Teaming Agreement?

Although the teaming agreement is not meant to be a substitute for the individual contracts that the members of the design-build team will eventually execute, it is a legal document and therefore carries legal consequences. While the specific subcontracts and contracts for services will outline defined scopes of work along with contract performance terms and conditions, the Teaming Agreement reflects the integrated nature of the design-build relationship and its associated task coordination. It also provides an excellent jumping-off point for discussions about a number of unique aspects associated with the design-build process such as:

:: Team makeup and selection
:: Core values and project goals

- Win strategies and collective advantage
- Legal issues and jurisdictional concerns
- Organizational structures and entity configuration
- Communication structures and technology
- Risk assessment and mitigation planning
- Incentives, shared savings, and rewards
- Quality management and value enhancement
- Financial and accounting issues
- Many other topics

As with all legal matters, ask your attorney and risk management counsel to review the teaming agreement, just to make sure that you have properly vetted all of the issues associated with the particular design-build project in question. This extra set of eyes can help corroborate expectations and obligations, so you can move forward with confidence.

Experienced design-builders establish standard teaming agreements that they can easily customize to fit the needs of any given project and circumstances. However, all parties still need to thoroughly discuss each item within the agreement to make sure that everyone understands its intent.

Implementing a design-build teaming agreement at the onset of a project pursuit garners substantial benefits. It is an opportunity to flesh out many of the issues and concerns that can plague a design-build team during project execution if they are not addressed before the work begins. The teaming agreement can set the stage for an open and honest relationship and establishes a solid foundation upon which the design-build team can successfully work together to achieve their project goals.

BUILDING TRUST

Teaming agreements are a great first step toward developing the type of relationship necessary for success in design-build. They help facilitate the sorts of discussions necessary to ensure that everyone agrees on each party's risks and accountabilities. Yet, although the teaming agreement will address many of the technical and logistical issues of project pursuit and execution, it is not enough to ensure that the team functions at its highest level. For this, there must be trust—it is the absolute cornerstone of the design-build relationship. Design-build is meant to operate in an **environment of trust** and, unfortunately, many project team members, including contractors, architects, and owners, are plagued by stereotypical attitudes and perceptions left over from past encounters in the design-bid-build, segregated services, "every man/woman for themselves" mentality. They must overcome this mistrust before a team can take advantage of the opportunities that design-build provides. Unfortunately, simply having the appropriate legal agreement in place will not achieve this goal. Bottom line, no trust—no team.

According to Patrick Lencioni, the author of *The Five Dysfunctions of a Team* (2002), "Trust lies at the heart of a functioning, cohesive team. Without it, teamwork is all but impossible."[7] He goes on to say, "In the context of building a team, trust is the confidence among team members that their peers' intentions are good, and that there is no reason to be protective or careful around the group. In essence, teammates must get

D-B DICTIONARY

Environment of Trust – A working environment in which there is an intentional commitment to practice trusting as part of the dynamic relationship associated with a design-build engagement.

In an environment of excessive government rules and regulations that ostensibly protect the government and the people's hard-earned tax money, we have allowed the pendulum to swing too far away from a place where common sense thrives, a place where a man's word is his bond, where public decency and open communication are expected, and where trust and value are the foundations of building good things together. This is how it was and should be. Design-build offers us an opportunity to return to these fundamentals and leverage the know-how and experience of the private sector to the advantage of the government/owner in a world of constrained public resources, hard choices, and even tough economic times. Risk is the new currency—how much to assume and how much to pass on (and pay for). It seems to me we would all be much better served as stewards of finite public funds if we could find that happy medium where trust reigns supreme and a spirit of collaboration ensues, leading to synergistic solutions and higher value for our investment. I believe this is achievable today, even with the rules and regulations that are in place, if we, on the government side of the equation, thought about risk and value in terms of what we are tasked to deliver (a quality project, on time, within budget, with optimized value) instead of all the other distracting reasons that typically rule, and then worked hand in hand with the best design-build contractor we can find to make it happen. In short, trust begets collaboration begets synergy, leading to value-based solutions (projects), and it all begins with the owner/agent who understands the two-step design-build process and leverages it to deliver quality. I have learned that design-builders want our business, like doing business with us, want to make a modest profit, but more importantly, want to uphold their reputation and build for us again. We should partner with them on these terms and do business together on the basis of trust.

Brig. Gen. John R. McMahon, PE
US Army Corps of Engineers
US Forces Afghanistan, Director of Engineering
San Francisco, California

comfortable being vulnerable with one another. This description stands in contrast to a more standard definition of trust, one that centers around the ability to predict a person's behavior based on past experience."[8]

If past experience is what we look to when trying to determine whether we can trust our design-build teammates or not, then we may already be fighting an uphill battle when it comes to building trust on the design-build team. However, for design-build to be successful, building trust is exactly what we must learn to do and that means that we must distinguish it in a way that makes sense and is useful.

Distinguishing Trust

All forms of trust involve relationships and interactions with people and necessitate some level of vulnerability and risk. Risk is only possible where disappointment, failure, and betrayal were possible, thus making us susceptible to danger and therefore suspicious of one another. This is why trust exists in all its different forms—everything from complete trust based on blind faith to total distrust based upon blind paranoia.

Figure 8.3

Trust spectrum

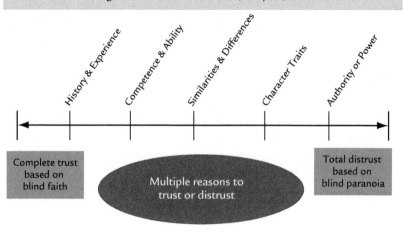

However, neither of these approaches seems logical when dealing with the design-build teaming relationship, although, more than one owner has commented unabashedly regarding their own design-build team, "I just don't trust those contractors." Trust is not merely a feeling or an emotion, although trust is often accompanied by feelings or emotions. Rather, trust is the result of a decision-making process—one in which we can consider numerous factors as a first step toward building a solid design-build relationship. See Figure 8.3.

Often we define trust in terms of reliability or predictability; we determine whether we can trust someone or not depending on how often they have delivered some spoken or unspoken expectations that we have of them. That's why familiarity often plays such an important role in our decision to trust or not. The more familiar we are with someone, the more likely it is that we will *decide* to trust them or not trust them in a given situation. Other factors will also influence our decision to trust or not to trust. Let us consider some of these factors.

Past History and Experience

Knowledge of a person's past performance usually has a direct relationship to our tendency to trust them or not. For instance, if your experience with someone has shown that the person is generally dependable and they have delivered on some commitment in the past, then you will most likely trust them again. On the other hand, if they have let you down in the past or broken a confidence, our tendency is to be cautious and not trust them in a new situation. However, in many instances there is no past history or experience upon which we can predict another's performance and reliability. Under these circumstances the best we can do is to check references and talk to other people who have knowledge of the team member's behavior and reliability. Obviously, as human beings we cannot meet every commitment that we make, but we must be careful not to overpromise and to openly communicate and acknowledge our failures when they do occur, without blaming someone else for our shortcomings. Our perceived dependability and consistency depends on how we manage our commitments, which in turn will influence whether our colleagues, partners, and team members will trust us.

Competence and Ability

Another factor that will influence our decision to trust someone is whether or not we perceive them to be competent and have the ability to carry out the tasks that assigned to them. Competency is often a critical factor in building trust in design-build simply because tasks are usually highly technical and carry a great deal of risk. We assess someone's technical competency using various criteria: We consider their years of experience, their education and licensing, and their past performance. However, just because someone is competent does not necessarily mean they are reliable or have the motivation necessary to deliver as expected and therefore should be trusted. We usually have to consider many factors when determining to what degree we will trust our fellow team members.

Similarities and Differences

It is generally easier to trust someone with whom we share similarities. There are both conscious and unconscious bonds at play when we work with people who are most like us and we develop loyalties whether we recognize them or not. Some of these bonds are due to organizational ties such as members of the same group or discipline; for example, it might be easier for architects to trust architects and contractors to trust contractors, or members of the same professional association to trust one another. Other connections are less obvious but influence our tendency to trust as well. For example, it may be easier for us to trust individuals who share our socio-economic level, or come from the same geographic area, or are the same age, gender, or ethnicity. Likewise, shared interests in sports, politics, or hobbies can result in a tendency to trust as well. Conversely, the more we are different, the more difficult it might be for us to trust. This is not to say that we cannot develop trust over time but it takes some effort and intention to learn about one another and understand how those differences can add value to the relationship.

Character Traits

Character traits like honesty, integrity, sincerity, authenticity, virtue, and honor are all pieces of the ethical pie that determines trust. Trust and truthfulness go hand in hand. In fact, the fastest way to destroy trust on a design-build team or in any relationship is to lie. The age-old adage often expressed by parents to their children that, "We don't care what you have done as long as you don't lie to us," carries a powerful message about trust. In fact, even in the face of severe consequences, making mistakes, falling short of expectations, failing to perform, or getting into trouble are all forgivable when you tell the truth about what you have done or have not done. Ultimately, trust can be restored and the relationship can be salvaged. But if you lie or withhold the truth from your teammates, trust will not have a chance, and your reputation will take a serious hit, potentially affecting your career for years to come.

Relative Power or Authority

Believe it or not, your standing in the company or on the team will often influence whether someone is willing to trust you or not. Many of you reading this book are probably quite familiar with the attitude of the 1960s expressed in the phrase "Don't trust anyone over 30," because they represented the "establishment." This attitude had nothing to do with competency, reliability, or character traits but rather had everything to do with perceived power, authority, and control. Company presidents, bosses, and team leaders are often not trusted simply because of their position. The fear is that if you disagree with your boss, question her thinking, or even suggest a different approach, there might be retribution. It is critical that every team leader understand this common human reaction when trying to establish team trust. If the members of your team do not believe that they can disagree with the team leader without retribution, they will not. You must be able to receive honest, reliable, and straightforward feedback from those in subordinate positions on your team. As a team leader, you should encourage questions and constructive dialogue in order to demonstrate trust while maintaining a commitment to open communication.

However, on a design-build team we often work with some individuals for the first time and do not have the luxury of a past history, or intimate knowledge of someone's competency. In these instances, we end up taking a chance, so to speak, usually based upon someone else's recommendation. In reality, knowing someone's track record may allow you to predict how they might perform on assigned tasks but it won't tell you about their ability to engage in a reciprocal relationship. In the book *Building Trust in Business, Politics, Relationships, and Life* (2001), Solomon and Flores refer to this type of trust as *simple* trust and distinguish it this way:

> *In simple trust, the focus tends to be on the high degree of confidence in the person trusted and the outcome. Thus, it takes the form of a focused optimism.*[9]

In the same book, the authors go on to distinguish an advanced notion of trust that might be far more useful in facilitating the design-build relationship and building an effective team. They describe something called *authentic* trust:

> *Authentic trust, by contract, is self-confident rather than simply optimistic. Its focus is on one's own responsibilities in trusting. Authentic trust is trust that is well aware of the risks, dangers, and liabilities of trust, but maintains the self-confidence to trust nevertheless.... Authentic trust differs*

from simple and blind trust in its willingness—indeed, in the necessity—to confront distrust. Practically, one does this by confronting the other person. So much of the literature on trust focuses its attention on trust as an attitude or as a social medium that the most important features of trust are ignored—conversation, communication, negotiation. Even in the most difficult cases of building trust, between age-old ethnic enemies or longtime warring factions, the key ingredient isn't some magical transformation of attitude so much as it is the possibly drawn-out tedium of bringing the sides together and making some mutual commitments, perhaps starting with small seemingly insignificant promises.[10]

In this context, "trusting, not trustworthiness, is the issue. The existential question is *how* to trust, not just who can be trusted."[11] With simple trust, failing to perform as expected, or disappointing those who have relied upon you, can mean the end of trust. But in the dynamic practice of authentic trust, breaches do not mark the end of trust but are a part of the trusting process. Recovery is possible when such failures, mistakes, disappointments, and even betrayals are discussed openly and frankly. However, to do so requires an intentional act of **committed openness** and transparency on the part of all team members, where honesty and truthfulness are essential. But recovery will not occur unless the team leader creates an environment in which people feel safe to say exactly what is so regarding any and all project issues or team members. No doubt, openness will result in the airing of complaints, accusations, and disagreements but believe it or not, such straight talk is the first step to building a truly trusting design-build relationship. As long as these doubts and concerns stay hidden, fear and resentment will rule, destroying any opportunity for real trust to develop. When properly managed, the contentious wrangling will eventually give way to mutual acknowledgement, shared commitment, and mutual respect. It will likely get messy before the air clears but it is a necessary step to developing true authentic trust—the kind that fosters real possibility and extraordinary results. Developing openness and trust takes know-how and practice.

> **D-B DICTIONARY**
>
> **Committed Openness** – The intentional practice of sharing concerns, admitting weaknesses, acknowledging mistakes, unveiling strengths, and revealing feelings in an open honest way in order to build trust in a relationship.

It's a lot easier to talk about trust than it is to actually demonstrate it in the light of day on a real project. Many owners and practitioners alike just can't make the leap because they are not used to sharing information, ideas, and especially cost data. But the more transparent you are, the more likely it is that you will be working in a trusting environment. Some team members claim to be transparent but only on a conditional basis. For example, they operate on a "need to know" basis, which is just playing games. Full disclosure, along with an open-book policy, will drive a spirit of trust that results in the open communication that is vital in design-build. Once each team member, including the owner, commits to operating in this fashion, the opportunities for success are endless.

Barbara Wagner
Senior Vice President
Clark Construction Group, LLC
Costa Mesa, California

Think of authentic trust as the type of trust that husbands and wives share in a successful marriage. The couple starts by making an intentional, conscious decision to trust one another. This doesn't mean that they will never have disagreements or question one another's motives or experience disappointment. They will. But by actively trusting first, they commit to open and honest communication and dialogue, which may require some messy discussions, discomfort, and possibly even some hurt feelings. But a willingness and commitment to talk openly and share what is bothering one another is the critical first step to building a trusting relationship. It is no different with design-build. The same commitment is required. Fears, doubts, and concerns must be freely expressed and reconciled before real trust can occur. It might take some practice and feel uncomfortable at first, but anyone who has experienced a successful marriage, a great friendship, or a reliable sibling relationship will understand how it works. The key is to be more committed to the relationship working and to positive results than you are to abandoning the relationship or getting your own way.

GETTING THE RIGHT PEOPLE ON THE BUS

Up to this point we have focused on teaming issues related to technical expertise and cultural fit. Now it is time to discuss human attributes that are more difficult to detect but that are equally, if not more, important to the success of the design-build team. Teams are made up of individuals with different personality types, character traits, temperaments, and talents. And although no single temperament is associated with individuals who thrive in a design-build environment, certain traits and talents seem to fit well with the design-build approach. As an analogy, many of us know how to read music well enough to play an instrument, but that does not mean that we can play jazz. Playing jazz, like design-build, is less structured and more fluid in nature and requires a great deal of improvisation and inventiveness. In other words, you must have the talent for it and quite frankly, not everyone does.

In his book *From Good to Great*, (2001) Jim Collins indicated that the leaders in his study who transformed their companies from good to great all indicated that it was far more important to figure out who they should have on their teams first and then what they were going to do second. They said, in essence, "Look, I don't really know where we should take this bus. But I know this much: If we get the right people on the bus, the right people in the right seats, and the wrong people off the bus, then we'll figure out how to take it someplace great."[12] Basically, the simple message is that people are the number one criteria for success.

If we adhere to these three basic principles as we build our design-build enterprise, then we must first ask, "Who are the right people?" Since early 2000 I have visited at least fifty or sixty companies across the country that have been ardent design-build proponents and who have enjoyed a significant amount of success in their design-build endeavors. I have interviewed well over 200 individual contractors, architects, engineers, subcontractors, and owners. Through these interviews and observations, I have identified a number of distinctive qualities that characterize the people who do best in design-build and who will likely be members of an exceptional design-build team.

Design-Builder Characteristics

True design-builders, not just people who do design-build, exhibit a number of characteristics and behaviors that are typically not seen in non-design-builders. The way people talk, listen, and engage one another provides the biggest clues to whether or not they have the right stuff to be a design-builder.

Letting Go of Ego

Ego—everyone has one and if you are a human being, you are never going to be able to let go of it completely. Having a realistic ego is a good thing—it gives us confidence. But too much ego can raise havoc on a design-build team. From an individual perspective, if you believe that you are the only one in the room or on the team that has the "right answer" and this keeps you from considering the opinions of others or you simply can't bear criticism of your ideas and become immediately defensive when someone starts to question your ideas or proposed solutions, then ego for you is going to be a problem. There is no room for individual "stars" on a design-build team.

On the other hand, there is plenty of room for "costars" who understand that the best solutions will come from the collective engagement of the entire team with its multiple perspectives, experiences, and expertise. People with a design-build mentality are far more committed to finding the best solution, regardless of where that solution comes from, than they are to getting their own way. This maturity characterizes a true design-build team player. It can take some time to develop this costar mentality if you are accustomed to "running the show" but ultimately the individual egos morph into a team ego and it is a powerful transformation to see—a group of individuals, all focused on the same goals, doggedly pursuing a level of excellence that they all recognize would not be possible by any one person alone.

Most of us are familiar with the expression "There is no 'I' in the word team," but I am always reminded about what basketball great Michael Jordan, one of the most respected team players of all time, said in response to that statement. He said, "It's true, there is no "I" in the word team but there is in win!" Performing well in design-build and learning how to win projects requires a team effort and the people who are able to set their individual egos aside for the sake of discovering superior solutions through dialogue and engagement will always come out ahead and better serve their clients.

Responsive in Nature

To visualize how design-build interpersonal relationships differ from those in design-bid-build, think of design-build as dancing and design-bid-build as marching. Although both dancing and marching represent coordinated efforts, dancing is fluid in nature and characterized by adaptability, flexibility, and responsiveness. And like dance partners, design-build teams need to be agile, nimble, and responsive—ready to move instinctively in any direction that leads the team toward their common goal. Responsiveness requires the ability to gauge the needs and reactions of other people so you can respond accordingly. Design-builders intentionally listen for what is missing so they can recommend and/or supply the resources or guidance necessary to move the project forward. They do not "fly off the handle" when things go wrong. Instead, they usually begin asking questions in a non-confrontational way—trying to figure out what happened so they can focus first on solving the problem. They don't waste energy by laying blame or looking to the past for excuses for why things went wrong. They remain "forward focused" and consistently demonstrate an emotional maturity that allows them to take challenges in stride while still recognizing the consequences and ramifications of the situation.

Sometimes you need a star on your team to win the project. For example, if signature architecture is required to win a project, then by all means, get that renowned person on your team if you can. But understand that with that "stardom" might come some personality characteristics and attitudes that the rest of the team is not going to appreciate. You will have to manage the relationship so the larger ego doesn't shut down the rest of the team and all of the other good ideas that would otherwise surface. It can be a challenge, but if you are aware going in you can usually find a way to make it work.

Getting Flying on the Way Down

As you have learned, the design-build project delivery model entails moving forward on a project before all of the details have been worked out. Likewise, one of the most important design-builder attributes is the ability to deal with uncertainty and the unknown. People who are particularly well suited to design-build cannot only deal with uncertainty, but they actually embrace it. I refer to this ability as "getting flying on the way down." Design-builders are not afraid to jump off into the unknown because they trust themselves (and their team) to be able to figure out whatever needs to be figured out once they are in the middle of the situation. In addition to their common sense, which allows them to instinctively apply their **knowledge-in-action** as they move forward in a challenge, they can also apply another kind of knowing that Donald Schön refers to as **reflecting-in-action**, the ability to think as they act to find the best solution. "Phrases like 'thinking on your feet,' 'keeping your wits about you,' and 'learning by doing' suggest not only that we can think about doing but that we can think about doing something while doing it."[13] Design-builders seem to feel right at home making on-the-spot adjustments and just-in-time decisions because they understand that, in dynamic processes like design and construction, you can't come up with the right answers until you have sorted through the right questions, which is exactly why the design-bid-build process is so flawed. In design-bid-build, architects and engineers are expected to come up with the "right design" before they have encountered all of the challenges—and not only that, they are expected to do it without engaging the other team players. And, most amazingly, people, clients in particular, actually expect there will be no design changes or adjustments after those 100% complete plans and specs are issued.

Design-builders generally spend far more time figuring out the right questions to ask then they do trying to come up with the "right" answer. They are very good at a skill that Marilee Adams calls **question thinking**, the ability to effect change by asking questions. Adams says that "the questions we ask will drive the results that we get and they virtually program the way we behave and what kinds of outcomes are possible."[14] Question thinking is an essential design-builder skill because it helps the team deal with the many unknowns initially associated with the design-build process. In their book *Embracing Uncertainty: The Essence of Leadership*, Clampitt and DeKoch put it this way:

> When you focus on getting all the answers correct, you are, in fact, limiting your knowledge... People who fixate on "getting all the facts" often forget the other essential component of knowledge: Learning the significant questions. If you want to embrace uncertainty, then you need to also focus on developing the right questions. The answers you discover are directly tied to the questions that you ask. If you are asking the wrong question, you may get the right answer but the wrong insight.[15]

People who are unable to function before they have all of the information and facts about a project should not be on a design-build team. They are not flawed or defective in any way, but they have too many

---

D-B DICTIONARY

Knowledge-in-action – Applying one's knowledge or know-how to a problem or situation without giving it much thought as you are engaged in the problem or situation. Common sense is a kind of knowledge-in-action that requires very little thinking before using it in a situation.

Reflecting-in-action – The act of thinking about what you are doing as you are doing it. Through observation and engagement, one's thinking may adjust and coordinate with what is happening in any given situation to come up with the most appropriate action.

Question Thinking – A term that describes a set of tools for transforming thinking, action, and results through skillful question asking.

doubts to be effective; their timid approach will hinder the team and may jeopardize the project's success. The best thing you can do for these individuals is to "get them off the bus"—and many of them will be relieved when you do. They can certainly play a supportive role to the team once the project is further developed, but you will be doing them, and your team, a disservice if you force people to participate who have no acumen for the process.

Possessing the Genius of AND

The first edition of the bestselling book *Built to Last,* by Jim Collins and Jerry Porras (1994), outlines a concept called The Genius of *AND*, which applies well to successful design-builders. In their study of visionary companies, Collins and Porras found that such companies:

> *…do not oppress themselves with what we call the "Tyranny of OR"—the rational view that cannot easily accept paradox, that cannot live with two seemingly contradictory forces or ideas at the same time. The "Tyranny of OR" pushes people to believe that answers must be either A OR B, but not both. It makes such proclamations as:*

- *You can have change OR stability*
- *You can be conservative OR bold*
- *You can have low cost OR high quality*
- *You can have creative autonomy OR consistency and control*
- *You can invest for the future OR do well in the short term*

> *Instead of being oppressed by the tyranny of OR, highly visionary companies liberate themselves with the "Genius of AND"—the ability to embrace both extremes of a number of dimensions at the same time. Instead of choosing between A OR B, they figure out a way to have both A AND B."*[16]

The concept applies not only to visionary companies but also to visionary individuals. The ability to embrace the "Genius of AND" is one of the most critical talents of a top-notch design-builder. It opens the possibility for solutions that otherwise could not be envisioned. In design-build, we are often challenged with requests that seem to demand options from two or more opposing dimensions such as high quality AND a reasonable budget AND delivered within a tight schedule. The automatic response from the "non-design-builder" mentality would be, "Sorry, it can't be done" or the well-known "Pick two of the three because you can't get it all" response. On the other hand, the natural reaction from someone with a design-build mentality is one of curiosity: "Hmmm, I wonder how we could deliver such a request." Not only do they willingly consider how the team might accomplish the task, but they delight in the challenge. The "Genius of AND" is a key indicator for selecting individuals who will be at home working in a design-build environment.

Even within an integrated firm, not all individuals have the right temperament, skills, and talents to perform on a design-build team, especially if the company is more oriented toward design-bid-build. Many companies today are intentionally identifying the employees who are best suited to design-build and working collaboratively on an integrated team. Putting the right people on your design-build teams is one of the best ways to mitigate risks and avoid costly problems down the road.

Having Emotional Intelligence

Finally, people with a high degree of emotional intelligence tend to do well with design-build. "Emotional intelligence taps into a fundamental element of human behavior that is distinct from your intellect."[17] It is a different way of being smart and many of the characteristics you learned earlier in this chapter are signs of high emotional intelligence.

Most of us are familiar with the notion of cognitive intelligence (IQ) but surprisingly, far fewer people are familiar with emotional intelligence (EQ), even though a great deal has been written about it and its relationship to overall success in life, both professionally and personally, since the 1980s. Your emotional intelligence centers around four basic skills: self awareness and self management, which are more about you; and social awareness and relationship management, which are more about how you are with other people.[18] The good news is that, unlike your IQ, which is relatively fixed at birth, your emotional intelligence can increase over time. Developing your emotional intelligence skills can greatly increase your effectiveness on a design-build team (or any team) and improve your ability to interact successfully with clients, associates, or any other social or professional interaction. "People who hone their emotional intelligence have the unique ability to flourish where others flounder. Emotional intelligence is the 'something' in each of us that is a bit intangible. It defines how we manage behavior, navigate social complexities, and make personal decisions that achieve positive results."[19]

As a prospective design-builder, you should become familiar with Bradberry and Greaves' concept of EQ, as described in their book *The Emotional Intelligence Quick Book*. You need to understand why it is so important. In addition, go online and investigate the many available EQ assessments. Consider taking one to benchmark your EQ skills to date. Once you are aware of your emotional competencies, you will be able to go to work building on the skills that you already have and put them to work on your next design-build team.

Peter Tunnicliffe, PE, BCEE, DBIA, CIRM
Corporate Director of Project Development
Chair of the Risk Management Committee
CDM
Cambridge, Massachusetts

Emotional intelligence is the centerpiece of collaborative accomplishment. It is a critical element of successful design-build delivery teams as schedule needs compress the essential activities needed to design, permit and deliver today's projects. Where yesterday's conventional design followed by bid construction was like the battle of the bands, today's design-build team is like a symphony orchestra; all synchronized and simultaneously performing a number of complicated movements under the guidance and direction of the maestro; today's design-build project manager. Not only must the multiple activities be synchronized, the multiple personalities of the team, the client, subcontractors and suppliers, and the permitting entities and regulators must be managed and harmony maintained for optimal performance.

Some of your characteristics are talents—the sorts of traits that you either have or you don't have. But many of them are skills that can be learned. As you have already observed, many of these characteristics overlap. Being a good listener goes hand-in-hand with being responsive. Listening to learn goes along with the ability to set your own ego aside. As you learn about these characteristics, you can begin to identify them for yourself and begin to recognize the affective strengths and weaknesses on your team. Then you can discuss and work on improving these skills. Determining a team's technical competencies is not nearly as difficult as discerning what affective skills might be missing.

ADVOCATING FOR THE PROJECT

Once you have the right people on the bus, with the right technical expertise and affective skills, and have made sure that you are taking advantage of each individual's strengths to support both the team and the project solution, you still need to check for one more characteristic in your team. You must now make sure that everyone on your team actually believes in what they are doing. Every team member *must* be an advocate for both the process and the project. In other words, they must be committed. It can be difficult to determine if someone is truly committed or if they are simply paying lip service to the project goals. Keep in mind that the way people talk and the language they use is a good indicator of how they really feel. According to Anne Donnellon, author of *Team Talk: The Power of Language in Team Dynamics*, analyzing the language used by team members can uncover whether true commitment exists.

> *For instance, when team members use the passive voice or third party pronouns to describe what they're doing—"That objective is yet to be realized"—that's a warning sign. Another sign of trouble is the use of abstractions to describe people: "Management wants us to do this." When team members are genuinely committed, they reflect this fact by personalizing the language that they use.*[20]

On a recent visit to an extraordinary design-build teaming session, the co-location office space had signs in every room that read "Advocate for the Project." When asked about the signs, one of the senior project managers said that everyone on the team was both completely on board for the project and for the team—or they were not on the team. Clearly the team leadership was serious. Everyone was aware that problems and hiccups were going to arise on this challenging project. But they were equally aware that no time or energy would be spent backpedalling or second-guessing the overall project delivery approach. Everyone had to be continually "forward focused," committed to the project's success, and willing to adjust and adapt to new circumstances without pointing fingers and trying to lay blame. There was absolutely no tolerance for naysayers or negative attitudes. At the same time, dialogue, constructive conflict, and

CASE STUDY
PRACTICING WHAT YOU PREACH

THE FACTS
Project Name:
Walker Brands
Project Type:
Office Building
Location:
Tampa, FL
Description:
Walker Brands is an award winning strategic brand agency. The project entailed an 8,400 square foot, two-story headquarters building
Cost:
$2,541,018
Duration:
10 months
Completion Date:
March 27, 2008

THE TEAM
Owner:
Walker Brands
General Contractor:
The Beck Group
Architect:
The Beck Group
Civil Engineer:
Atwell Hicks, LLC
Structural Engineer:
Structures One PL
MEP Engineer:
H.C. Engineering for Architecture
Specialty Consultant:
Holmes Hepner & Associates Architects

Courtesy of the Beck Group.

THE LEAD-IN

When you sell "strategic branding" programs to premier destination real estate developments, attractions, retailers, and health care facilities

Courtesy of the Beck Group.

nationwide, your "brand" and philosophy better be well represented in your own work environment and facilities. So when Walker Brands decided to design and build a new 8,400-square-foot corporate headquarters near downtown Tampa, they knew the facility had to celebrate the company's distinct brand personality of "collaboration and innovation." It also needed to demonstrate their understanding of how a place can create a custom stage for employees, customers, and visitors to have "on-brand" experiences, thereby increasing the company's value. Likewise, to be true to their own values, they would have to select a project delivery model built on collaboration and innovation. Design-build was the logical choice—especially when you consider Walker Brands' own commitment to providing completely integrated solutions to its own clients.

THE STORY

Since 1992, Walker Brands has been working with clients to develop integrated brand packages that combine identity, marketing, public relations, placemaking, and customer experiences in an effort to increase their long-term value and their bottom line profits. Walker believes that clients should experience a company's brand throughout every interaction with the company. So when Walker Brands went looking for a company to design and construct their new headquarters office, they needed a firm that could help them achieve this "brand diffusion" throughout their entire facility. In addition to embodying Walker's reputation for collaboration and innovation, the facility also had to represent Walker's commitment to the environment and sustainability. President Nancy Walker considered the new facility to be "an opportunity for us to tell

Courtesy of the Beck Group.

our story with an innovative design that also demonstrated environmental responsibility."

As a small business, Walker Brands did not have a dedicated facilities staff or the time to invest in understanding all the nuances of planning, architecture, and construction. One of the mandates from Walker Brands was that they find a team that would truly partner with them in achieving their vision of building a brand—a firm that could assist them from start to finish, create a real estate asset that could be completed on time and on budget and also be a source that they could trust. They needed a partner that had the interest and capability to become an extension of their own team.

Nancy Walker selected the Beck Group for the design and construction of her new headquarters, in part, because it could use in-house architectural and construction expertise in an integrated way, from the earliest conceptual stages of the project. An integrated team could consider and address all issues and opportunities while the project was being designed.

In addition to incorporating many "brand" features into the overall project design, they also introduced their client to the LEED® green building rating system certification program through the United States Green Building Council (USGBC) and set out to help the new Walker Brands facility become the first privately funded LEED certified building in Tampa, which was a perfect illustration of Walker's commitment to express both collaboration and innovation in all that they do.

By becoming completely immersed in the Walker culture, the Beck team committed to Walker's vision, their budget, and their LEED goals from day one. Once on board, Beck assumed risk for design, budget, quality, and schedule and delivered as promised. However, the project was not without its challenges. After all, it can be difficult to hold to a fixed, hard budget when you are working with such a creative and

Courtesy of the Beck Group.

free-thinking organization as Walker, not to mention the tight project time line.

It was no small feat to deliver the project on budget. As a creative organization, Walker wanted every design detail to be a manifestation of their brand. Beck worked diligently to make sure that every activity and space within the building exuded Walker's brand. The architecture, interior design, and environmental graphics combine to create a memorable stage that tells the Walker story.

The building was designed from the inside out—centered on a core collaboration and information exchange zone. Walker's signature brand color is incorporated in an orange exterior wall that bisects the building's midsection as a dramatic, confident, and memorable gateway entrance. A magnificent kinetic mobile, designed by nationally recognized sculptor Mel Ristau, hangs twenty feet into the building's atrium, adding an element of public art for employees, clients, and guests to enjoy. The windows are purposefully designed in a rhythm of random sizes and shapes punched into the building façade to reflect the company's creativity, energy, and passion for distinction. Even the bathrooms are designed to transport visitors to other destinations, such as the Trevi Fountain in Italy and Times Square in New York City, through environmental graphics. Both the interior and exterior signage throughout the building tells the Walker Brand story and educates clients regarding the various "green" elements and "placemaking" features throughout the facility.

Beck optimized these design elements while maintaining the budget by working with all their subcontractors throughout the design stage to find areas to scale back so that the most important aspects of the building maintained their integrity. Beck kept Walker's vision in mind, while also considering cost, which required the integrated team to work closely to identify the most cost-effective methods to deliver the project. They helped their client make good decisions on modifications that ensured that everyone would be satisfied with the both the design results and the bottom line. Although the client's involvement in the details of construction sometimes created changes and challenges, according to Nancy Walker "Beck was always flexible, accommodating, and customer-focused."

With similar design-build finesse, Beck offset any schedule delays by capitalizing on efficiencies in other areas. Walker commented that they eventually stopped worrying about issues that were out of their control, knowing that Beck would oversee and overcome any challenges. In the end, the job was delivered on time as promised from day one.

One physical challenge that Beck had to address in the very beginning of the project was the placement of the building itself. Aerial electrical and communication lines were in the way of construction. In order to meet the schedule, the masonry and structural steel had to be completed before the lines were moved, so Beck built the structure around the overhead lines until they could be relocated.

Achieving LEED certification is a challenge on any project and it was important that Beck keep the entire project team motivated to focus on the goals and point credits that needed to be achieved. LEED was such a new concept at the time that Beck intentionally kept it simple and didn't overwhelm the team with complex strategies for achieving points; instead, Beck achieved the LEED certification by incorporating a variety of recycled materials that minimized environmental impacts. Additionally, over 95% of the building is exposed to natural daylight, reducing the need for artificial lighting. Reduced water usage, energy efficiency, community connectivity, sustainable features and materials, and access are just some of the other components and elements of the design that helped meet the certification criteria. The MEP systems were also a key element of the project, especially because of the LEED certification goal. The air conditioning systems had to be redesigned to be more efficient. Rather than using two units, Beck had the mechanical engineer redesign the entire plan to accommodate six units for increased efficiency.

By all accounts, the Walker Brands project stands out in terms of innovation, creativity, and particularly for the speed at which it was designed and built, the partnering effort required of all stakeholders, the budget constraints, the complexity of the LEED certification process and the site requirements. The new headquarters was purposefully designed to leverage the Walker brands to create engaging and memorable customer experiences. This project demonstrated the power of design-build and Beck's ability to overcome challenges, devise innovative strategies to difficult problems and meet tight deadlines.

THE CLOSE

Walker Brands reports that their business has reaped many benefits from their "incredible facility," including reduced utility costs, increased recruiting opportunities, greater productivity and employee satisfaction, heightened pride in their firm, an abundance of media coverage, increased client interest and opened doors with audiences they would have never known otherwise. But more than anything else, it is quite evident that the building represents the Walker brand in every detail and demonstrates that as a company they truly practice what they preach.

From the Owner

The whole Beck team took the time to really listen, understand our business, incorporate our ideas and partner with us. Beck not only designed and built a facility that represented our brand in terms of the bricks and mortar, but time and again came to the table with ways to incorporate our brand into the details of the space. Our goals became Beck's goals and it was a partnership in the truest sense of the word. Beck provided an invaluable service by committing to our vision, our budget and getting it done. Beck not only introduced us to LEED, but at each step of the process we learned more and more about the benefits of design-build. We participated in planning meetings, review stages, and progress updates and watched them orchestrate the entire project. Not only did Beck come together internally, but also with their subs, the City, and our team. They provided the transparency that brought out the best in all parties. In the end, we were the beneficiaries of the incredible collaboration and this building stands as a testament to their efforts. For anyone who would argue that design-build lessens the owner's leverage, I would offer that design-build plus integrity is a winning combination.

Nancy Walker

President

Walker Brands

[Case study adapted from The Beck Group's submission for the Design Build Institute of America 2008 Design-Build Award Competition, by permission.]

communication, along with collaboration and integration, were fully encouraged. The team clearly practiced the three essential elements of successful design-build teaming, communication, collaboration, and integration.

COMMUNICATION, COLLABORATION, AND INTEGRATION

Communication, collaboration, and integration are three words that are often used in the design-build and integrated project delivery environments. But unless we can properly define what they mean in practical terms, we can't properly assess our mastery of any one of them—and mastery is what an effective design-build team needs.

Communication

We all talk about communication—almost like we discuss the weather—but nobody seems to know how to improve their communication skills,

Figure 8.4
Communication model

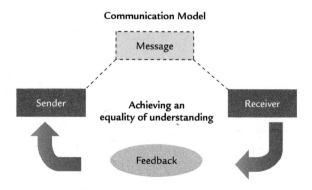

which are vital to all relationships. To communicate means to achieve an "equality of understanding." But just because you speak and someone else listens does not mean that any communication has occurred. The only way to verify that you communicated your message is to purposely check for understanding. That requires soliciting feedback. See Figure 8.4.

You must take the time to give and receive feedback to verify that the message you intended to send was understood the way you meant it. Paraphrase and ask probing questions that will help you discern the level of understanding that has occurred—questions like:

:: "What do you think about what I just said?"
:: "Tell me what you are thinking."
:: "Does that make sense to you?"
:: "Let's recap what we have said so far."
:: "Let's make sure we are on the same page."
:: "Tell me what you've heard so far."

It's also important to understand that communication is not all about talking, a fact summed up by the expression "If you think communication is all talk, then you haven't been listening." Listening is by far the most important component of communication; unfortunately it takes practice and intentionality to get really good at it. I refer to this type of listening as **committed listening**, the ability to listen as if what the other person has to say actually matters. Committed listening is listening to learn. It requires suspending your opinions and judgment of what the other person is saying. It does not matter whether you agree or disagree with the other person's words. In committed listening, the purpose is to simply accept it as the other person's truth—regardless of what you think. Most of us are lax about the words we choose in our efforts to communicate with others, so we have to be diligent when listening. We have to dig a little to get at the

---

D-B DICTIONARY

Committed Listening – Listening intently to another person to make sure that you clearly understand what they mean by the words they are speaking—without judging, assessing, or negating what they are saying.

true meaning of what people are saying—to gain clarity and achieve an equality of understanding.

On the other hand, learning how to use **committed speaking**, speaking clearly, with intention and purpose, to enhance the probability of the listener "getting" your message as you intended, is a practice worth honing as well. How we talk influences the thoughts and feelings of others and we have to become more aware of the power that our speaking has on others—particularly in a team setting. Donnellon asserts:

> *For teams, the work they do as a unit is conversational. The conversations may be face-to-face, over the phone, or through email, but regardless of the channel used, teams do their work through language. That is, the work of most professional or managerial teams is to reconstruct new meanings—in the form of new product developments, enhanced processes, or the solution to a vexing problem—by sharing and integrating their knowledge. Team work is essentially a linguistic phenomenon.[21]*

To improve the effectiveness of your design-build team, provide training that will teach members how to listen and speak in ways that will empower them. By doing so, you will enhance the contributions of all individual members and increase the likelihood of a successful outcome. This is a far better option than letting them unknowingly continue to converse in ways that sabotage team efforts, and therefore team results, without anyone actually knowing why.

Every individual member of the team is responsible for ensuring that real communication occurred in any given interaction. It will do no good to blame other team members for their failure to get your communication when you have done nothing to help them. Fortunately, there are countless techniques and tools to do just that.

Some Useful Techniques and Tools for Communication

Communication is a field of study unto itself; its breadth should not discourage us from learning more about it. You can benefit from just a few specific skills you can learn and actions you can take that will improve communication on your design-build team:

Suspending Assumptions

One of the most powerful communication techniques available to any team is the disciplined practice of intentionally suspending your assumptions and opinions. Learning the difference between dialogue and debate is a good place to start. Dialogue focuses on learning from one another, whereas debate focuses on making your case.

> *The purpose of dialogue is to go beyond any one individual's understanding. There is no opposition or debate in dialogue. No one is trying to win. In dialogue, individuals gain insights that simply could not be achieved individually. Instead the group explores complex difficult issues from*

D-B DICTIONARY

Committed Speaking – Speaking with intention and purpose, using words specifically to clearly convey your exact meaning.

many points of view. Individuals suspend their assumptions but they communicate their assumptions freely. The result is a free exploration that brings to the surface the full depth of people's experience and thought, and yet can move beyond their individual views. The purpose of dialogue is to reveal the incoherence in our thought."[22]

In other words, dialogue helps us find the holes in our own thinking and fill them. After all, what is the point of a diverse, multidiscipline design-build team if we are not going to communicate in a way that takes advantage of the team's collective genius?

What makes dialogue difficult is that we seem to be easily drawn to defend our thoughts and assumptions as if they are the truth. David Bohm, the author of *The Special Theory of Relativity*, puts it this way: "The mind wants to keep moving away from suspending assumptions...to adopting non-negotiable and rigid opinions which we then feel compelled to defend."[23] We have to practice not getting sucked into debate, argument, and confrontation. Once we begin to dig in our heels around an issue before it has been fully vetted, it is difficult to reverse the damage done to the team dynamic, and trust often takes a hit. Instead, we need to translate our dialogue to discussion, so we can make decisions about the project in a systematic, intentional manner. When a team unconsciously slips into debate mode, communication quickly deteriorates, causing individuals to become detached and discouraged without understanding why. To manage the communication among team members, you need to purposely direct the discussion to optimize solutions and make good design and construction decisions.

Exercising Cascading Communication

How many times have you led a meeting where you thought your team reached agreement on an important project decision, only to find out several days later that one attendee, who seemed to offer an affirmative nod at the time, vehemently denies agreeing to such a decision? This is a common occurrence on teams. A technique called **cascading communication**, described by Patrick Lencioni in his book *Overcoming the Five Dysfunctions of a Team: A Field Guide for Leaders, Managers, and Facilitators*, is "…an activity following a meeting in which team members go to their respective departments and report on the agreed-upon decisions and outcomes of the meeting. It is designed to take place within 24–48 hours following the meeting, and occur face-to-face or by phone—no email—so people can have a chance to ask questions and fully understand the decision and the reasoning behind it."[24] The technique encourages team members to publicly acknowledge their concerns and get their questions answered, because they know they will be responsible for communicating the decision down the line and therefore cannot "hide out" at meetings.

Practicing Positive Intent

The way we listen to other people reflects how we perceive their intent. If we automatically assume that others have negative intentions

---

D-B DICTIONARY

Cascading Communication – A concept developed by Patrick Lencioni in which team members, following a meeting, return to their departments and report, via phone or personal communication, the meeting outcomes and decisions. This reporting gives others a chance to ask questions, gain understanding, and enhance their ability to communicate it to others.

Cascading communication is a great way to keep the team aligned and help people clarify and confirm their understanding. It is a simple technique but one that requires significant discipline to maintain. At first it may seem too time consuming. But it does not take half as much time as the rehashing and backtracking caused by people who do not voice their concerns or confirm their understanding of decisions already made—or the havoc caused by frustrated and discouraged team members.

behind their questions, comments, suggestions, or actions, we can impede real communication. Our need to be right makes us approach communication with fear and distrust. On the other hand, if we assume positive intent and deliberately assume that other people have good reasons for doing and saying the things they do, we foster productive communication. When you practice finding positive intent, you intentionally think of a good reason why the other person might be questioning your approach, or making a suggestion, or taking some action that otherwise might be upsetting to you. This attitude will defuse any knee-jerk reaction you might have and avoid creating a destructive environment of mistrust and skepticism. If you have a team composed of the right people, who are as committed to the project as you are, you can use these techniques to improve communication to avoid sabotaging the team dynamics.

Studying communication can make a huge difference in your effectiveness in leading a team, delivering a project, achieving your goals, securing an agreement, or generally getting things done. While it can be difficult to carve out the time you need to improve these skills, the effort is well worth the cost in both time and dollars. A project environment with good communication results in less conflict and litigation, and fewer claims, headaches, and ulcers—and that's priceless!

TAKEAWAY TIPS

Turn the Radio On

Consider requiring your project teams to start carrying radios again to communicate with one another, instead of depending solely on email or cell phones. The radios allow you to hear other people's conversations. Having access to this "team chatter" helps keep everyone in the loop instantaneously, while the conversation is happening, preventing a lot of mistakes and miscommunications.

Collaboration and Integration

It might seem that collaboration, or working together, is the key to successful design-build teaming. But collaboration alone is not the magic pill that allows ordinary design-build teams to perform in an extraordinary fashion—that role is reserved for integration, the blending into a whole. While both collaboration and integration are vital aspects of design-build teaming, integration is what gives design-build its real power. But the two terms are often misused, so it can be difficult to differentiate between them and assess their application and effectiveness within the context of the design-build teaming commitment.

Simply speaking, design-build collaboration can occur without integration, and often does. But design-build integration takes the intent of collaboration to a whole new stratum to delivers projects at the highest level. How does this happen?

Distinguishing Between the Two

Collaborating requires cooperation and collegiality. Integrating requires something more—it requires an attitude of engagement, inquiry, and analysis. Collaboration and integration both require

> True design-builders love integration because they know that the collective genius of the team can only be realized when team members share their knowledge, expertise, and experience along with their differing frames of reference and engage in a mutual exploration to discover the absolute best solution that none of them could have achieved alone. They are completely comfortable with not having all the answers because they know that no single person or discipline could ever have all the answers to such complex problems. This fact does not scare them—instead it invigorates them. It is another characteristic of the truly gifted design-build professional.

sharing, but what gets shared and how it gets shared differs in the two approaches. According to Peter Beck, "Collaboration is a *data-centric* activity wherein each discipline contributes information to other disciplines for processing to achieve common objectives."[25] This sharing of information happens easily on good design-build teams. "By contrast, integration is a *knowledge-centric* activity and relies on participants sharing their knowledge, experience, and expertise to solve problems." See Figure 8.5.[26]

Another way to distinguish between collaboration and integration is to consider that collaboration cobbles independent thoughts, the fragmented pieces of information that are patched together to create a solution. Michael Turner described the process as "tantamount to sprinkling in the ingredients without mixing."[27] On the other hand, integration is a process that melds information to "discover" solutions as a part of the engagement itself. Collaborating to share information is still necessary, but the information is synthesized to create an integrated solution through a mutual exploration of the problem at hand.

In reality, even fully integrated firms report that they often miss the mark in trying to achieve team integration, acknowledging that many procedural barriers can get in the way; these barriers must be addressed at the corporate level. The resistance appears to be tied to the segregated services mentality associated with traditional project delivery. This resistance can affect every aspect of the AEC business, from hiring practices to accounting methods. As long as contractors, architects, engineers, and owners see themselves as competing entities (even within the same firm), achieving full team integration will continue to be a struggle. But as more and more team members are selected based upon the specific

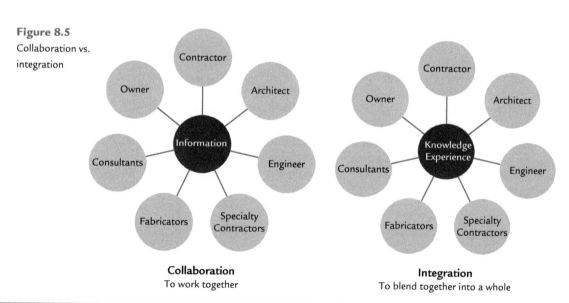

Figure 8.5
Collaboration vs. integration

Collaboration
To work together

Integration
To blend together into a whole

skills and talents mentioned earlier in the chapter, and each discipline routinely practices suspending their assumptions and actively listening to learn from one another, integration will occur. However, it cannot happen without effective leadership.

LEADING THE DESIGN-BUILD TEAM

Back in Chapter 7, you learned how design-build PMs need to step back from their discipline training and look at the bigger project picture from the client's point of view, to get perspective on the team's various discipline interests. However, the concept of leadership on a design-build team does not stop with the project manager or even the project director. Leadership should extend across the entire design-build team, and it is easy to recognize when it exists and when it does not. There are hundreds of books on leadership style, but the one that best describes the ideal form of leadership for a design-build team is **servant leadership**.

Robert K. Greenleaf coined the phrase in his 1970 essay "The Servant as Leader." In his 1977 book entitled *Servant Leadership: A Journey into the Nature of Legitimate Power and Greatness*, he stated that servant leadership begins with "the natural feeling that one wants to serve. Eventually people who possess this servant-first attitude aspire to the leadership position, but it initiates from their desire to first serve other people—never from a desire to be in command or power."[28] This leadership style lends itself perfectly to the design-build project delivery model.

> The traditional "control and command" leadership style associated with design-bid-build is not effective in the collaborative, integrated design-build environment—not if you want to achieve the extraordinary results that are possible with the process. Team members, associates, and clients alike will trust design-builders who exhibit a servant leadership style far more than they will trust those who operate under a power-based leadership model.

D-B DICTIONARY

Servant Leadership – A leadership philosophy, first articulated by Robert Greenleaf in the 1970s, that focuses attention on serving others and building trust.

> It's hard to describe the difference in working within a group or organization that believes in and practices servant leadership. There is a level of trust like that of a close-knit family. In fact, those of us that work at TDIndustries often refer to ourselves as the TD family. Of course, at times, there will be disagreements, differing viewpoints, and unexpected challenges (opportunities) as there inevitably are within any group of motivated people. Under servant leadership, those times are always handled with respect, compassion, and trust in one another. In our meetings, all are equal and all have the opportunity to be heard. An outside observer can see the interaction of such a team and may participate in the group's activities, even if he does not understand how it works. He may not even realize what he is seeing. Anyone can certainly reap the benefits of the groups' performance, whether or not they actively participate. They can see the trust within the group, and they quickly learn to trust the group and the individuals in it. So this is why I cannot imagine any better approach for performing design-build with an integrated team.

Drew Yaggy, PE, DBIA, LEED AP
Chief Engineer and Preconstruction Manager
TDIndustries, Inc.
Dallas, Texas

> When I hear architects and contractors wrangle over who should be in control on a design-build team, I already know that they don't get it. They speak about the leadership role as some kind of static position that flows from some justifiable superiority bestowed upon them by their education, status, rank, or discipline. They don't understand that design-build leadership is about serving others and creating an environment in which every individual on the team can excel.

Whether they are aware of it or not, the most successful design-build individuals (and firms) use the servant leadership approach. Their actions speak far louder than words. They hold the interests of their clients, end users, and teaming partners in high esteem and work diligently to create a win-win situation for everyone involved, without ever sacrificing the interests and needs of their own firms. Because they understand and use servant leadership, in addition to the "genius of and," their project results uniformly reflect their commitment to excellence and satisfaction.

The behaviors and actions of these service-oriented leaders are noticeably different from power-oriented leaders; servant leaders are not interested in the power of a traditional "control and command" position common under the design-bid-build model. They are more interested in results. Greenleaf differentiated between those who lead with a servant-first mentality and those who lead with a leader-first mentality.

> *The servant-first lives the service model of leadership. The leader-first lives the power model of leadership. The power model assumes a hierarchy shaped like a pyramid. Only a few people have power—those at the top of the pyramid. In the service model, the hierarchy doesn't matter. That's because anybody in a family, organization, or community can be of service. Anybody can identify and meet the needs of others. Anybody can be a servant-leader. The paradox is that a servant leader can gain power without seeking it. People trust servant-leaders, and give them power, because they know that servant-leaders use power to benefit everyone.*[29]

This is exactly the mentality that should be nurtured on a design-build team. Leadership, along with every other aspect of the design-build process, should be delegated and shared as appropriate to produce optimal value for the project. For this reason, individual egos must be set aside on a design-build team. The design-build team leader should take full advantage of the diverse experience and expertise available on her team and view every member as a valuable resource and contributor to the overall solution and entrust others to take the lead when it will best serve the project—regardless of rank, position, or discipline.

The servant leadership model is so well suited to design-build because it naturally encompasses another leadership concept known as **co-leadership**, in which more than one person is accountable for a team's success. The design-build relationship has leaders from two distinct but essential disciplines—design and construction—coming together, leveraging each discipline's unique contribution, to achieve results that neither could accomplish individually. In this relationship, one discipline cannot have "power" over the other.

While a team cannot operate effectively without a skipper at the helm, that skipper should focus on being a good steward and sharing leadership responsibilities with the rest of the team. Peter Block defined stewardship as "the willingness to be accountable for the well-being of the larger organization by operating in service, rather than in control, of those around us."[30] Such a model is well worth considering on a

D-B DICTIONARY

Co-leadership –
A leadership arrangement that embraces differing frames of reference and is based upon mutual respect and trust, and in which more than one person is accountable for directing the team.

design-build team that brings together so many professional, supportive partners to take on the huge challenges confronting our industry.

> We know that every successful organization has, at its heart, a cadre of co-leaders—key players who do the work, even if they receive little of the glory... Power and responsibility are dispersed, giving the enterprise a whole constellation of costars—co-leaders with shared values and aspirations, all of whom work together toward common goals... Successful costars are consummate team players and, thus, valuable models for everyone interested in effective collaboration. Usually servant-leaders, they tend to be self-reliant, yet committed to organizational goals... Outstanding co-leaders know that they don't have to be at the top of the organizational chart to find satisfaction—that exercising one's gifts and serving a worthy cause are far more reliable sources of satisfaction than the title on one's office door.[31]

Those who love the design-build process express over and over the professional success and reward they feel when led by a truly integrated leader. You must take off your specialty "hat" (contractor, architect, engineer) and represent all members fairly and respectfully. You must know the personalities, culture, and "ways of thinking" of each member of your team. You can't overemphasize budget at the expense of design or compromise engineering in exchange for schedule. I have asked many architects why they have not yet embraced design-build and far too often they say that it's because the "contractor" team leader is biased toward the construction perspective and does not represent others' interests equitably. But you cannot serve the owner's best interest if you don't unify the team around common goals and objectives that focus on the overall best design and construction solution.

John Alley, DBIA
Vice President
Layton Construction
Co., Inc.
Sandy, Utah

LEARNING HOW TO DANCE

The traditional model for practicing architecture, engineering, and construction management is to focus and protect by position, by discipline, and by segregated interest, regardless of the impact on the project as a whole. Doing design-build this way, with a traditional design-bid-build mentality, is like trying to dance without any music—you can follow all the steps but somehow the rhythm is all off. A design-build team is not merely a group comprised of contractors and designers joined to a client by a contract. Delivering complex projects in an interdisciplinary team format is a completely different way of doing business for most of us and it requires a completely different way of thinking. No matter how successful you have been in delivering design or construction under the segregated services model, your discipline-specific education and skills alone is insufficient to perform well on a design-build team. To plan, design, execute, and deliver design-build projects requires an entirely new battery of tools, techniques, and teaming skills. To obtain them, a company must instigate advanced training in such areas as communication, facilitation, collaborative negotiation, conflict resolution, commitment management, action language, and integrated design-build leadership. Companies that master these new skills will be in a league of their own.

Not for Everyone

Even with proper training, the success of any team still depends on the character, commitment, and talents of its individual team members. It might seem that just about any contractor, architect, or engineer can learn to be a design-builder. Most likely, any AEC professional can understand how design-build is structured contractually and how the two-step procurement process works. But understanding the design-build process from a technical perspective does not automatically give individuals the affective skills they need to perform effectively on a design-build team. Some people simply do not belong on a design-build team—not because they are flawed human beings, defective personalities, or bad people—but because they are uncomfortable working in the uncertain, dynamic environment of design-build. They often become paralyzed when they do not have all the details worked out before moving forward and must have every "t" crossed and "i" dotted before they can trust in the relationship—and even then they remain skeptical and wary. It is not in their nature to engage in dialogue and healthy conflict and debate on their way to building trust. Therefore, every company that pursues a design-build project must hand-pick its internal and external team members and invest in the training necessary to help them perform at their highest level. No other action that you can take will better mitigate your risks and ensure your success.

> **CHAPTER HIGHLIGHTS**
>
> 1. Transitioning from design-bid-build to design-build involves making a mental shift from old mental models that influence how we behave and respond to one another as team members. Many people in the low-bid mentality of design-bid-build view people in other disciplines with mistrust and skepticism. In design-build a new mental model is needed to create a sense of collaboration and full disclosure.
>
> 2. Building the right team takes an intentional blending of diverse backgrounds, experience, and viewpoints along with the discipline knowledge required to address a client's project needs at the highest level.
>
> 3. The three major parties in the design-build team are the owner, the contractor, and the primary designer. Many additional people will be woven into the team (for example, subcontractors and vendors), and the sooner they are involved, the better.

4. In design-build you can customize your team by pulling in the right people at the right time. Some of these people may be strategic partners who are third party associates and their participation and input are critical to the success of the project. Strategic partners might include fire marshals, code and permitting officials, regulatory agencies, and community organizers.

5. After selection of the teaming partners, a teaming agreement should be formed to establish a framework for defining the roles, responsibilities, and risks during both the proposal development phase and the post-award phases of the project.

6. At the proposal development phase, the teaming agreement should answer these key questions and others: Who will be responsible for what activities? Who will be responsible for final review and approval of the proposal before submission? Who will provide the technical oversight for the design? How will the costs associated with the proposal development be shared? How will confidentiality be maintained regarding proprietary information and strategies?

7. The teaming agreement must also address the following questions during the post-award phase: Who will be responsible for what tasks during the contract performance phase? How will the design and construction work be coordinated? How will design and construction errors be handled? What will be the protocol for engaging the client during design reviews and approvals? What will happen if team members do not honor the commitments spelled out in the teaming agreement?

8. The teaming agreement sets the stage for the integrated nature of the design-build relationship and the associated coordination of tasks by addressing: team selection; core values and project goals; win strategies and collective advantage; legal issues and jurisdictional concerns; organizational structures and entity configurations; communication protocols and technology; risk assessment and mitigation planning; incentives, shared savings and rewards; quality management and value enhancement; and financial and accounting issues.

9. Design-build is meant to operate in an environment of trust. Many factors influence the decision to trust or not to trust, including: past history and experience, competence and ability, similarities and differences, and character traits. On a design-build team, trust is demonstrated by committed openness and transparency on the part of all team members; honesty and truthfulness are essential.

10. A successful design-build team is made up of individuals with a set of traits and talents suitable for the design-build approach. Being a design-builder requires shifting individual ego to a collective ego, being responsive in nature, being able to deal with uncertainty and the unknown, possessing the genius of AND, and being emotionally intelligent.

11 For a team to be successful, every member must be an advocate for both the process and the project. One indicator of a team member's commitment can be found in the way they talk and the language they use.

12 Effective communication is vital to the success of the design-build team. Communication involves reaching an "equality of understanding" and this can be achieved through committed speaking and committed listening. A few effective techniques and tools are: differentiating dialogue, cascading communication, and practicing positive intent.

13 It is important to understand that collaboration and integration are not the same things. Collaboration means to work together and integration means to blend into a whole. Collaboration operates from a more independent thinking perspective, where information is shared to arrive at a solution. Integration operates from a more synergistic perspective and solutions are "discovered" as the team thinks through a problem together.

14 The leadership style that epitomizes the design-build leader mentality is servant leadership. Effective design-build team members who use this style hold the interests of their clients, end users, and their teaming partners in high esteem and work diligently to create a win-win situation for everyone involved, without sacrificing the interests and needs of their own firms.

RECOMMENDED READING

- ***The Fifth Discipline, The Art and Practice of the Learning Organization,*** Peter M. Senge, Currency Doubleday, 1990
- ***Sway, The Irresistible Pull of Irrational Behavior,*** Ori Brafman and Rom Brafman, Doubleday, 2008
- ***The Five Dysfunctions of a Team: A Leadership Fable,*** Patrick Lencioni, Jossey-Bass, 2002
- ***Building Trust in Business, Politics, Relationships, and Life,*** Robert Solomon and Fernando Flores, Oxford University Press, 2001
- ***Good to Great,*** Jim Collins, Harper Business, 2001
- ***The Reflective Practitioner,*** Donald Schön, Basic Books, 1983
- ***Change Your Questions, Change Your Life,*** Marilee Adams, Berrett-Koehler, 2004
- ***Embracing Uncertainty, The Essence of Leadership,*** Phillip G. Clampitt and Robert J. DeKoch, M.E. Sharpe, 2001
- ***The Emotional Intelligence Quick Book,*** Travis Bradberry and Jean Greaves, Fireside Simon and Schuster, 2005
- ***Team Talk: The Power of Language in Team Dynamics,*** Anne Donnellon, Harvard Business School Press, 1996

- **Servant Leadership: A Journey into the Nature of Legitimate Power and Greatness,** Robert K. Greenleaf, Paulist Press, 1977
- **Stewardship: Choosing Service Over Self-Interest,** Peter Block, Barrett-Koehler Publishers, 1993
- **Co-Leaders: The Power of Great Partnerships,** David Heenan and Warren Bennis, John Wiley and Sons, 1999

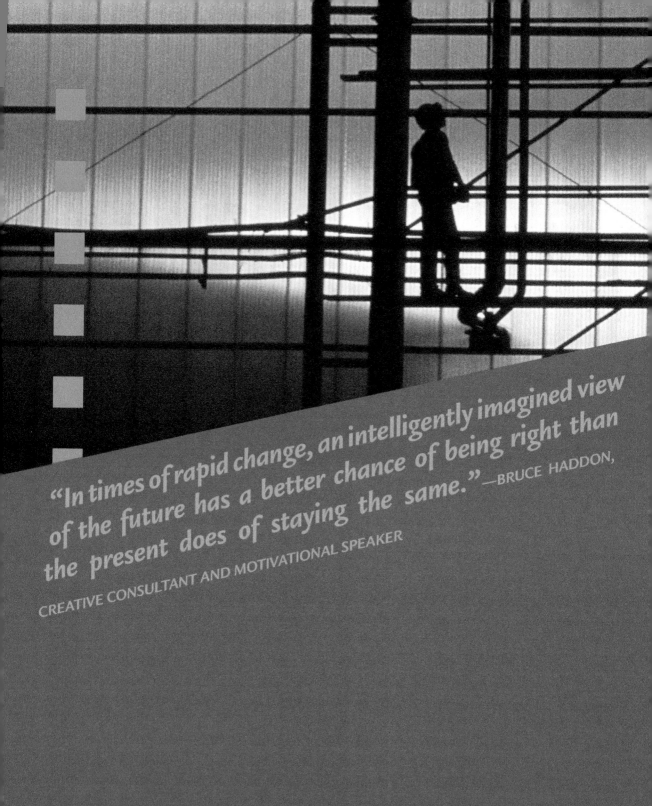

9
IMPLICATIONS FOR THE FUTURE

I started this book by describing how rapidly our industry is transforming—whether we are ready for it or not. This transformation requires a shift in the way we think, the way we relate, and the way we behave. It will dramatically affect the way we perform and the results we produce. The demand for more collaborative approaches to planning, designing, and construction means that alternative project delivery methods like design-build are no longer optional. They are mandatory if you want to remain competitive.

Owners have already discovered that they do not have an option. As their budgets diminish and their facility needs increase, they can no longer afford the mediocre and costly results associated with the segregated services model of design-bid-build. As technology makes it easier to share our information, knowledge, and resources, there are no excuses for lumbering along with old project delivery techniques that are too slow, too burdensome, and too contentious. This final chapter explains how the integrated design-build process, along with a number of new initiatives and technologies, will complete this transformation once and for all.

THE PERFECT STORM

The phrase "perfect storm" has come to represent the simultaneous occurrence of circumstances or events which, if considered individually, may be interesting and useful, but far less potent than when they arise concurrently. In the world of commerce, a perfect storm is rare, but we are definitely in the midst of one in the AEC industry. This storm will tip over the traditional design and construction boat once and for all. The world, the United States included, is experiencing significant economic challenges on multiple fronts. Interestingly, these challenges have coincided with the emergence of several trends and advanced technologies. In particular, these four forces have tremendous transformational potential, especially when applied to design-build:

:: Integrated Project Delivery (IPD)
:: Building Information Modeling (BIM)
:: Sustainable Design
:: Lean Construction

Let us take a look at each of these potential industry changers.

INTEGRATED PROJECT DELIVERY

The term **integrated project delivery (IPD)** is not a new term, although it only really gained acceptance after the American Institute of Architects (AIA) started to promote it as a preferred project delivery model at its 2008 National Conference. The California AIA, the organization that jump-started the national AIA's interest in the IPD model in 2007, defines IPD as follows:

> IPD is a project delivery approach that integrates people, systems, business structures and practices into a process that collaboratively harnesses the talents and insights of all participants to optimize project results, increase value to the owner, reduce waste and maximize efficiency through all phases of design, fabrication and construction. IPD principles can be applied to a variety of contractual arrangements and IPD teams will usually include members well beyond the basic triad of owner, architect, and contractor. At a minimum, though, an integrated project includes highly effective collaboration between the owner, the architect, and the general contractor ultimately responsible for construction of the project, from early design through project handover.[1]

Simply stated, IPD is any project delivery method that engages the project contractor as part of the design and construction team at the beginning of the design process. Under this definition, design-build, CM-at-Risk, design-assist, and other approaches could all be considered integrated project delivery models. However, design-build is the most effective integrated project delivery approach because it contractually combines responsibility for the design function and construction function under a single source contract. Other IPD approaches still operate under a separate contract model, with each party responsible only for its part of the solution, with no accountability for the project as a whole.

You should also be aware of another contracting model that takes collaboration even one step further. The **Alliance Contract**, also referred to as a **Project Alliance Agreement**, not only links the contractor and the designer contractually, but also obligates the owner to share in the gain or pain proposition associated with the contract execution. The alliance participants collectively establish the performance goals and targets against which the project's success will be measured. The three primary parties (owner, contractor, and designer) all assume responsibility for the project and jointly share in the profits (gain) or losses (pain)

Sidebar:

The term integrated project delivery is not new. In 2000, California Polytechnic State University in San Luis Obispo established an Integrated Project Delivery (IPD) minor for construction management, architecture, or engineering majors. This popular program was nicknamed the design-build minor. However, the more generic official title, integrated project delivery, emphasizes the integrated, multi-discipline, team approach to design and construction. At the same time, this title de-emphasizes the controversy and misunderstanding surrounding the design-build project delivery concept. At the time, design-build was still considered a fad by many faculty members and industry practitioners. However, in reality, design-build is the original integrated project delivery model, using highly collaborative teams to deliver fully integrated solutions.

D-B DICTIONARY

Integrated Project Delivery (IPD) – A project delivery model that emphasizes the integration of design and construction personnel, along with owner representatives, at the earliest stages of the project initiation to maximize efficiency and reduce waste through all phases of design, fabrication, and construction.

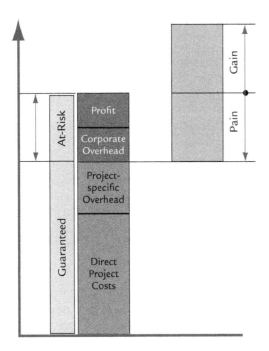

Figure 9.1
Alliance contracting

D-B DICTIONARY

Alliance Contract –
A contractual arrangement intended to create an alignment between the primary participants of the project, including the owner, contractor, and architect, and possibly others. The alliance contract typically includes a shared savings-loss feature intended to provide incentives for the team. Also referred to as a **Project Alliance Agreement**.

Target Price –
An estimated cost established jointly by a project team that serves as a benchmark for project performance.

Gainshare –
In an alliance contract, the amount of money saved on the project when the actual project costs are less than the targeted costs. This savings is shared among the alliance participants.

Painshare –
In an alliance contract, the amount of money lost when the actual project costs are more than the targeted costs. This loss is shared among the alliance participants.

resulting from the project's performance. (In some instances the major specialty contractors and subconsultants are also brought into the initial alliance.) As you can imagine, this arrangement dramatically aligns the interests of all parties and provides a real incentive for collaboration. (See Figure 9.1.)

In an alliance contract project, optimal project results occur when the project cost (including both design and construction) is less than the **target price;** the residual is shared as a **gainshare**. Less than optimal results occur when the project cost is more than the target price; in that case, the overrun is shared as a **painshare**. The painshare penalty is usually capped at the contractor's or design-builder's overhead and profit. When costs exceed this point, the owner/client accepts liability for all direct costs until the project is completed. Obviously, it is in everyone's best interest to set reasonable targets to begin with, and to work together to add efficiency, reduce waste, and save dollars so all parties can benefit from the gainshare.

The sharing of gains and losses in an alliance contract project has a tremendous effect on how team members behave and significantly influences the parties' performance under the contract. When the team does a thorough risk analysis and mitigation plan, and sets realistic price targets and contingencies, the resulting strategic alliance can produce remarkable results. This shared risk model removes the fear of project unknowns that often adds thousands of dollars to the project cost when risk is expected to be absorbed by a single party.

Although still somewhat rare, these types of arrangements are generating more interest among owners who desire a collaborative effort with a team that has many incentives to perform at its best.

D-B DICTIONARY

Building Information Modeling – The process of compiling and managing data associated with a building's design as a digital representation of the physical and functional characteristics of the project. The data includes information pertaining to the building's configuration and geometry, spatial relationships, geographic information, and material properties.

BUILDING INFORMATION MODELING

The second major factor shaping the future of design-build is **building information modeling (BIM)**. This expanding technology is probably the main reason that groups such as the Construction Users Round Table (CURT) and the AIA have moved toward project delivery models that incorporate collaboration and integration.

According to an article published in the Fall 2008 issue of the AIA's electronic newsletter, *Practice Management Digest*:

> *In its most basic form, building information modeling (BIM) is the move from analog to digital design and construction. It is a model-based technology linked with a database of project information. And it is poised to fundamentally change the way projects are built and the way project stakeholders communicate with each other. Radically transforming the way designs are created, communicated, and constructed, BIM is not just the electronic transfer of paper documents. It greatly increases the ability to control and manipulate data and information in an unprecedented way and in an interoperable format. The move from paper-centric information to parametric, model-based information means that the digital design can be used for cost estimations, simulations, scheduling, energy analysis, structural design, GIS integration, fabrication, erection, and facilities management. These models are not just the electronic drafting tools that firms now think of as digital practice, nor are they three-dimensional renderings with separate construction documents. The move to an integrated, parametric, and object-based system should lead to dramatic changes in design and construction as well as, possibly, compensation and risk allocation.[2]*

MENTORING MOMENT

Julie Stites, AIA, LEED AP
Architect
The Haskell Company
Jacksonville, FL
Education: Cal Poly State University, San Luis Obispo - Graduated 2005
Major: Architecture

To BIM or Not to BIM

Generally, younger employees aren't given much responsibility because they lack the experience and the know-how that more seasoned architects and contractors working in the design-build arena have—and rightly so. There's an awful lot of talk about how this generation is impatient and how we "want it all right now." But the truth is, we do understand that we must work our way up through the ranks of the traditional career ladder. Our impatience is more about our deep desire to contribute and not so much about entitlement or arrogance.

And in a design-build environment, there are many opportunities to contribute at a younger age—especially if you have trained in BIM or LEED.

This younger generation has one advantage that the more mature generation does not—an aptitude for technology. This aptitude can provide direct access to projects and project teams in which you might otherwise not be able to participate until you are much older. Knowledge of advanced technologies puts you in the big tent, so to speak, where you can listen and learn from some of the most experienced project people in the company. It's very humbling when you understand how much you still have to learn but at the same time, rewarding to know that you can bring something of real value to the team that will contribute to the overall success of the project.

Although I learned all of the basic software programs in college, at the time there wasn't that much being taught in terms of building information modeling—certainly not as much as some universities offer today. However, I picked up training wherever I could, and was fortunate enough to be selected as part of a company team to receive more immersive training in BIM even though I had only been with the company for a few years. Those of us that were on the BIM pilot team are now mentoring our colleagues as they, too, make the transition.

As I see professionals struggling to make the transition to the new technologies on the fly, I realize more and more how valuable the technology training available at the universities is; it is a true luxury to learn it over an extended period of time. Every student should take full advantage of the opportunity to learn these advanced technologies that are transforming our industry. It will certainly give you an edge and may even open up some doors that otherwise might take you years to enter.

From a practical standpoint, the goal of building information modeling is to "build a building virtually, prior to building it physically, in order to work out problems, and simulate and analyze potential impacts."[3] Whereas CAD (Computer Aided Design) vastly improved our ability to produce 2-D drawings much more efficiently, BIM provides 3-D information. The addition of this third dimension allows all parties to visualize the building of a project from the foundation up, along with all the parts and pieces that interface with it and information associated with those parts. At the same time, BIM allows us to create a model-based cost estimate and detailed project schedule instantaneously. This allows all members of the design-build team to quickly test the effects of every component of the design. This, in turn, supports the team in making crucial, informed decisions long before the first shovel of dirt is ever disturbed on the building site.

Within the past five years, technology costs have dropped and capabilities have increased, leading to the ability to deploy BIM on any project. This tool creates a tremendous opportunity to integrate the work of all participants in the process. BIM's ability to visually display the project while incorporating a wealth of data behind the scenes allow us to truly see the real design, including cost of elements, program verification, systems coordination, and the scale of spatial progression.

Very early in the design process, the contractor can provide meaningful constructability input. Column spacing trade-offs, floor-to-floor heights, duct distribution, all become discussion points. As a result, the design evolves in a much more integrated fashion, and increases its effectiveness in balancing all the factors that bear on a project's development. The design-build approach coupled with BIM technology helps us design in the truest sense of the word.

Jay Whisenant, AIA, CCS
Principal, COO
NTD Architecture
San Diego, California

As you can see, BIM allows you to integrate and analyze every type of decision-making and planning scenario, taking into account the operational and commercial potential of any project. Among other things, you can use BIM for overall facilities planning, productivity analyses, and staffing scenarios.

Kimon Onuma, FAIA
President and Founder
Onuma, Inc.
Pasadena, California

From an architectural perspective, the information in BIM is the true value. Architects produce an immense amount of information that is under-utilized. Information that is not accessible will become irrelevant in this information-centric age. Architects that understand this will flourish like the Internet and those that do not will fade like Betamax. At the end of the day, it is not the technology or tools that matter, but the ability to translate our knowledge into a format that is relevant and valuable. BIM is just a container for data. A changed process with new tools gives us broader choices on how data is used. Technology is not the barrier to change; the cultural shift in how we collaborate is the challenge.

Many design and construction firms have already embraced the BIM technology. Some have all aspects of the BIM capabilities up and running, while others are only using the 3-D visualization component to detect system clashes and address constructability issues. However, many companies and individuals still question the viability of the investment. Granted, at this point in time, it may not be appropriate to consider BIM technology for every project. As one contractor told me, "We still have smaller subcontractors out there that we work with everyday who are just now getting on board with CAD." Nevertheless, it is only a matter of time before BIM does become mainstream and many believe the pace of its acceptance can and should be accelerated.

One BIM proponent in particular believes that, because BIM is first and foremost information-based, the key to its widespread acceptance is simplifying the process of adding information into the model, regardless of the contributor's technical skills. Generally, you have to know how to design in BIM before you can input data into the model. Unfortunately, many of the people who have the information needed to build a credible model (subcontractors, vendors, regulatory agents, etc.) do not know how to use the complex BIM technology, and have no reason to learn it. This creates a communication challenge between the information providers and the designers who need the information, and makes BIM more difficult to optimize.

Kimon Onuma, whose architecture firm Onuma Inc., of Pasadena, California, has been involved in using BIM since the mid-1990s, has focused on the information in BIM. He realized that the industry could be transformed by the development of an interface that allowed designers, contractors, and other participants to focus on the information exchange process, rather than on the details of making the BIM software do what they want. To encourage collaboration and to make the model easy to access, he envisioned a Web-based interface.

The buildingSMART™ alliance referenced in Figure 9.2 was created to spearhead technical, political, and financial support for advanced digital technology in the real property industry—with applications ranging from concept, design, and construction through operations and management. The building SMART alliance operates within the independent non-profit National Institute of Building Sciences (NIBS). You can learn more about this organization at http://www.buildingsmartalliance.org.

Putting his money where his mouth is, Onuma developed a web-based BIM interface called the ONUMA Planning System (OPS™) that uses the **open standards** of the web to make the model accessible to users, without requiring them to download the entire BIM. You access OPS via an online platform that allows participants to interact with the model and input information in a way that is most familiar to them, whether the preferred interface is a spreadsheet, a word document, a plan, or even a sketch. Onuma calls an online OPS collaboration a BIMstorm™ session, with each project having its own BIMstorm site. Using OPS, an entire design-build team can hold a virtual design charrette, sharing ideas, information, and drawings in real time, without ever leaving their offices.

In an effort to demonstrate the power of OPS technology, Onuma developed several BIMstorm planning exercises that solicited voluntary participation from all parts of the globe. The first BIMstorm exercise, called BIMstorm LAX and held in January 2008, focused on a sixty-block area east of Dodger Stadium in Los Angeles. (See the Case Study "BIM and Beyond," later in this chapter.) Since then, BIMstorm collaborations have focused on several other cities, including Boston, New Orleans, London, Vancouver, Washington DC, Pasadena, and Tokyo. OPS has allowed architects, planners, contractors, vendors, end-users, stakeholders, students, and the business community from all over the world to collaborate live, generating design schemes from any location with internet access.

BIM is one of the most powerful forces at work in the perfect storm that is altering our industry. To demonstrate the revolutionary spirit associated with Onuma and other proponents of BIM, participants in the Pasadena BIMstorm signed a "Declaration of Information Independence" April 23, 2008, as shown in Figure 9.2

D-B DICTIONARY

Open Standards – Non-proprietary formats and databases that support the exchange or joint use of various types of information by differing software tools.

Figure 9.2
Declaration of Information Independence. Courtesy of Onuma, Inc.

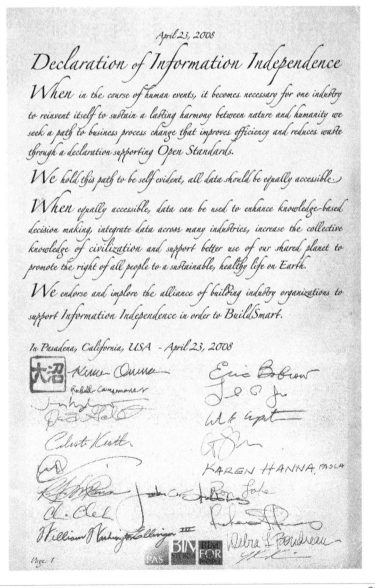

BIM AND BEYOND

CASE STUDY

THE FACTS
Project Name: BIMStorm
Description: In 2008, 14 BIMStorms, 2,458 sites, 485,883,746 square feet of space, 3,964 buildings, 493,914 rooms

THE TEAM
3,000 participants, 14 countries

Pennsylvania State University, iCON Lab during the 24 Hour BIMStorm LAX, John I. Messner, Ph.D.

THE LEAD-IN

In the late 1980s, Kimon Onuma regularly flew back and forth between his architectural office in Pasadena, California, and his father's architectural office in Japan. The two were collaborating on a number of United States government projects. The travel alone was a challenge. Onuma sometimes made the trip twice a month. Another major challenge was trying to manage data from the field and get it into the proper hands for evaluation and synthesis. On top of that, team members had to transfer huge CAD and 3-D files between the offices. At the time, Onuma was using FileMaker® to capture the project data and was forced to train bilingual personnel in both offices on how to use it. Furthermore, the only option for transferring the files was a direct, expensive dial-up service that clients were not interested in paying for. Onuma was weary of all the travel and frustrated with the cost and inefficiency associated with cumbersome and expensive techniques. He knew there had to be a better way.

THE STORY

When Onuma discovered building information modeling (known as virtual building models in 1993), he saw that the tool had far more potential than simply serving as 3-D modeling software. He decided to focus on the information component of BIM. He began to develop a program that would help him manage complex project information over long distances.

Courtesy of ConnectivityWeek

Over the years, he created customized computer files for ArchiCAD® filled with valuable information. His team eventually developed a way to use the Web to organize and store this information for use by both offices. Finally, they developed a server program called the Onuma Planning System (OPS) which allowed clients from around the world to access information about a particular BIM via the Web. OPS also operated as an editor, allowing users to modify a BIM in real time and have the changes made available to other people working on the same BIM.

Initially, the point of OPS was to allow Onuma to communicate and collaborate with his father's firm in Japan more efficiently and without all the travel. However, it soon became clear that the OPS system had far more potential than just aiding one company in the delivery of its architectural services. The program captured architectural knowledge. This knowledge base was expanded with each project, making it possible to automate mundane tasks. This, in turn, left more time for design. Large scale planning projects could be produced with fewer people and more accurate information. This added efficiency led to higher profits, which were reinvested into more research and refinement of the software.

Initially OPS was only used within his firm; clients were unaware of the powerful tool that allowed Onuma to be so efficient. The architectural firm delivered standard CAD drawings to their clients just like everyone else; the clients remained unaware of the rich databases associated

BIMStorm LAX Tower Study, Design by: BIMStorm LAX Teams, 3D Model and Rendering by Acusourcing

with the projects and the BIM models that were being developed in the background.

This all changed in 2003, when the United States Coast Guard set out to overhaul its facilities. The Coast Guard planned to establish thirty-five new Sector Command Centers that would combine its group commanders and port operations into unified facilities. Having worked with Onuma before, the Coast Guard learned about the OPS system and wondered if it could be employed to capture and catalog all of its facilities data. Before this could happen, Onuma's team needed to make sure the Coast Guard personnel could easily interact with the OPS program. The result was a powerful Web-based interface that allowed project stakeholders to enter information into the model easily.

According to Onuma:

Making a meaningful building information model can be as easy as making reservations for a plane ticket, hotel, rental car, and dinner on the Expedia.com web portal. On Expedia.com you don't have to know XML or HTML programming. You don't have to know what software programs are working together to optimize your search for a hotel and restaurant near the convention center. With an "Expedia®" type web portal, you can make a building information model without even knowing what "BIM" means. The complex integration of relevant information happens in the background and the user is presented with a customized decision-making tool in an interface that is familiar to that particular user's expertise."[4]

In other words, no special expertise or technology is required. The Coast Guard field operators didn't want to hear about interoperability [how one system works with another system] or BIM. They just wanted tools to plan for the command centers with a simple Web-based interface that allowed them to log in, see the pieces they needed to work on and start using the tool. OPS was a perfect fit. As they worked on the project, the system constantly collected site-specific data to populate the database for later use.[5]

It would have taken the Coast Guard approximately 350 months to complete the planning for the thirty-five Command Centers. By utilizing OPS and BIM, they were able to reduce that time to only six months.

In the end, the USCG's entire thirty-three-million square-foot facility portfolio was inputted into OPS, allowing for at least low-level creation of BIMs for each facility. Nearly three-million square

BIMStorm LAX Tower Study, Design by: BIMStorm LAX teams, Composite view of rendered tower on Google Earth, 3D Model and Rendering by Acusourcing.

feet of space was rendered in high detail with ArchiCAD…[The team] created a calculator to allow planners to quickly understand personnel and equipment demands at each facility. Tools were also developed to allow blast threat analysis (BIM Bombs™) and other security tools.[6]

In January of 2008, Onuma decided to push the OPS technology to its limits and develop a grassroots exercise that would demonstrate the program's full capabilities. He knew that the Web-based tool had the capacity to engage project participants from every area of the globe, incorporating information from numerous perspectives in real time. The exercise, known as BIMstorm LAX, was pitched as the "Woodstock of BIM." Even its promotional poster was designed to mimic symbols associated with the rebellious counterculture music festival of the 1960s. Writing for AIArchitect, Michael Tardif described the event like this:

> Developed as an online design charrette, the exercise engaged 133 design professionals from 11 countries over a 24 hour period with 700 additional people participating as observers. The group was loosely organized into 25 teams worldwide and the entire group went to work to develop plans for large sections of the city of Los Angeles, California, creating massing models and schematic designs for 420 buildings totaling over 55 million square feet. Buildings ranged in size from a 600 square-foot house to a proposed 2 million square-foot bank tower.[7]

In order to encourage efficiency and autonomy, Onuma and his team hoped to eliminate communication outside of the BIMstorm. Instead, they wanted to encourage participants to focus only on the information entered into OPS.

> We tried to minimize or eliminate completely the need to communicate by traditional methods—e-mail, voice, or documents—and focused on

communicating with data. For example, a team in California submits a building into OPS. It gets downloaded in Honolulu, where they design the structural system and submit it back. A team in Manila continues with the structural design; a team in the Netherlands picks up the building and designs the mechanical system; a team in the U.K. runs the same building through Ecotect for energy analysis. Most of these teams did not communicate at all other than through the model, or did not even know the other team existed. They just had to log on and view the reality of what was being developed and run with it.[8]

After BIMstorm LAX, another thirteen events took place in 2008, focusing on a variety of cities. In 2009, BIMStorms focused on low-carbon collaboration, starting with the University of Southern California. According to Onuma: "Every BIMStorm is unique and organic; some are sixty minutes in length and others last more than twenty-four hours. BIMStorms can be stand-alone events or work in conjunction with other BIMStorms. They are, for the most part, not real-world projects, but driven by real-world needs, so the data generated by BIMStorms are not owned by any entity. Think of them as a BIM sandbox for the AEC community. Unlike a project, the results of the BIMStorm are available to all who participate."[9]

At the 2008 Associated General Contractors' conference in Tahoe, Nevada in June 2008, teams from the AGC participated in a BIMStorm that took the tower design from the LAX event and ran through a series of constructability reviews, including detailed analysis of the structural, MEP and curtain wall systems. The team created a number of "what-if" scenarios for consideration. "The two-day BIMStorm generated data in hours instead of the typical weeks or months, and demonstrated how collaborating at these early stages utilizing open standards can provide huge benefits."[10]

THE CLOSE

Kimon Onuma believes that the "I" in BIM is the most important initial in the acronym. Using information wisely through the life of a building is what will allow our industry to experience increased productivity gains and get the real payback of true integration. Open standards are at the core of the Onuma OPS program. [The term open standards refers to non-proprietary formats and databases that support the exchange or joint use of various types of information by differing software tools.] They clearly illustrate how data can be captured and used not only to build a BIM, but to go beyond the building model to manage and make decisions about all kinds of planning and project scenarios. "Open standards allows participants in a planning and design process to use whatever tool they choose to complete a particular task. Participants might not even know which tools other participants are using. The tools used for BIMStorm LAX included paper-and-pencil overlays for early concept planning and design; ArchiCAD by Graphisoft, Revit Architecture by Autodesk, and VectorWorks by Nemetschek North America for building design; Ecotect by Square One Research for energy analysis; Elite CAD by Elite Software for HVAC systems design; MARS Facility Cost Forecast System by Whitestone Research for operations and maintenance cost forecasting; Google Earth for placing buildings on a virtual representation of Los Angeles; Sketch Up, Green Building Studio, Tekla, Navis Works, and Solibri; and, of course, the OPS model server and editor for generating a variety of data and graphical reports, including spatial data summaries and construction cost estimates."[11]

There is no question about the power of BIM to accommodate collaboration and facilitate the integration of a design-build team, but the process associated with OPS takes BIM to a whole new level of collaboration. OPS enables architects to perform rapid prototyping of multiple scenarios and early testing of conflicting requirements to determine whether they can be resolved, and if so, how. It allows many parties with a high degree of professional knowledge—architects, owners, engineers, security analysts, contractors, and others—to interact in real time and incorporate their specialized knowledge into the project data. It allows project teams to visualize all types of data, document the decisions they make, and continually eliminate ambiguities as decisions are made. Best of all, viewing the data requires little or no professional knowledge or computer skill. "The goal of the entire system," says Onuma, "is to allow users having little knowledge and only minimal technical skill to get to the knowledge." With tools like this, there is no reason for anyone on a design-build team to be left out of the design process. Everyone can contribute information that will ultimately create exceptional results. (Visit http://www.bimstorm.com/ to learn more.)

From the Owner:

OPS shifted the reality of true scenario-based planning to an enterprise level, enabling us to link shore facilities and infrastructure to mission execution and strategic Coast-Guard-wide outcomes. This better enabled us to allocate infrastructure resources to the most important mission execution and outcomes. The integration of BIM, geospatial data, real property data, and mission requirements supports the need of a common operational picture for the United States Coast Guard. This common operational picture can include real-time tactical information, as well as longer-term strategic information, which is enabled by the ONUMA Planning System.

David Hammond, Chief, SFCAM Division Commandant, (CG-434) United States Coast Guard

▪▪ SUSTAINABLE DESIGN

So far we have looked at two major forces that will affect the future of our industry—integrated project delivery and building information modeling. Now we come to the third, sustainable design.

In 1987, the United Nations Brundtland Commission defined sustainability as "meeting the needs of the present without compromising the ability of future generations to meet their own needs."[12] According to Dr. William Rees, a bio-ecologist by training and former Director of the University of British Columbia's School of Community and Regional Planning, "For humans to live sustainably, the Earth's resources must be used at a rate at which they can be replenished. However, there is now clear scientific evidence that humanity is living unsustainably, and that an unprecedented collective effort is needed to return human use of natural resources to within sustainable limits."[13] For this collective effort to succeed, the worldwide design and construction industry must play a part. The contribution our industry can make in developing sustainable communities is significant. By all accounts, the architecture, engineering, and construction industry is embracing this responsibility. In fact, this is one of the areas of common purpose that seems to be uniting the disciplines in a collaborative effort.

Andrea Murray, AIA, NCARB, LEED AP
Architect
Bread Loaf Corporation
Middlebury, Vermont

Design-builders are in a terrific position—when compared to separate, highly fragmented architecture and construction firms—to deliver sustainable building solutions for a greater value. This is the case because we often have multiple disciplines working together on a project from the onset. We can consider alternative design and building strategies and look for creative ways to include them in our projects. We can weigh the costs and benefits of such strategies and seek out a range of solutions that allow each building component to have more than one purpose and work synergistically to maximize the benefits realized.

As mentioned in Chapter 1, architects, engineers, and contractors have all jumped on the LEED bandwagon in response to the growing demand by clients and communities for design and construction solutions that promote economic and social development and that avoid environmental degradation and exploitation of resources. As of March 18, 2009, 17,434 individuals had achieved the designation of LEED AP (80% in the United States and 20% internationally). Furthermore, eight to ten times more LEED projects are now in the works than in the entire sixteen years that the USGBC has existed. Keep in mind that this is happening during a very severe market downturn for new construction.[14]

At California Polytechnic State University alone, approximately 20% of the currently enrolled construction management students have already achieved their LEED AP status or will complete it before graduation.

Furthermore, the topic of sustainability has become an integral part of the Construction Management program, across all courses in the curriculum. Several other universities have done the same. Student interest in sustainability education should come as no surprise—after all, the future we are trying to sustain is theirs.

According to an article by Dane Kose and Dan Sisel of Mortenson Construction, PricewaterhouseCoopers (PWC), one of the big four accountancy firms, proclaims that sustainability has now "reached the tipping point." They cite a 2006 global survey of their clients in which 70% of chief executives of mid-sized to large companies believe that sustainable strategies are directly linked to their profitability and two-thirds believe such strategies will remain a high priority going forward."[15] With such demand for sustainable projects, it only makes sense to take advantage of the sustainability advantages offered by design-build. Kose and Sisel go on to suggest that design-build is perhaps the delivery method best suited to minimizing "the adverse environmental impact of any project." They explain:

> *Each member of the project team has unique knowledge of alternative sustainable strategies as well as the corresponding relationship with other elements of the program. Engaging all parties early in a collaborative manner maximizes the likelihood that the best strategies emerge and that execution closely follows strategic intent. [It is no surprise that] some of the most successful sustainable construction projects have used the design-build project delivery method... Collaborative design-build project teams are uniquely positioned to deliver a broader set of cost-effective sustainable strategies. [Because the builder is engaged early], the builder becomes a more proactive asset to the project team. Accordingly, the evaluation of alternatives can be considered fully in the context of all project filters: programmatic requirements, constructability, economic viability and sustainability. The result is faster decision making and more environmentally friendly outcomes.*[16]

People interested in sustainable building projects naturally turn to design-build. After all, sustainable building projects require intensive communication, coordination, and collaboration, all of which are hallmarks of the design-build process. Greg Gidez, Corporate Design Manager with Hensel-Phelps Construction, and one of the leaders in both design-build and sustainability shares these observations:

> *The ability to accelerate the construction and delivery of the project through an integrated design-build process allows us to make critical decisions earlier and with greater knowledge. The shortened schedule should also reduce general conditions costs and construction loan costs. More money can then be used to enhance the sustainable characteristics of the building and increased performance. More money goes into project betterment and less into overhead and carrying costs. For example, on a project I was involved with, by shortening the design-construction schedule from the owners' thirty-six months down to the twenty-six months actually required to complete the project, the project went from*

a LEED "certified" level to a "Gold" level. The project out-performed the energy model, resulting in a Platinum LEED for existing buildings. Over 110 strategies were employed in the building to help it achieve the rating, and many of the innovative solutions were brought to the table by design-build subcontractors and material suppliers who would not have been at the table under a traditional design-bid-build model. Concurrent to this project, there were four other buildings built by the owner at the exact same time—all based on the same bridging documents. Although all were design-build contracts, the other four buildings approached it with a design-bid-build mentality and strategy and let the market control the outcome. Those four buildings barely reached a certified LEED level, and suffered from performance problems. The one that took the fully integrated design-build approach obtained a Platinum LEED level and out-performed expectations, paying dividends for years to come. The owner contends that productivity has gone up and call-backs and warranty issues have gone down in the Platinum building, all thanks to collaboration.

There is no question that design-build is the perfect collaborative structure for systematically integrating the various project objectives into a holistic sustainable solution—and when combined with BIM technology, the efficiencies and effectiveness of the process are even greater.

LEAN DESIGN AND CONSTRUCTION

The term lean manufacturing was first introduced by Jim Womack, Dan Jones, and Dan Roos in 1990 in their book *The Machine That Changed the World*. They define lean manufacturing as a process that focuses on "doing more of everything with less of everything." In their book, they describe a production system that is better, faster, and cheaper than traditional manufacturing processes. Lean manufacturing requires less space, less inventory, fewer labor hours, and avoids wasteful practices. The book introduced American manufacturers to the tremendous performance capability of the Toyota Production System (TPS).[17]

One of the most fascinating features of the book is that only one chapter actually focuses on manufacturing. The book is really about lean enterprise, and the application of lean principles to every aspect of an organization, including sales and marketing, manufacturing, engineering, human resources, finance, design, product planning, and purchasing.[18] The authors' point is that, if you only focus on the manufacture of a product, you can squeeze only so much waste out of the labor and material. By broadening your focus to include every other facet of an operation, you can achieve far greater efficiency.

Lean enterprise principles can be applied to design-build with extraordinary results. Lean design and construction focuses on eliminating waste in the project delivery process by continuously getting rid of any and all non-value-adding activities, minimizing uncertainty,

The basic process of assembling an automobile and creating a building are similar. Influenced by the Toyota Production System, our approach to design-build focuses on five Ps—philosophy, process, people, partners, and problem solving. We question and challenge how things are done and why, with a view toward improving and making them better. We continually focus on quality design and construction and encourage active improvement feedback regularly from the entire team, including our subcontractors and suppliers, to help us improve efficiencies, add value, and reduce wasted effort at every step of the process.[19]

Jim Gray
Chairman and Chief Executive Officer
Gray Construction
Lexington, Kentucky

and decreasing variability. This continuous improvement philosophy, referred to as *kaizen* by the Japanese, works hand in glove with integrated design-build, which flourishes in a teamwork and relationship environment.

Despite the advantages associated with lean enterprise principles, it can be difficult to get your whole team on board with the philosophy. Having your major subcontractors and suppliers team up to

The big thing with lean applications in design-build is the streamlining of the process and the elimination of waste. I believe that packaging the design and construction under one contractual responsibility gives the design-build entity lots of opportunities to remove waste. Lean principles can be applied from the very beginning of the process. For example, a typical cause of waste in the traditional process is the "over design" that happens on projects because the architects and engineers don't know who is going to install the work; they include significant safety factors in the designs as a protection against faulty work. Furthermore, a huge collection of drawings is required for the same reason. But a design team that knows, from day one, that it will be designing the project, and that knows who will be installing the work, can produce only the drawings needed by that team.

Other, more obvious efficiencies can be realized in terms of both materials and methods. For example, applying lean principles to the production side of the equation can produce a great deal of time and dollar savings through the prefabrication of components, which eliminates multiple on-site steps that require a great deal of manpower and equipment. When your workforce can assemble parts and pieces at waist height, instead of over their heads, they can produce the work faster and more safely. My own studies have found that design-build can reduce costs by 6% and overall time for design and construction by as much as 33%. In my experience, adding lean principles to the equation increases savings by least half in both categories.

Victor Sanvido
Senior Vice President
Southland Industries
Garden Grove, California

deliver their work based upon the most efficient means and methods is effective only if the general contractor supports them in their efforts. However, the value that such partnerships can bring to any project are tremendous.

You have just read about two industry practitioners who have embraced various lean practices, utilizing *kaizen* as a means of improving their design and construction processes. Other firms are following suit. Lean manufacturing is one more industry-improving initiative that seems to be a perfect fit with design-build project delivery and a logical complement to efforts associated with sustainable goals.

THE COLLABORATION ECONOMY

The fourth industry-changing factor is the rise of the collaboration economy. Why is the idea of collaboration so important? The design and construction industry is notorious for keepings its information, ideas, and resources as secret as possible. Because we are so accustomed to competing on a low-bid, every-firm-for-itself basis, the sharing of anything does not come easily. But that attitude will not work in today's environment. According to Don Tapscott and Anthony Williams, the authors of *Wikinomics: How Mass Collaboration Changes Everything*:

> *Companies can no longer depend only on internal capabilities to meet external needs. Nor can they depend only on tightly coupled relationships with a handful of business partners to keep up with customer desires for speed, innovation, and control. Instead, firms must engage and co-create in a dynamic fashion with everyone—partners, competitors, educators, government, and, most of all, customers.*[20]

According to Tapscott and Williams, advances in technology, demographics, and the worldwide business economy have brought us into a new age in which people participate in the economy like never before.[21]

Some examples of the networking that is the hallmark of this new, collaborative economy include Facebook, MySpace, Flickr, Second Life, and YouTube. According to Tapscott and Williams, "The Web is no longer about idly surfing and passively reading, listening, or watching. It is about peering: sharing, socializing, collaborating, and, most of all, creating within loosely related communities."[22] How does this translate to the world of design and construction? The BIMstorms described earlier in this chapter epitomize the use of peering to allow people from all over the world and from every discipline to participate in solving the same problem. But you have probably used peering in your own work to solve problems. For example, asking your subcontractors to help you think through the best approach to a challenging construction issue is common form of peering. Now, consider how the internet could take this practice to the next level.

We all work in a traditionally reserved industry, where trade secrets and business strategies are held close to the vest. Times are changing and there is a great opportunity out there for the next generation to network and collaborate through improved technology and communication—bringing about the transformation of our industry from yesterday's paper, fax machines, and file cabinets into today's digital, web-based environment, where trades people can discuss cutting-edge building techniques, students can access the knowledge and wisdom of industry leaders, or a builder can sell excess materials for a profit instead of dumping them at a landfill—all without ever leaving the office. This future is now and a whole generation out there is ready to embrace it.

Josh McGarva
Vice President
Western Water
Constructors, Inc.
Santa Rosa, California

Although the design and construction industry has been slow to take advantage of these mass collaboration opportunities, the tightly guarded boundaries that have separated design and construction professionals for so long are beginning to open up a bit. For example, the new networking site http://www.constructionexchange.com/, launched recently by the youthful professionals at Western Water Constructors in Santa Rosa, California, is, as far as I know, the only professional networking site exclusively for the construction industry. See Figure 9.3.

Figure 9.3
Construction Exchange website

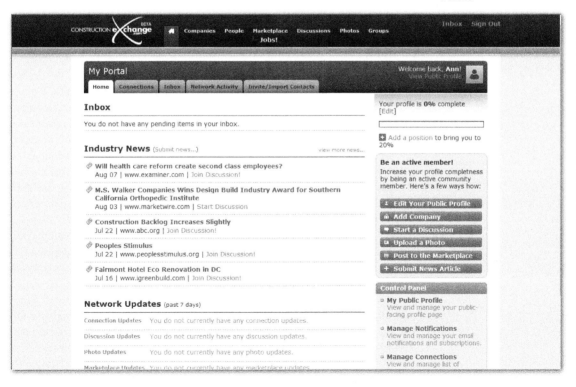

Jeffrey L. Beard
ACEC Vice President
Director, Institute for
Business Management
American Council of
Engineering Companies
Washington, D.C.

A Bright Future Before Us

During summers in the late 1960s and early 1970s, I worked for a general contractor in Southern Maryland, doing day labor and carpenter assistant jobs on custom houses, commercial buildings and Navy projects including barracks and aircraft hangars. The summer employees coveted the work at Patuxent Naval Air Station (described by Tom Wolfe in *The Right Stuff*) because of the prevailing wage rates. It was one of the few sources of good pay and helped toward college savings.

After obtaining a degree from Rutgers College, I returned to Maryland and became an estimator/project manager for a construction firm while working nights and weekends as a sports editor of the local weekly paper.

I eventually began to work exclusively for the construction firm, spending five years there before accepting a job in downtown Washington, DC as an office engineer for a Pittsburgh-based steel firm. The company, H.H. Robertson, produced rolled shapes for everything from steel deck for commercial and institutional buildings, to guardrail, to curtainwall systems for high rise offices. A regular exposure to architects and engineers was inherent in the job as I accompanied two seasoned sales engineers on calls throughout the mid-Atlantic region, and did some minor detailing and bills of materials back in the office.

While working in steel, I kept up contacts with a few Rutgers alums. One day I received a call from John Berard, who was public relations director for the Associated General Contractors of America (AGC). AGC was looking for a couple of people to staff their market divisions, and I subsequently became Associate Director of the Heavy-Industrial Division. It was my first stint as an association staffer, but I enjoyed the national conferences and the opportunity to make a difference in the business conditions for an entire industry.

By 1983, the 64-mile commute was getting to be too much, so I decided to look for a job closer to home. I went to work as Director of Contracts and Field Services for Harms Associates Engineers/ Wheeler Goodman Masek Architects in Pasadena, California and Annapolis, Maryland. The projects were marvelous: public schools, historic preservation for the City of Annapolis and hospital and health care projects. My job allowed me to see the tension between designers and constructors from the A/E perspective. With every RFI, change order and battle over scope, I kept thinking that there must be a better way, perhaps something new and innovative, perhaps an older method that had been eclipsed by technology or business customs.

Reading in the library one day, I stumbled across a passage about the project approach pioneered by Stone and Webster nearly a century before, as the two engineers sought to satisfy the burgeoning demand for power plants and electrical distribution facilities in late nineteenth- and early twentieth-century America. Here was a concrete example of a firm that included both designers and constructors; it took responsibility for conceiving and delivering a fully functioning project to an owner without the ongoing divisiveness apparent in many of the sequential design-bid-build projects that I had experienced.

While I was extremely grateful for four years in the A/E world, I missed the non-profit world's focus on the industry's big issues, and in 1989 accepted a position at the American Society of Civil Engineers (ASCE) as one of two Government Relations professionals. By this time, I was brimming with ideas about how to help the design and construction industry, and constantly urged my boss to allow me to pursue some new initiatives. At least some of the ideas came to fruition, such as my idea for a nationwide infrastructure report card, which ASCE later produced. But the singular most important role for me was as a staffer for ASCE's Design-Build Task Force in 1991. Our blue-ribbon panel met three times over the course of a year, and I drafted a final report that seemed to pave the way for engineers to again embrace integrated delivery, provided that certain caveats were met in the application of design-build. The ASCE design-build task force report's first printing was quickly snapped up by interested members and others, such as reporters for Engineering News Record (ENR) and other industry publications.

Later that year, I received a call from Preston Haskell, Chairman of the Haskell Company of Jacksonville, Florida, requesting multiple copies of the report and offering to pay for a second printing. He also flew to Washington to meet with me and I remember being more than impressed by the gravity of his voice, the logic of his ideas and his can-do spirit for large-scale pursuits. We agreed that I would ask ASCE if the leadership would be willing to carve out a place for design-build practitioners, and failing that, if I could make discrete inquiries at other national non-profit organizations (such as AIA, AGC) to gauge their degree of support for this trans-disciplinary group of individuals and firms. Suffice it to say that the requests were not warmly received.

Instead, Preston Haskell, through the force of his personality and depth of his convictions, decided and found the Design-Build Institute of America (DBIA). My background role during the spring of 1993 was to draft the incorporation papers and bylaws, and apply to the Department of the Treasury for non-profit status. By June 1993, we were open for business at 601 Pennsylvania Avenue, midway between the Capitol and the White House.

All of the early DBIA chairs were phenomenal leaders, and each contributed mightily of their time and ideas. Preston Haskell personally drafted many chapters in the Manual of Practice. Rik Kunnath also contributed sections to the Manual and championed the research arm of DBIA. Grant McCullagh insisted that the organization needed succinct, streamlined contracts. Don Warren provided strategic leadership that permitted growth of the National Design-Build Conference and the educational programs. Kraig Kreikemeier understood the technical side of design-build delivery and led the way into a performance-based specification system (not a commercial success, but a brilliant idea somewhat ahead of its time).

I salute all those who spend their lives in design and construction. Design-build and integrated delivery have a bright future because they are driven by the spirit of collaboration and the democratization of information. New technologies, such as BIM and embedded sensors, serve to increase the design and construction practitioners' ties to each other and to their work. And although great strides have been made over the years in refining and improving the design-build process, the early adopters of these innovative tools will help take design-build to the next level and carry it into the future.

This useful website provides a simple environment with powerful tools for sharing ideas, disseminating information, and promoting growth through networking. Josh McGarva, one of the creators of the Construction Exchange is hoping the free-access site will grow through the collaborative efforts of its users. People are invited to engage in discussions, participate on projects, and join in on problem-solving endeavors associated with any type of construction or design topic. It is a great place for AEC professionals and students alike to share ideas and get to know each other.

You can do a lot of things at the Construction Exchange site, including:

:: Creating a public profile that lists your professional history, interests and specialties
:: Create a company profile that lists your company's specialties, past projects, and contact information
:: Start or participate in an online discussion about industry topics
:: Post an ad regarding jobs, or the sale or purchase of materials and equipment
:: Post photos of current or past projects

It is encouraging to see a new generation of builders, who have grown up online, embrace this technology and use it to serve the AEC industry at large. It epitomizes this generation's ethic of openness, participation, and cooperative engagement. The implications for design and construction collaboration and integration are enormous.

THE NEXT GENERATION

Many industry veterans believe that we are closer then we have ever been to a tipping point in our industry, and that this next generation of architects, engineers, and constructors will provide the impetus required to bring about wholesale change. If we are smart, we will let them—but with guidance, of course. I am a professor who works with young construction management, architecture, and engineering students every day. In my opinion, they will either transform this industry or leave it altogether. This generation is as committed as any before them, but their focus is on working smarter, not harder. Fortunately for us, they will be equipped with the tools and technology required to do just that. Socially and ethically they are most like their grandparents' generation, showing the same can-do attitude attributed to the veterans of World War II. Furthermore, they are without a doubt the most collaborative and connected generation in history. They are by all accounts born for design-build.

Neil Howe and William Srauss refer to this generation as the Millennials. Their book, *"Millennials Rising: The Next Great Generation,"* describes them as follows:

As a group, Millennials are unlike any other youth generation in living memory. They are more numerous, more affluent, better educated, and more ethnically diverse. More important, they are beginning to manifest a wide array of positive social habits that older Americans no longer associate with youth, including a new focus on teamwork, achievement, modesty, and good conduct. Only a few years from now, this can-do youth revolution will overwhelm the cynics and pessimists. Over the next decade, the Millennial Generation will entirely recast the image of youth from downbeat and alienated to upbeat and engaged—with potentially seismic consequences for America.... Over the coming decade this rising generation will introduce itself to the nation and push the nation into a new era. Once this new youth persona begins to focus on convention, community, and civic renewal, America will be on the brink of becoming someplace very new, very "millennial" in the fullest sense of the word. That's when the "end of history" stops, and the beginning of a new history, their Millennial history starts.[23]

For some people the potential impact of this next generation on the AEC industry may be unsettling. But for those who are excited about a transformed design and construction industry, the thought of their influence is an invigorating source of hope. The key is to lay a solid foundation upon which this next generation can build the future—a future that not only embraces technology but also embodies an attitude of collaboration, integration, and endless possibility.

Many young people graduating from construction, architecture, and engineering programs today have an intense desire to make a difference. Those who have been exposed to design-build and want to make that their career path seem to fully embrace the notion of collaboration and understand the importance of having all the right tools to accomplish the mission at hand. They have a deep desire to create value for their team and company the very first day on the job, which can actually become a challenge for their managers. The exceptional new hires quickly identify the opportunities associated with making an extra effort, usually on their own, to become certified Design-Build Professionals, LEED Accredited, and achieve professional licensure at an accelerated pace. Quite frankly, they are an inspiration, and we have to be careful not to dim their aspirations but instead find ways to acknowledge and encourage their contributions.

Coleman Walker
Director of Talent Acquisition
The Haskell Company
Jacksonville, Florida

This group, born in or after 1982, first started entering the university around 2000. According to Howe and Strauss, they will still be flowing through our colleges until at least 2020. That means that we have already seen some of these young adults graduate from our architecture, engineering, and construction programs and that our classrooms are filled with them right now. This provides the educational community a

The Design-Build Educator Workshop is held each year in conjunction with the DBIA Annual Conference. You can learn more about this workshop, and find out how your construction management, architecture, or engineering program can get involved in the DBIA University Program by visiting the DBIA website at http://www.dbia.org/. You can learn more about the DBIA Certified Design-Build Professional credential at the same website.

tremendous opportunity to influence an entire generation that already possesses an integrated, collaborative attitude and have actually been working as teams since kindergarten.

Implications for Education

Education is the key to completing the transformation of the AEC industry from a segregated services model to an integrated services model. Unfortunately, most architecture, engineering, and construction programs still teach the traditional segregated process of design and construction along with the traditional roles for architects, contractors, and engineers. We need to encourage traditional construction management, architecture, and engineering programs to find ways to incorporate more integrated classes, with participation from multiple AEC disciplines. Such a change will serve our industry as well as our students. Therefore, our industry may have to exert significant influence, in the form of time and dollars, to make this change happen.

Expanding the Design-Build Curriculum

As a leader in undergraduate design-build education, California Polytechnic State University joined forces with the California Center for Construction Education (CCCE) to conduct the first Design-Build Educator Workshop in 2006 for faculty interested in teaching design-build at their universities. The workshop was funded by a grant from the Charles Pankow Foundation (a private industry foundation) which was specifically earmarked to encourage universities to start teaching fundamental design-build principles to undergraduates in construction management, architecture, and engineering students. With support from the DBIA, CCCE has held the Design-Build Educator Workshop each year since 2006; approximately sixteen to twenty-five professors per year have attended the program. Several of the workshop attendees have gone on to participate in the DBIA University Program. Through this program, individual faculty can qualify as approved instructors, giving them the right to use the DBIA's core curriculum to supplement their existing AEC courses. This, in turn, has allowed their students to jumpstart their pursuit of the Certified Design-Build Professional™ credential available through the DBIA Certification Board. (See Figure 9.4.) This opportunity is timely, as we are seeing more and more owners

TAKEAWAY TIPS

Let's Go for a Ride

Consider pairing up young interns and new graduates with your senior managers and executives. Because Millenials are most like their grandparents, they often prefer being mentored by folks from the "veteran generation" (those born from 1922 to 1945). Consider finding ways to blend the seasoned experience of your veteran employees with the open mindedness, creativity, and technological savvy of these young minds. Millennials are very respectful and appreciative of the experience that senior people have and covet the opportunity to hear about their experiences. These mentoring moments can occur as scheduled lunch dates, impromptu invitations to ride out to a job site, or a special invitation to sit in on a high-ranking meeting. These senior mentors do not have to be associated with a project or involved with the young constructor or engineer on a day-to-day basis. As a matter of fact, it is probably better if they are not. You should assign other technical, on-the-job mentors as well. These mentoring opportunities have everything to do with integrating the wisdom that can only come from years of experience with the fresh thinking of young minds. Together these dynamic duos can learn from each other and create a future that most of us can only imagine.

Figure 9.4
Certified design-build professional

listing design-build education as part of their RFP evaluation criteria. In one recent RFP, the owner allotted twenty points for teams with members who possessed certification as a Design-Build Professional. Young constructors, architects, and engineers who graduate with a fundamental education in design-build will have a hiring advantage if this trend continues.

GETTING OWNERS ON BOARD

Design-build and other integrated project delivery methods have grown in popularity largely because of owner demand. Ironically, owners are often reluctant to make the mental shift themselves and fully embrace the collaborative model. They want their architects and contractors to work well together, but they are unwilling to join the party themselves. However, if owners really want to lead this drive toward more open communication, collaboration, and integration, they too must choose to trust the teams they hire and create an environment in which the process can unfold as it is supposed to.

Long ago we referred to individuals who were capable of managing the design and construction process from a holistic, comprehensive perspective as master builders. Today, the master builder is not a recognized profession in the industry. However, as design-build gains prominence, it would benefit the industry if individuals who possess the interdisciplinary fluency necessary to manage projects from multiple perspectives, along with the affective skills associated with integrating a team, were recognizable in the marketplace. Perhaps a distinct body of knowledge focused on these capabilities will be developed and the master builder of old will evolve into the design-build project integrator of the future.

Often it's easy for the contractor and the architects to point fingers at each other because they are afraid to offend the owner (for the obvious reasons), when in fact it's often the owner that creates the situation through mismanagement of the process, underfunding, lack of timely decision making, changing the design late in the game, etc. and as a result sets the project up for inter-relational challenges.

Greg Gidez, AIA DBIA LEED AP
Corporate Design Manager
Hensel Phelps Construction Co.
Greeley, Colorado

Many owners are still developing the designs until they are 40% to 70% complete before releasing the RFP. They continue to use low price as the primary evaluation criteria, even when one of the most beneficial opportunities of design-build is to add value and enhance project performance and results. They often press for unreasonable changes regarding who bears the risk for the project, thinking that design-build releases them from any and all liabilities. They often hold fast to rigid design-bid-build protocols and practices such as "liquidated damages" and others that destroy any incentive for innovation and creative problem solving. In these circumstances, there's little or no chance for AEC practitioners to actually function as a collaborative, cohesive unit and provide the value that only a multi-disciplined, integrated design-build team can do. As owners demand more communication, transparency, collaboration, and integration from design-build teams, they themselves remain skeptical and distrustful—undermining the very collaborative process that they are trying to propagate.

However, we have to remember that much of the skepticism and distrust that has developed over the years from past dealings has stemmed from the segregated services design-bid-build model, which by its very nature, pits one party against another and diminishes opportunities to create value and optimize results. The bottom line is this: owners need to fully commit to design-build to make it work. The trust required in the design-build model must come from the owner as well as from the other project participants. Unless owners are willing to give design-build a fair chance of working, they will never realize the extraordinary results it is capable of producing.

Of course, owners will not be inclined to trust the design-build process without encouragement. As industry professionals, we need to educate our clients. For my part, I hope owners will embrace the lessons within these pages as much as contractors, architects, engineers, and every other service provider and vendor associated with the built environment.

THE TIPPING POINT

Our world faces enormous challenges associated with health care, education, infrastructure, and energy resources, among others. The design and building industry is poised to play a major role in addressing these challenges. However, we must focus on a collective, synergistic contribution and not simply on the independent parts that we can play as individual disciplines.

I believe with all of my heart that contractors, architects, and engineers are some of the most creative people on the face of the planet. Unfortunately, for far too many years they have been working under a project delivery model that has minimized the effects of their talents, resources, and collective genius. Imagine what will happen when the barriers that have kept us working in isolation come down, so that communication, collaboration, integration, and trust become the new cornerstones of their relationships. What would happen then? What problems could we solve? What new technologies could we develop? What extraordinary outcomes could we expect when we collaborate in an integrated fashion?

Far-fetched and idealistic—I don't think so. I am by nature an optimistic integrator. I am by nature a thinker in pursuit of possibility. I am by nature a person who never gives up, and never takes no for an answer. I can't help it—I *am* a design-builder. What about you?

CHAPTER HIGHLIGHTS

1. Contractors, engineers, and architects who use communication, collaboration, and integration in conjunction with the proper use of design-build project delivery will have an enormous impact on a project's success.

2. The challenge for companies participating in design-build projects is to make sure everyone in the organization knows how design-build should be done. In addition to this, a design-build team requires the right aptitude and temperament, along with the commitment to fully communicate and collaborate across the entire team.

3. Four forces that are converging to propel the industry forward at an unparalleled pace are: 1) integrated project delivery (IPD), 2) building information modeling (BIM), 3) sustainable design, and 4) lean construction.

4. IPD is any project delivery method that engages the project contractor as part of the design and construction team, at the beginning of the design process. Design-build, CM-at-Risk, and design-assist can be considered IPD models, with design-build being the most effective because it contractually combines responsibility for the design function and construction function under a single source contract.

5. The alliance contract is a model that takes the single source concept further. The three primary parties—owner, contractor, and designer—assume responsibility for the project and jointly share in the profits (or losses) resulting from the project's performance. A target price is set with all parties sharing optimal results (gainshare) or less then optimal results (painshare).

6. BIM provides 3-D information that allows all parties to visualize the project from the ground up. An instantaneous model-based cost estimate can be created as well as an automatically generated detailed project schedule. The BIM model includes information about every piece and part so that the design-build team can witness the cost and schedule impacts of every component of the design, quickly allowing the testing of various approaches.

7. The USGBC's LEED Accreditation program and the implementation of it by AEC practitioners is a response to the growing demand by clients and communities for design and construction solutions that promote

economic and social development in ways that avoid environmental degradation, exploitation of resources, or pollution.

8. Design-build, especially when combined with BIM technology, can systematically integrate various objectives (cost, schedule, quality) into a holistic sustainable solution by synchronizing the complete interdependencies between the performance requirements, material options, and various building systems.

9. Lean design and construction focuses on eliminating waste in the project delivery process by continually getting rid of any and all activities that do not add value, minimizing uncertainty, and decreasing variability.

10. The Internet has led to an increase in virtual networking opportunities. Tightly guarded boundaries in the industry can begin to open up through collaborative online peering, which enables participants to share ideas, solve problems, socialize, and disseminate information.

11. The generational group entering colleges around 2000 or later have been termed Millennials. Young adults from this group who are entering the industry are bringing with them a strong sense of collaboration gained through working as teams throughout their school years.

12. Education is the key to completing the transformation of the AEC industry from a segregated services model to an integrated services model. Expanding design-build curriculums will take the commitment of both industry and faculty.

13. DBIA created the Designated Design-Build Professional program in 2002 to classify individuals who demonstrate, by experience, education, and testing, the qualifications to be recognized as a Design-Build Professional (DBIA). This credential is available to any AEC practitioner or associate who completes the educational requirements, documents design-build experience, and is able to pass the designation exam.

14. Successful design-build requires the full participation of the owner. The owner must set up the design-build process correctly and let it function as the collaborative, integrated process it is meant to be. Owners must play their part in creating the environment of trust that is vital to design-build success.

RECOMMENDED READING

- **Wikinomics: How Mass Collaboration Changes Everything,** by Don Tapscott and Anthony D. Williams, Portfolio, 2006.
- **The Toyota Way, by Jeffrey K. Liker,** McGraw Hill, 2004.
- **Millennials Rising: the Next Great Generation,** by Neil Howard and William Strauss, Vintage Books, New York, 2000.
- **Think Better: Your Company's Future Depends on It,** by Tim Hurson, McGraw Hill, 2008.

ENDNOTES

CHAPTER 1

1. Gary Hamel, *Leading the Revolution* (Boston: Harvard Business School Press, 2000), 13–15.

2. Jeffrey L. Beard, Michael C. Loulakis, and Edward C. Wundram, *Design-Build: Planning through Development* (New York: McGraw Hill Professional, 2001).

3. Davis and White, "How to Avoid Construction Headaches," *Harvard Business Review,* March–April, 1973.

4. American Institute of Constructors. http://www.aicnet.org.

5. Ibid.

6. Beard, et al.

7. Ibid.

8. The Collaborative Process Institute, "Collaboration in the Building Process," (1996), http://www.dprconstruction.com/news-and-events/technical-papers/tech-collabor.cfm.

9. Construction Users Round Table, http://www.curt.org/.

10. Construction Users Round Table, "Collaboration, Integrated Information, and the Project Lifecycle in Building Design Contracting and Operation," CURT Document WP 1202, August 2004.

11. Ibid.

12. Construction Users Round Table, "Optimizing the Construction Process: An Implementation Strategy," CURT Document WP 1003, July 2006.

13. National Building Information Model Standard Committee (NBIMS), http://www.buildingsmartalliance.org/nbims/about.php.

14. McGraw-Hill Construction, *A Working Definition, Integrated Project Delivery* (AIA California, 2007).

15. Lean Construction Institute, http://www.leanconstruction.org/.

16. Ibid.

17. United States Green Building Council, http://www.usgbc.org/.

18. Green California, "Governor's Green Building Action Plan," http://www.green.ca.gov/GreenBuildings/leed.htm.

19. James P. Cramer and Scott Simpson, *The Next Architect: A New Twist on the Future of Design* (Osberg: Greenway Communications, 2007).

CHAPTER 2

1. ZweigWhite, *2007–2008 Design/Build Survey of Design and Construction Firms* (Natick, MA: Zweig White Information Services LLC, 2007).

2. Ibid.

3. Brenton E. Diller, Daniel Kerr, P.E., and Dr. David R. Riley, "Study Shows Why Design/Build is Just Plain Better," *Contracting Business* (April 27, 2005), http://contractingbusiness.com.

4. Jeffrey L. Beard, Michael C. Loulakis, and Edward C. Wundram, *Design-Build: Planning through Development* (New York: McGraw Hill Professional, 2001).

5. ZweigWhite, *2007–2008 Design/Build.*

6. Dean Kashiwagi, "Bond Programs: Accountability and Best Value," *DesignShare,* http://www.designshare.com.

7. Carol Eaton, "Design-Build Gains Ground in Healthcare Market," *California Constructor,* Volume 39, Number 2 (February 2009), 8.

8. Victor Sanvido and Mark Konchar, *Selecting Project Delivery Systems* (Fairfax, VA: The Project Delivery Institute, 1999).

9. Ibid.

10. Ibid.

11. Ibid.

12. Ibid.

13. Ibid.

14. *The Wizard of Oz* (1939; Burbank, CA: Metro-Goldwyn-Mayer, Warner Home Video).

CHAPTER 3

1. "US Navy's Submarine Rescue Team," Dateline, August 17, 2000, http://usgovinfo.about.com/library/weekly/aa081700a.htm.

2. Ibid.

3. ZweigWhite, *2007–2008 Design/Build Survey of Design and Construction Firms* (Natick, MA: Zweig White Information Services LLC, 2007), 87.

4. Ibid.

5. Ibid.

6. Answers.com, http://www.answers.com/topic/specification?cat=technology.

7. Design-Build Institute of America, *Design-Build Contracting Guide*, DBIA Manual of Practice, Document Number 510, November 1997.

8. Ibid.

9. Ibid.

10. Nadine M. Post, "Sutter Health Unlocks the Door to a New Process," *Engineering News Record*, November 2007.

11. Ibid.

12. Diana Hoag and Nancy Gunderson, "Contract Incentives and Design-Build: Re-Thinking Acquisitions Strategies," *Design-Build DATELINE*, February 2005.

13. Lee Evey, President 2005–2009 Design-Build Institute of America, Keynote Address, Florida Regional Chapter Conference, April 2007.

CHAPTER 4

1. Frank Cushman and Michael Loulakis, *Design-Build Contracting Handbook*, 2001 (Aspen Publishers, 2001), 321.

2. Office of Management and Budget, *OMB Circular No. A-11* (2006).

3. Ibid.

4. Laurie Sorten, "Krispy Kreme Doughnuts On the Double," Baking & Snack, December 2002.

5. Michael C. Loulakis, *Design-Build Lessons Learned: Case Studies from 2003* (Wickwire Gavin, P.C. and A/E/C Training Technologies, 2004).

6. Design-Build Institute of America, *Design-Build Contracting Guide*, DBIA Manual of Practice, Document Number 510, November 1997.

7. Diana Hoag and Nancy Gunderson, "Contract Incentives and Design-Build: Re-Thinking Acquisitions Strategies," *Design-Build DATELINE*, February 2005.

CHAPTER 5

1. Joe Calloway, *Work Like You are Showing Off* (Hoboken, New Jersey: John Wiley & Sons, 2007), 2.

2. *Miracle* (Walt Disney Pictures, 2004).

3. The Collaborative Process Institute, "Collaboration in the Building Process," (1996).

4. Ibid.

CHAPTER 6

1. Jim Crockett, "Inside Design-Build: Conceptual Estimators," *Design-Build*, June 2001.

2. *UniFormat™: A Uniform Classification of Construction Systems and Assemblies*, The Construction Specification Institute (2001).

3. Ibid.

4. Funding Universe, Hilton Hotels Corporation, Company History, http://www.funduniverse.com/company-histories/Hilton-Hotels-Corporation-Company-History.html.

5. Port of San Diego, Port of San Diego Overview, Real Estate, Development Projects, Hilton San Diego Bayfront Hotel, http://www.portofsandiego.org/hilton-san-diego-bayfront-hotel.html.

6. "San Diego Celebrates 'Topping Off' of New Hilton San Diego Bayfront," Reuters, http://www.reuters.com/article/pressRelease/idUS198204+25-Jan-2008+BW20080125.

7. Crockett, "Inside Design-Build."

8. Office of Management and Budget, *OMB Circular No. A-11* (2006).

9. Crockett, "Inside Design-Build."

10. Phillip G. Clampitt and Robert J. DeKoch, *Embracing Uncertainty—The Essence of Leadership* (M.E. Sharpe, 2001).

11. Margaret Wheatley, "The New Science of Leadership: An Interview with Margaret Wheatley," Interview adapted from the radio series *Insight & Outlook* hosted by Scott London (2006).

CHAPTER 7

1. Gary Hamel, *The Future of Management* (Boston: Harvard Business School Press, 2007).

2. New York City Transit, "Subway and Bus Ridership Statistics 2008," http://www.mta.info/nyct/facts/ridership/index.htm#chart_s.

3. Al Baker and Richard Pérez-Peña, "With Terrorism Concerns in Mind, Police Prepare to Guard a Shuttered System," *New York Times*, December 20, 2005, http://www.nytimes.com/2005/12/20/nyregion/nyregionspecial3/20security.html.

4. New York City Transit, "2008 Subway Ridership," http://www.mta.info/nyct/facts/ridership/ridership_sub.htm.

5. Thomas, S. Kuhn, *The Structure of Scientific Revolutions*, 3rd ed. (Chicago: University of Chicago Press, 1996), 7.

6. John Steele, ADePT Awareness Workshop, Rancho Cucamonga, CA. 2008.

7. Cheryl Allen Richer, "The Case for Performance Standards," U.S. Department of Transportation, Federal Highway Administration, Public Roads, May/June 2004, Volume 67, No. 6, http://www.tfhrc.gov/pubrds/04may/03.htm.

8. Hamel, *Future*, 21.

9. Adrian Gostick and Chester Elton, *The Invisible Employee* (Hoboken, New Jersey: John Wiley and Sons, 2006), 33.

10. Ibid., 13.

11. ADePT Design Software Suite, "Essential Tools for Managing Design in Design-Build Projects," AMI Technologies, the TechnoCentre, Puma Way, Coventry, CV1 2TW, United Kingdom.

12. Joe Flores, Personal interview, July 18, 2007, The Beck Group, Dallas, TX.

CHAPTER 8

1. Peter M. Senge, Art Kleiner, Charlotte Roberts, Richard Ross, and Bryan Smith, *The Fifth Discipline Fieldbook* (Currency Doubleday, 1994), 236.

2. Peter M. Senge, *The Fifth Discipline: The Art and Practice of the Learning Organization* (Currency Doubleday, 1990), 174.

3. Senge, et al., *The Fifth Discipline Fieldbook*, 236.

4. Senge, *The Fifth Discipline: The Art and Practice*, 254.

5. Ori Brafman and Rom Brafman, *Sway: The Irresistible Pull of Irrational Behavior* (New York: Doubleday, 2008), 70.

6. Design-Build Institute of America, "Design-Build Teaming Agreement Guide," *Design-Build Manual of Practice*, Document No. 308, September 2000.

7. Patrick Lencioni, *The Five Dysfunctions of a Team: A Leadership Fable* (San Francisco: Jossey-Bass, 2002), 195.

8. Ibid., 195.

9. Robert Solomon and Fernando Flores, *Building Trust in Business, Politics, Relationships, and Life* (Oxford University Press, 2001), 92.

10. Ibid., 92-94.

11. Ibid., 14.

12. Jim Collins, *Good to Great: Why Some Companies Make the Leap... and Others Don't* (New York: HarperBusiness, 2001), 41.

13. Donald Schön, *The Reflective Practitioner* (Basic Books, 1983), 54.

14. Marilee Adams, *Change Your Questions, Change Your Life* (Berrett-Koehler, 2004), 25.

15. Phillip G. Clampitt and Robert J. DeKoch, *Embracing Uncertainty: The Essence of Leadership* (M.E. Sharpe, 2001), 110.

16. Jim Collins and Jerry Porras, *Built to Last: Successful Habits of Visionary Companies* (New York: Harper Collins, 1994), 43-44.

17. Travis Bradberry and Jean Greaves, *The Emotional Intelligence Quick Book* (San Diego: Fireside Simon and Schuster, 2005), 25.

18. Ibid., 23.

19. Ibid., 24.

20. Anne Donnellon, *Team Talk: The Power of Language in Team Dynamics* (Boston: Harvard Business School Press, 1996).

21. Ibid., 6.

22. Peter Senge, *The Fifth Discipline, The Art and Practice of the Learning Organization*, 241.

23. David Bohm, *The Special Theory of Relativity* (New York: W.A. Benjamin, 1965).

24. Patrick Lencioni, *Overcoming the Five Dysfunctions of a Team: A Field Guide for Leaders, Managers, and Facilitators* (San Francisco: Jossey-Bass, 2005), 143.

25. Peter Beck, "Collaboration vs. Integration: Implications of a Knowledge-Based Future for the AEC Industry," http://www.di.net/articles/archive/2437/.

26. Ibid.

27. Michael Turner, "Strategic Teaming and Its Impact on AEC Integration," http://beckgroup.com/content/about/features/WhitePapers/StrategicTeaminganditsImpactonAECIntegration.pdf.

28. Robert K. Greenleaf, *Servant Leadership: A Journey into the Nature of Legitimate Power and Greatness* (Mahwah, New Jersey: Paulist Press, 1977), 13.

29. Kent M. Keith, *The Case for Servant Leadership* (The Greenleaf Center for Servant Leadership, 2008), 26-30.

30. Peter Block, *Stewardship: Choosing Service Over Self-Interest* (Barrett-Koehler Publishers, 1993).

31. David Heenan and Warren Bennis, *Co-Leaders: The Power of Great Partnerships* (New York: John Wiley and Sons, 1999), 3–15.

CHAPTER 9

1. AIA California Council, *Integrated Project Delivery: A Guide*, ver. 1 (AIA National, 2007).

2. "Preparing for Building Information Modeling," AIA Practice Management Digest, Fall 2008.

3. "An Introduction to Building Information Modeling (BIM)," Journal of Building Information Modeling, Fall 2007, 12.

4. Kimon G. Onuma and Dianne Davis, "Integrated Facility Planning using BIM Web Portals," The Federal Facilities Council of the National Academies, Washington, D.C., October 9, 2006, http://onuma.com/BIM/FFC_BIM_Portals_Onuma.pdf.

5. McGraw Hill Construction, "Interoperability in the Construction Industry, Smart Market, Design and Construction Intelligence," 2007 *Interoperability Issue*, 28–29.

6. Ibid.

7. Michael Tardif, "BIMstorm Hits LA," *AIArchitect*, Volume 15, February 15, 2008.

8. Kimon Onuma, "Engineering a BIMStorm," *Consulting-Specifying Engineer*, March 2009, 31–32.

9. Ibid., 32.

10. Tardif, "BIMstorm Hits LA."

11. Ibid.

12. World Commission on Environment and Development, *Our Common Future* (Oxford: Oxford University Press, 1987).

13. Michael Gismondi (2000). Interview with Dr. William Rees, *Aurora Online*, http://aurora.icaap.org/index.php/aurora/article/view/18/29.

14. Ever Blue LEED News and Updates, LEED Exam and Certification Information, http://www.everblueenergy.com/blog/2009/03/18/how-many-leed-aps-how-many-projects/.

15. Dana Kose and Dan Sisel, "Design-Build and Sustainable Construction: Success Now for our Future," Mortenson Construction, www.mortenson.com/files/DBIA.SustainableConstruction.pdf.

16. Ibid.

17. James M. Morgan and Jeffrey K. Liker, *The Toyota Product Development System: Integrating People, Process, and Technology* (New York: Productivity Press, 2006), 3–4.

18. Ibid.

19. "Success Means More than Being Dependable," *ENR Magazine*, March 5, 2007, 29.

20. Don Tapscott and Anthony D. Williams, *Wikinomics: How Mass Collaboration Changes Everything* (New York: Portfolio, 2006), 20.

21. Ibid., 10.

22. Ibid., 46.

23. Neil Howe and William Strauss, *Millennials Rising: The Next Great Generation* (New York: Vintage Books, 2000), 4–6.

GLOSSARY

A

Agency CM Under this model the construction manager acts as an agent or advisor to the owner and provides pre-construction and oversight services for a fee. It is a service added to a project delivery method such as multiple prime or design-bid-build. The CM agent or advisor is not liable for either the design or the construction.

Alliance Contract A contractual arrangement intended to create an alignment between all the primary participants of the project which typically includes the owner, contractor, and architect, but may include others as well. The intent is to align the parties' cultures, behaviors, systems, and individual commitment in order to achieve superior results on a project. The alliance contract typically includes a shared savings-loss feature intended to incentivize the team's behavior and project performance.

Alliance Contracting A cooperative contracting model that is characterized by a collaborative process establishing an alignment of values, goals, and objectives among the project owners, designers, and contractors. This contracting approach is meant to promote openness, trust, and risk sharing among the project participants.

Alternative Project Delivery Any project delivery method that is not design-bid-build. Design-build and CM-at-risk are examples of alternative project delivery methods.

Annotated Drawings Project drawings that reflect the adjustments and modifications made to the proposal design immediately after contract award at the design validation stage.

Award Celebration An organized and intentional acknowledgment and celebration of the whole team for putting together a successful proposal resulting in a contract award for a project.

Award Fee An incentive mechanism applied to target cost contracts that triggers a financial reward when the design-builder meets or exceeds certain performance standards set forth at the onset of the contract. It is typically used to incentivize noncost performance items such as safety, schedule, communication, and customer satisfaction.

B

Background of Relatedness The idea that a diverse group of individuals will also share an array of common interests, aspirations, goals, and objectives.

Best and Final Offer (BAFO) A term used to describe the final revised proposal submissions in an Equivalent Design/Low-bid design-build competition. Also referred to as *final revised proposals*.

Best Practice Capture Log A tool used to document the most effective methods, techniques, and means associated with pursuing and managing a design-build project.

Best Value An approach that considers both quantitative (price) and qualitative (design, schedule, past performance, management plan, etc.) factors to deliver a project solution that meets or exceeds the owner's expectations.

Bridging A process in which the owner directly hires a designer (bridging architect or engineer), independent of the design-build team, to develop the project design prescriptively as part of the requirements of the RFP.

Building Information Modeling (BIM) A software tool that generates and manages building data such as spatial relationships, building specifications, geographic data, and the quantities and qualities of the individual building components. This data is used to produce a three-dimensional, dynamic, real-time model that can be utilized during both the design and construction phases to preview constructability challenges and meet budget and scheduling goals.

Business Enabler An individual or a team whose contribution helps their clients improve their overall business performance by helping them achieve their goals and objectives in a more efficient manner.

C

Cascading Communication Following a meeting, team members report outcomes and decisions to their departments via phone or personal communication. This reporting methods gives others a chance to ask questions, gain understanding, and enhance their ability to communicate that understanding to others.

Change Order The way in which work can be added or removed from the original scope of work agreed upon in the contract. In typical delivery systems these changes will alter the original estimate and schedule.

CM-at-Risk Performed by contractors who can offer pre-construction services. The contract has two parts. Part A includes pre-construction services that conclude when the design is progressed to the point where the project price can be established and negotiated. Part B includes the final cost estimate and the traditional construction services to build the project.

Co-Leadership A leadership arrangement that embraces differing frames of reference and is based upon mutual respect and trust where more than one person is accountable for directing the (design-build) team.

Co-Location A space or office arrangement that has contractor and designer members of the design-build team stationed in close proximity to one another on a temporary basis in order to work closely together on a specific project.

Commitment Management System A formal process for tracking the promises and commitments made between members on a design-build team.

Committed Listening Listening intently to another person to make sure that you clearly understand what they mean by the words they are speaking without judging, assessing, or negating what they are saying.

Committed Openness The intentional practice of sharing concerns, admitting weaknesses, acknowledging mistakes, unveiling strengths, and revealing feelings in an open honest way in order to build trust in a relationship.

Committed Speaking Speaking with intention and purpose, using words specifically to clearly convey your exact meaning.

Competitive Range A term that describes those proposals that are reviewed and found to contain no fatal errors or deficiencies and therefore deemed responsive in certain low-bid types of design-build competitions.

Computer Aided Drafting (CAD) A computer technology that aids in drafting part or all of a project. It is a combination of a visual drawing and a symbol-based method of communication.

Conceptual Design A preliminary stage of the design process that addresses the owner's initial program and focuses on the broader, overall design parameters such as total square footage, number of stories, and building mass and configuration. Conceptual design typically includes diagrammatical sketches and drawings without detail, outline specifications, and other documentation that illustrates the scale and relationship of the project.

Conceptual Estimating Systems-based estimating that utilizes parametric formulas to determine project costs when design is less than 100 percent complete.

Constructability Reviews Design reviews typically performed by contractors on a design-build project to consider how the design may impact construction means, methods, productivity, fabrication, schedule, cost, etc.

Construction Users Roundtable (CURT) A nonprofit organization whose primary goal is to strive towards continued improvements and changes in the construction industry.

Contingency An amount of money added to an estimate to account for various unknown factors at the time the estimate is initially developed. There are construction contingencies, design contingencies, and owner contingencies. As the design develops, the contingencies generally decrease.

Contingency Round Table A model for conducting a joint risk assessment and contingency development session where the various disciplines and associated parties to the project can voice their concerns for the project from their subjective perspectives and agree to realistic contingency amounts to include in the estimate.

Contract Payment Structure Payment methodology decided upon during the procurement process when the owner chooses the type of contract to be used. See: Lump Sum, Cost-Plus, Guaranteed Maximum Price, Unit Price.

Contractor-Led Design-Build Most prevalent design-build entity model where the construction firm serves as the prime contractor with the owner and the contractor hires an independent design firm to provide design services

Controlling Party The party holding the prime contract position in a multiple firm design-build contractual relationship. Also referred to as *managing entity* or *controlling entity*.

Cost Model A framework for identifying and quantifying all project costs relative to a conceptual design in a systematic fashion. Once the model is completed it is used to guide the completion of the design by setting the boundaries and limitations for costs and quantities originally defined, allocated, and proportioned for the project.

Cost Model Briefing Session A formal meeting that takes place with all members of the design-build team shortly after the model is created to explain how the cost model will be used to manage the design to budget throughout the entire design process.

Cost-Plus Contract payment structure where the design-builder is reimbursed for actual cost incurred to perform the design and construction services plus an agreed upon fixed fee or percentage.

Cultural Fit Compatibility between the personality of the firms or individuals working together. Firms should be compatible in terms of their ways of doing business, values, attitudes, and behaviors.

D

Design-Bid-Build A project delivery method in which the owner holds and is responsible for managing two separate contracts for design and construction. The process is linear and each function typically completes before the next one begins.

Design-Build A project delivery method in which design and construction services are provided under a single contract using a team approach to deliver an integrated project solution, which is in contrast to the two-contract method of design-bid-build in which designers and contractors deliver their independent services under separate contracts.

Design-Build Charrette A collaborative design session in which multiple disciplines associated with the design-build project gather together in the same space to conduct a very engaging brainstorming event to flesh out multiple design options quickly. Ultimately, the multi-disciplined design-build team will emerge with one design concept on which it will build its response to the RFP.

Design-Build Entity An organizational unit comprised of design professionals and construction professionals paired together to provide design and construction services. The DB entity may be organizationally configured in several different ways.

Design-Build Institute of America (DBIA) A unique interdisciplinary organization comprised of architects, engineers, contractors, public and private owners, specialty contractors, vendors, university students and faculty, and many others affiliated with the AEC industry.

Design-Build Presentation A design-build team puts forth their proposal to be evaluated by a jury or panel of qualified individuals who represent the interests of the client. The presentation is part of the competitive design-build process.

Design-Build Plus A term used when services are added to the design-build contract beyond design and construction, such as financing, operations, or maintenance.

Design-Build Project Manager (DB-PM) A unique title that distinguishes the design-build project manager role from the conventional construction project manager role. The DB-PM is responsible for managing the design and construction of a project as an integrated process. Also referred to as a *design-build process manager*.

Design Commitment Design commitment is made when a design element is verified to comply with the requirements of the RFP by the designer-of-record and has been priced and confirmed "within budget." Only then can purchasing decisions be made and construction planned.

Design-Construct Coordinator This person is often responsible for managing both the design-cost interface and the design-construct interface tasks.

Design-Construct Interface The relationship between the design and carious aspects of construction, such as the availability of quality labor, fabrication efficiencies, schedule constraints, timely delivery of materials, and other constructability considerations.

Design-Cost Interface An aspect of the design-build management process associated with reconciling the design with the budget throughout the design phase of the project. Management task primarily applied during the design-development stage.

Design-Cost Reconciliation A design-build management task associated with the design-cost interface that entails reviewing the various elements of the design as it progresses and confirming that it continues to stay within budget and within the original scope.

Design Evolution Log A document that tracks the changes in the design throughout the design-build process and the logic associated with them.

Design Impact Risks These are the types of risks that influence the design direction and final design details, things like loading requirements, environmental studies, seismic issues, regulatory restrictions, and others.

Design Manager Also called a *design coordinator*, this person is responsible for managing the design work flow, scope maintenance, and seeing that the target design schedule is met. The design manager may also be responsible for managing the design to budget.

Design Parameters Information that sets the boundaries of the design and guides the direction relative to scope, size, cost, quantities and any other types of constraints and limitations created by the owner's program and economic requirements or the design-build team to help manage the design process.

Design-Performance Interface The relationship between all of the elements of the design and the performance requirements for the project.

Design Process A series of information inputs and outputs that result in a completed project design.

Design Specifications Detailed descriptions of how a product or system must be manufactured and installed or constructed. Also includes identification of specific products, materials, manufacturers, and model numbers. Similar to a prescriptive specification.

Design Stability Refers to the state of completeness of any given design element at time of pricing. Considers what the likelihood is that the design will change between the conceptual stage and the construction-document stage of design. The more likely it is that the design will change, the less stable the design is.

Design Validation The process of reconciling the proposal design with the owner's expectations. Design validation meetings should occur shortly after contract award before moving ahead with any further design efforts.

Design Work Breakdown Structure An organizational chart in which the critical design work elements, called design tasks, of a project are illustrated to portray their relationships to each other and to the project as a whole.

Designer Led Design-Build The design firm holds the prime contract with the owner and hires out the construction services to a general contractor.

Designer of Record The DOR is the designer who stamps the drawings and is professionally liable for the design. In design-build this individual is also responsible for managing the design-performance interface and confirming that all design details comply with the requirements spelled out in the RFP and proposal. This person may also be responsible for design coordination issues such as de-conflicting all drawings.

Diagnosis Bias As human beings we have a tendency to label people, ideas, and things based upon our first impressions. These biases can influence how we behave and interact with one another, even when there is evidence that contradicts our initial opinions or opportunities to see a person, situation, or thing differently.

Discussions One-on-one meetings between the proposal reviewers and the design-build team regarding deficiencies, mistakes, errors, misinterpretations, and needed clarifications in the technical aspects of their submission.

Draw-Build A variation of design-build in which the owner has completed the design to such an advanced degree that the design-builder's role is limited to completion of the detailed construction drawings and specifications and construction, thus diminishing the opportunity to impact the design and add value.

E

Early Release Package A bidding package or scope of work that includes all necessary design and specification criteria needed to secure pricing and contract commitment earlier than would be normally achievable under the design-bid-build model.

Emotional Intelligence (EI) Involves four basic skills: self-awareness, self-management, social awareness, and relationship management. EI is usually measured as an Emotional Intelligence Quotient (EQ).

Engineer's Estimate This estimate is usually prepared by someone on the owner's staff or by an independent consultant working for the owner or designer. The purpose of the estimate is to check for initial project feasibility and establish a budget. The timing on this estimate is usually many months and even years before the actual bidding occurs. Also may be referred to as *architect's estimate*.

Enhancement Criterion An evaluation item associated with a design-build RFP involving some component of a project design that the owner wishes to see improved or enhanced.

Environment of Trust A working environment in which there is an intentional commitment to practice trusting as part of the dynamic relationship associated with a design-build engagement.

Estimating Churn The progressive estimating process that tracks the design from the schematic stage through design development and finally to construction drawings.

Evolutionary Specification Development Approach Utilized to develop the project requirements. The owner begins the process by preparing the general functional descriptors for the project, followed by the more results-oriented performance requirements associated with the various project systems. The design-builder then develops design specifications to meet the performance requirements.

F

Full Disclosure The act of revealing all relevant information about a design-build project and the processes by which it is estimated, designed, and constructed.

Functional Requirements Broad descriptions of what a facility or structure needs to be and how it will be used. Also includes an expression of project needs, goals, constraints, or challenges.

G

Gainshare The amount of money saved on the project in an alliance contract, when the actual project costs are less than the targeted costs. This savings is shared among the alliance participants.

Go-or-no-go Decision An analytical approach to determine whether to pursue a project or not based upon information provided in the RFQ and RFP documents and other project circumstances and conditions. Also referred to as *pass or play decision*.

Guaranteed Maximum Price (GMP) Combines features of both the cost-plus and lump-sum contract methods. The design-builder is reimbursed for actual cost incurred to perform the design and construction services, plus an agreed upon fixed fee or percentage up to a guaranteed maximum price.

H

Hard Bid A contract method whereby contractors submit bids to owners based on the design team's plans and specifications and the owner awards the contract to the contractor with the lowest price. No other criteria are considered.

High Performance Contract A contract that results in lower costs, faster performance, higher quality, less litigation, and more satisfied customers.

I

Information Interviews These are conversations that are conducted in a question-asking interview format where the intention is to gain specific insight, knowledge, and understanding.

Information Outputs These are the design deliverables associated with a design-build project work plan. They may include the drawings and specifications but are not limited to them. In addition to drawings and specifications, information outputs may include narratives, reports, calculations, memos, sketches, or any other form of documented communication that supports both the design and construction processes. Design information may be released to the construction team before it is incorporated into a formal drawing or specification.

Integrated Design-Build Firm Also referred to as *full service design-build firm*. This is a multidisciplined firm that has its own in-house designers and contractors. Other disciplines may also be associated with the firm, such as planners and interior designers.

Integrated Partnering Program A teaming process that focuses on building trust and encouraging collaboration across an interdisciplinary project team through a formal set of practices to build productive working relationships resulting in improved results.

Integrated Project Delivery A project delivery model that emphasizes the integration of design and construction personnel, along with owner representatives, at the earliest stages of the project initiation to maximize efficiency and reduce waste through all phases of design, fabrication, and construction.

Integrated Project Leader (IPL) A title intended to clearly distinguish the primary accountability associated with leading a design-build team. Although every organization is different, an IPL may take the place of a traditional project manager in design-build, or the project manager may be subordinate to the Integrated Project Leader. Also referred to as *integrated team leader*.

Integration Manager A recommended title describing the distinct accountability associated with managing the team relationships and communications to ensure high levels of collaboration, cooperation, and integration. This person may be responsible for more than one project team at a time.

Interdisciplinary Fluency The ability to think across multiple disciplines when addressing problems and not just from one's own single discipline perspective.

Interoperability The ability of a system or a product to work with other systems or products without special effort on the part of the user.

Invitation for Bids (IFB) The most common solicitation instrument in design-bid-build, in which the contract is usually awarded to the contractor with the lowest bid.

J

Joint Risk Assessment A multidisciplined, comprehensive risk identification, evaluation, and analysis session that involves representatives from the entire design-build team.

K

Kaizen A philosophy that, when applied to business, means the continual improvement of all functions associated with a process to improve production and eliminate waste.

Knowledge-in-Action Applying one's knowledge or know-how to a problem or situation without giving it much thought as you are engaged in the problem or situation. Common sense is a kind of knowledge in action that requires very little thinking in advance of using it in a situation.

L

Leadership in Energy and Environmental Design (LEED) Rating system developed by the U.S. Green Building Council that provides a group of standards for environmentally sustainable construction.

Lean Construction This type of construction is focused on the continuous improvement of all aspects of the built environment in order to provide maximum value with minimum cost while meeting the customer's needs.

Low-bid Design-build Any design-build procurement process that stipulates "price" as the predominant evaluation criterion and selection determinant.

Lump Sum Contract payment structure where the design-builder agrees to complete the design and construction services as spelled out in the contract for a fixed price regardless of how much it actually costs to perform these services.

M

Managing Risk Process of assessing and mitigating uncertainties associated with the design and construction of a project.

Master Builder Historically, this term was applied to an individual who was responsible for the design, construction, and all other related services necessary to deliver a project.

Mental Model An established way of thinking and looking at the world that directs our behavior and actions.

Multiple Prime Project Delivery Method Owner has a separate and distinct contract with the architect for traditional services but opts to forego the services of a general contractor in favor of contracting directly with each specialty-trade contractor needed to complete the project.

N

Non-linear Innovation Innovation that goes beyond predictable improvement to encompass whole new ways of doing business or activities.

O

One-on-One Briefing One-on-one meetings conducted by the design-build project manager with individual team members to clarify expectations and communicate how vital the team member's role is in achieving success on the project.

Open Book Policy A philosophy that encourages transparency regarding all costs associated with a design-build project.

Open Standards Nonproprietary formats and databases that support the exchange or joint use of various types of information by differing software tools.

Outline Specifications A brief narrative outline corresponding to the design-build estimate describing the range of materials, products, options and quality that each cost category in the estimate represents. Also called a ***design narrative*** or ***project scope definition***.

P

Painshare When the actual project costs are more than the targeted costs in an alliance contract arrangement, the losses are shared among the alliance participants.

Partnering Program (Teaming Session) A structured process of collaborative activities and procedures that helps build stronger team relationships and trust. It allows groups to achieve measurable results through productive work interactions. This process provides structure for teams to establish a mission by using common goals and shared objectives.

Parametric Formulas These are derived from data collected from past projects, and they help an estimator quantify and estimate certain elements of the design before the design details have been worked out. Some are informal rules of thumb, and others are more elaborate.

Performance Criteria The means and measures used to verify design and construction performance for compliance with the stipulated performance requirements in an RFP. Performance criteria are comprised of industry and/or manufacturers' standards, tests and testing, calculations, on-site mock-ups, and any other methods used to measure performance.

Performance Requirement A design requirement expressed in terms of an expected outcome or final result as opposed to a prescriptive specification that dictates specific products, manufacturers, and techniques.

Performance Standard A written expression of the performance threshold(s), requirement(s), or expectation(s) that must be met to be appraised at a particular level of performance. In order to be useful, the performance standard should identify what methods of measurement and/or tests will be used to verify compliance to the performance standard.

Platform Skills Specific publicspeaking and presentation techniques and skills designed to improve the effectiveness of professional communications, presentations, and interviews.

PreConstruction Services Usually entails services such as design and drawing reviews, scope-definition assistance, preliminary and schematic budgeting, constructability reviews, and value engineering. Early contractor selection allows for these services to be incorporated into the design-build process.

Preemptive Marketing An intentional and directed effort to target specific projects and clients in advance of an RFP release or procurement effort on the part of the client. The directed effort entails specific project preparation and due diligence activities to gain information, knowledge, and insight about the client and the project. Some companies refer to this as "pre-winning a project."

Prescriptive Specifications The traditional method for stipulating materials, products, manufacturers, and techniques in design-bid-build. These stipulations are detailed and explicit.

Price Proposal The portion of the design-build proposal that includes the price for the design and the construction of the project. The price may or may not be a part of the RFP request.

Prime Contract The entity that holds the contract has the overall responsibility for managing the design and the construction of a project.

Procurement Acquisition of goods and services.

Procurement Workshop An organized (usually public) briefing session in advance of an official RFP release, where design-builders are invited by the owner to voice their opinions about an upcoming project.

Programming The decision-making process and research that identifies the scope of work and the functional and operational requirements of a building project.

Progressive Design-Build A qualifications-based design-build procurement method where the design-builder is selected primarily based on qualifications and then the design-builder and the client enter into a progressive two-part contract that is implemented in stages.

Progressive Estimating Describes the design-build estimating process. The estimating task begins at the earliest stages of design with feasibility or concept estimating, and progresses as the design develops through the schematic estimating stage, the design-development estimating stage, and finally the detailed estimating stage.

Project Characteristics Workshop A meeting attended by all primary DB team members where participants discuss all aspects of a project, including risks, concerns, characteristics, challenges, etc., and everyone gets to hear the different perspectives.

Project Debriefs A process in which the entire DB team reviews project performance at the end of the job and makes note of what worked well and what did not. Lessons learned and best practices are noted and captured for future use.

Project Delivery The comprehensive, start-to-finish process for organizing, implementing, executing, and completing the design, construction, and start-up of a building, facility, or structure.

Project Director (Project Executive or Project Advocate) This person usually comes from the business-development or sales side of the organization and selects the design-build projects that the firm will pursue. He or she usually initiates the proposal-development process and continues to participate in a leadership role on the project after contract award.

Project Enhancement Approach A hybrid evaluation method that has the design-builders compete on the basis of the enhancements that they propose to the bridging design included in the RFP and overall proposal price. The total cost is divided by the total enhancement points, called quality points. The team with the lowest cost per quality point wins the competition.

Proposal Response Team The group of contractors, designers, and consultants who put together the proposal package in response to a project RFP. These same individuals usually stay with the project after contract award if they win the job.

Q

Qualifications Based Selection (QBS) A selection methodology traditionally used by owners to select architects and engineers based on the design professionals' qualifications in relation to the project. In some states, procurement laws now allow the QBS method to be used to select contractors as well. Price is not a competitive criterion in a qualifications based selection method.

Qualifications Package A proposal developed by a design-build team in response to a Request for Qualifications that identifies and highlights the unique qualifications and characteristics of the design-build entity in regard to their experience, past performance, and capabilities as they relate to the specific project in question.

Quantity Take-Off Activity performed during the estimating process involving counting the number of material and work items needed to complete a construction project. The material and work items are then priced to formulate a bid.

Question Thinking A term that describes a set of tools for transforming thinking, action, and results through skillful question asking.

R

Redundant Contingencies When duplicate contingencies are included in a design-build estimate. The desire is to prevent redundant contingencies.

Reflecting-in-Action The act of thinking about what you are doing as you are doing it. Through observation and engagement, one's thinking may adjust and coordinate with what is happening in any given situation in order to come up with the most appropriate action.

Relational Contracting A contracting methodology that focuses on the relationship between the parties to the contract, as well as the transactional aspects of the contract, such as products and services. Responsibilities, risks, and benefits are meant to be apportioned fairly and transparently, establishing a basis for trust and cooperativeness.

Request for Information (RFI) Used in the construction industry when it's necessary to confirm interpretation of details or specifications from the architect or client.

Request for Proposal (RFP) The solicitation document that an owner releases to elicit design and construction services in a design-build procurement process. The document typically becomes a component of the initial contract.

Request for Qualifications (RFQ) Used to gather information about a project team. It can produce a short-listed group of candidates to be considered for a project.

Responsibility Matrix A tool used to outline the various design and construction tasks necessary to prepare the proposal and perform the contract obligations post-award, along with assigning accountability for each task and a due date for completion.

Risk Assessment The determination of both quantitative and qualitative impacts associated with various known and unknown factors affecting the design and construction of a project.

Risk Mitigation Plan A structured approach to managing the uncertainty associated with a design-build project through calculated decision-making and management strategies.

Risk Register A document that tracks the decisions made in relation to the identified tasks of a design-build project. It is a dynamic tool designed to capture the logic associated with the decisions made.

S

Scope Used in project management to describe the totality of work involved in completing a project. The main tool for describing scope is the work breakdown structure.

Seamless Team Used to describe a fully integrated team and its ability to perform and communicate in a holistic, comprehensive fashion to respond to the project and its solution as a whole unit.

Servant Leadership A leadership philosophy, first articulated by Robert Greenleaf in the 1970s, that focuses attention on serving others and building trust.

Shortlisting A method used in design-build procurement to narrow the field of potential RFP responders to only the highest-ranked three or four teams based on qualifications submitted in response to a Request for Qualifications.

Signature Architect An architect of renowned stature with significant name recognition and notoriety.

Single-source Responsibility A procurement concept that places the responsibility for both the design and the construction of a project with one party under a single contract.

Site Analysis Gathering of basic data about a specific project site. This data can involve location, surrounding neighborhoods, human-made and natural features, zoning laws, and climate considerations.

Source Selection A formal process employed in the procurement of design and construction services. It entails making decisions about the form of solicitation, the evaluation and selection process, and the contracting method.

Standard of Care and Warranties Under a separate contracts model, designers are professionally liable to uphold a generally accepted standard of care relative to the design they deliver to the owner, and the contractor is required to warrant the construction work. In design-build, the design-builder is required to warrant the construction work as well as the design developed under the direction of the contractor post award.

Stipend (Honorarium) A stipulated amount of money paid to the unsuccessful proposers by the project owner in a design-build competition. The money is intended to help defray proposal development costs.

Strategic DB-Teaming Initiative A plan to identify teaming partners in advance of a project pursuit.

Strategic Partners Individuals who are identified as having potentially significant influence on a project's outcome. These team members often represent third-party entities, such as end user organizations, special permitting agencies, code enforcement offices, political associations, community groups, and others. It is recommended that these strategic partners be engaged as early in the design-build process as possible.

Subcontractor-Vendor Workshop A planning session involving the design-build team along with its subcontractor and vendor partners, specifically designed to evaluate and assess various procurement options as they relate to the construction needs and design constraints associated with a project.

Submittal Shop drawings, samples, and material data used mainly by engineers and designers to ensure that the correct products are used on the project.

Surrogate Owner When a team does not have access to the actual project owner, it may initiate contact with a substitute or alternate owner who is in the same or similar business in order to gain information and insight into the needs, goals, and objectives of its actual client.

Systems Thinker Someone who can think through solutions from a broad perspective while considering many viewpoints at the same time and has the ability to leverage information and options to come up with a resolution that serves multiple interests.

T

Target Cost Contract An incentivized cost-reimbursement-type contract that sets a target price as a goal for project performance and holds both the design-builder and the owner responsible for cost overruns and savings.

Target Design Schedule A schedule for design delivery that has been systematically coordinated and aligned with the procurement and construction needs of a project through joint discussions and agreements made with various members of the design-build team.

Target Price An estimated cost that has been established jointly by a project team and that serves as a benchmark for project cost performance.

Team Alignment The compatibility of personal and/or corporate values, beliefs, and principles as well as project goals and objectives.

Team Integration The state at which a design-build team moves beyond the normal sharing of information and cooperation associated with a project and into a synergistic, highly engaged collaboration where integrated solutions are discovered through the sharing of knowledge, expertise, and experience.

Team Readiness Assessment A formalized process that profiles team member compatibility, alignment, and fit. It consists of both individual and team assessments accompanied by a proposal briefing session.

Teaming Agreement A vehicle for structuring the relationship between the various team members associated with a design-build project pursuit. The agreement establishes and defines the roles, responsibilities, and expected performance of the team members.

Teaming Partners Various companies brought together to complete all aspects of a project. In design-build it is important to identify companies as potential partners by looking for companies that have a reputation for technical knowledge along with cultural fit.

Teaming Profile A formalized process that profiles team member compatibility, alignment, and fit. It can include both individual and team assessments accompanied by a proposal briefing session.

Technical Leveling A technique utilized in the Equivalent Design/Low Bid evaluation method whereby the technical components of the design-build proposals are reviewed by the owner's evaluation team in an effort to equalize or level the competitive playing field.

Technical Proposal The portion of a design-build proposal that contains the non-price evaluation components such as the design. Other non-price components may also be requested, such as the project schedule, quality assurance plan, or overall management approach.

Technical Review Team A group of individuals who are uniquely qualified, based on their background and experience, to provide a review of one or more elements of the project at various stages of design development. The members of this review committee are typically not associated directly with the project.

Technical Transfusion Disclosing technical information from one design-build proposal during discussions that results in the improvement of a competing proposal.

Trend Management A design-build management technique that tracks and documents cost fluctuations between milestone estimates as the project design progresses in an effort to keep the project on target with the budget.

Two-Envelope Proposal A proposal process that requires that the technical components of a design-build proposal be submitted separately (as in a separate envelope) from the price component of the same proposal, with the price component remaining sealed until all of the technical factors have been evaluated and scored.

Two-Phase Procurement A selection process that begins with a shortlisting procedure conducted through a qualifications evaluation process, and concludes with a best value assessment requiring the preparation and submission of a complete design-build proposal from each of the shortlisted teams. Also referred to as *two-stage procurement*.

U

Unifying Mechanism Any activity, device, technique, symbol, slogan, or tool that serves to facilitate and encourage team cohesion.

V

Value Engineering A method for evaluating the cost effectiveness and value of the various methods, materials, and means by which we design and construct projects. The analysis considers the initial cost, maintenance cost, energy usage cost, replacement cost, and life expectancy.

W

Weighted Criteria Matrix A tool used to assess a best-value selection in which evaluation criteria and weights are displayed and scores are distributed to determine the winner in a design-build competition.

Win Strategy An organized, well-thought-out plan that focuses on highlighting the design-build team's strengths, competencies, and capabilities relative to the client's specific goals and objectives for the project.

Work Breakdown Structure A tool used to define and group a project's discrete work activities in a way that helps organize and define the total scope of work of a project.

Z

Zero Dollar Change Order A document used to modify the project's contract based on design or scope changes made at the design validation stage. Zero dollar change orders so not result in an increase in the contract price.

CONTRIBUTORS

John Alley, DBIA
Vice President
Layton Construction Co., Inc.
Sandy, Utah

John Alley was the first DBIA-designated constructor in the state of Utah. His career has been focused on preconstruction services using integrated design-build delivery methods. Mr. Alley is currently Vice President of Layton Construction Co., Inc., and has more than twenty-three years of construction experience. He has been part of a number of projects, including 201 Commerce Center, a 110-acre master-planned industrial park; River Park, an 89-acre master-planned Class A office space; and the Westminster College Performing Arts Center. Mr. Alley holds a B.S. in construction management from Brigham Young University, as well as a B.S. and an A.S. in building construction from Salt Lake Technical College. He currently serves as chair of the Utah chapter of the Rocky Mountain Region Design-Build Association. Mr. Alley also serves as chair of the Brigham Young University Construction Management Industry Advisory Committee.

Jeffrey L. Beard
ACEC Vice President
Director, Institute for Business Management
American Council of Engineering Companies
Washington, D.C.

Jeffrey L. Beard is Vice President of the Education Division of the American Council of Engineering Companies (ACEC), Washington, D.C. He recently conceived and implemented RCEP.net, a nationwide portal and educational management system (in partnership with NCEES) for engineers and related professionals. RCEP.net honors accredited continuing education programs (for professional licensure and other purposes) and saves the PDHs/CEUs of course participants in a secure database. In addition, Mr. Beard oversees ACEC's leadership, management, and business education programs, including the award-winning Senior Executives Institute (provided by ACEC and the Brookings Institution). Previous to his work at ACEC, Mr. Beard was president of DBIA, where he served from DBIA's founding in 1993 until 2003. He holds a bachelor's degree from Rutgers University, a Master of Planning from the University of Virginia, and is a doctoral candidate at the Georgia Institute of Technology, with a major in integrated facilities delivery and a minor in the history of technology.

Peter Beck
Managing Director
The Beck Group
Dallas, Texas

Henry C. (Peter) Beck III has been associated with The Beck Group and its affiliates since 1978. During this period, he has held various positions within the firm, and has served as Managing Director since 1991. The Beck Group is an integrated builder offering project finance, design, construction, and development services through nine offices across the United States and Mexico. The firm completed transactions in excess of $707 million in 2007. Among other distinctions, The Beck Group has been recognized as one of the "100 Best Companies to Work For" by *Fortune* magazine. Mr. Beck's mission is to integrate the building disciplines through unique processes and technologies, resulting in order-of-magnitude improvements in design, cost, and schedule. He received his B.S. in civil engineering from Princeton University in 1977 and his M.B.A. from the Stanford Graduate School of Business in 1981. Mr. Beck currently serves on the Board of Trustees of the Southwestern Medical Foundation, The Dallas Foundation, Dallas Citizens Council, Design Futures Council, and Stanford GSB Management Board. He also chairs the Civil and Environmental Engineering Advisory Council at Princeton University. He has formerly served on the board of the Greater Dallas Chamber (which he chaired in 2001), the Texas Parks & Wildlife Commission, and the Stanford Business School Trust.

Erin Brozovich
Project Engineer
CDM Constructors Inc.
Rancho Cucamonga, California

Erin Brozovich is a Project Engineer at CDM, an integrated consulting, engineering, and construction firm. She attended Ohio State University and graduated with a civil engineering degree in 2003. She began working in the environmental consulting division at CDM in 2004. After two years of engineering design, Ms. Brozovich transferred to CDM's construction division to work on a $250-million design-build program for the government. She has helped to manage various water and wastewater design-build projects, including pump stations, well replacements, pipe line rehabilitation, and treatment plant upgrades. She continues to assist with managing design-build projects on Camp Pendleton and is working toward becoming a project manager.

Pat Burns, PE (Brig. Gen., United States Air Force, retired)
Vice President
Federal Contracting Group
Mortenson Construction
Alexandria, Virginia

Since 2005, Pat Burns has been employed by Mortenson Construction as a vice president in their Federal Contracting Group. Mortenson is the 28th-largest general contractor in the United States, and is well known for complex construction such as the Walt Disney Concert Hall, Denver Coors Field, and, most recently, the Denver Metropolitan Museum of Art. Mr. Burns is responsible for Mortenson's federal program project pursuits, which accounted for $350 million in company revenue in 2007. He retired from the United States Air Force in 2005, where, from 2001, he served as the Director of Installations, Headquarters Air Combat Command, Langley Air Force Base, Virginia. He was responsible for engineering programs totaling $2.4 billion annually. He served in a similar capacity in charge of engineering for all Air Force bases in the Pacific from 1999 to 2001. He has received the Air Force Distinguished Service Medal from the President of the United States for his achievements at Air Combat Command. Buildings magazine named his organization one of the "Top Twenty-Five Innovative Building Organizations in the U.S. for 2004," the first time that a military engineering organization achieved that recognition. Engineering News-Record magazine named him one of the "Top 25 Engineer Newsmakers for 2005." And he was named the "2005 Outstanding Eminent Engineer" by the University of Wyoming College of Engineering.

Michael Cebulla, LEED AP
Superintendent
Ryan Companies US, Inc.
Minneapolis, Minnesota

Michael Cebulla approaches each new project anticipating its success and his customer's satisfaction. As a superintendent, Mr. Cebulla oversees the entire construction process, paying particular attention to budget, schedule, and safety issues. He is responsible for monitoring, mobilizing, and managing the Ryan field team while coordinating subcontractor activities. He excels at building a project team and a work force, fostering communication and collaboration in delivering a product of the highest quality. His dependability, people skills, and a genuine pride in his work are just a few of the reasons his customers are continually pleased with his results. He has been involved with a number of projects in Minnesota, including: Lexus of Maplewood, a 10,000 square-foot dealership; Lexus of Wayzata, a 75,000 square-foot dealership; and The Aberdeen, a 105,000 square-foot condominium complex and parking garage. He is also a LEED-certified professional.

Dave S. Crawford, PE, DBIA
President and Chief Operating Officer
Sundt Construction, Inc.
Tempe, Arizona

President and Chief Operating Officer of Sundt Construction, Inc. and member of the Board of Directors of the Sundt Companies and Sundt Construction, Inc., Dave S. Crawford is responsible for the acquisition, administration, and execution of work. Sundt is active in the civil, commercial, military, institutional, educational, residential, hotel, medical, mining, industrial, and manufacturing markets of the United States, primarily the southwestern and western states. Mr. Crawford has been in the construction industry and affiliated with the Sundt family of companies since 1968. He served as Executive Vice President and Manager of Operations and Vice President and Arizona District Manager, and he was elected to the Board of Directors in 1992. Mr. Crawford was an active participant in drafting, lobbying, educating, and passing legislation permitting alternate project-delivery methods and qualifications-based selection for public construction in the state of Arizona. He is also actively involved in this effort in other states and on a national basis. Mr. Crawford received a B.S. in Civil Engineering from the University of Arizona. He is a member of a number of committees and organizations, including The Arizona Builders' Alliance (Government Relations Committee), The American Concrete Institute ("Tolerances" Committee), the Design-Build Institute of America, the Alliance for Construction Excellence (Chairman of Legislative Committee), and the Associated General Contractors of America (Project Delivery Committee).

Jeffrey Cupka, AIA
Design Principal
Ryan Companies US, Inc.
Minneapolis, Minnesota

Jeffrey Cupka directs the design effort for Ryan's integrated full-service Architecture & Engineering team. He is responsible for design leadership, direction, quality, and coordination of the entire team. With more than twenty-five years of experience, Mr. Cupka has particular expertise in understanding how to be an effective participant in a design-build team and how all design disciplines support the overall goals of any given project. His customers appreciate his ability to take their goals, business needs, and aesthetic tastes and transform them into beautiful, functional designs that reflect their organizations. He has an expansive portfolio and has worked on many diverse types of projects, including: Retail, Office, Medical Office/Clinic, Industrial, Manufacturing, Tenant Space Planning, Speculative Projects, and Distribution Facilities. He also has extensive experience in Site Planning, Multi-use Business Parks Planning, Master Planning, Urban Planning, and Branding Exercises. Mr. Cupka led Ryan Companies' initiative to introduce and utilize Revit Software (BIM) into the standard practices. He is a registered architect in Minnesota, New Hampshire, Arizona, Wisconsin, and South Dakota. He is a member of the California Center for Construction Education (CCCE) Curriculum Advisory Board at California Polytechnic State University at San Luis Obispo. Mr. Cupka is also an active member of AIA Minnesota, AIA National, DBIA Midwest, Urban Land Institute, and NAIOP. He received a degree in architecture from the University of Minnesota.

Matt Dahlberg, AIA, DBIA, LEED AP
Architect
The Haskell Company
Jacksonville, Florida

Matt Dahlberg joined Haskell in August 2004 as a construction assistant project manager. He worked on several education-facility projects before transferring to the architect/engineering group in October 2005. As an associate architect, he has been a part of many successful projects, from campus renovations to new university housing. He has supplemented his education by becoming a LEED Accredited Professional and a DBIA Designated Design-Build Professional. His projects include the Osprey Fountains Student Housing, Titusville High School renovations/additions, and the Astronaut High School renovations/additions. Mr. Dahlberg received a Bachelor of Architecture from California Polytechnic State University at San Luis Obispo, where he minored in construction management and environmental design. He is a member of the United States Green Building Council (USGBC), North Florida Emerging Green Builders, the American Institute of Architects, and the Design-Build Institute of America (DBIA).

Ryan DelaRiva, LEED AP
Field Engineer
Pankow Builders, Ltd.
Honolulu, Hawaii

Ryan DelaRiva is a Field Engineer at Pankow's Hawaii office, a design-build general contractor. He recently finished a project in the Central Valley of California for Pankow's San Francisco office. When Pankow built an aquatics complex for Chevron Energy Solutions and the Dinuba Unified School District, Mr. DelaRiva took an active role in the project's structural redesign at the midpoint of the project, detailing and working side by side with the structural engineer, architect, and owner. He coordinated field activities, trades, project documentation, and all Division of the State Architect inspections. Prior to the Dinuba Aquatics Complex project, he was working in Pankow's Oakland office performing preconstruction estimates and modeling future projects in Revit. Still at Pankow, Mr. DelaRiva has recently moved to Oahu, Hawaii to work at the Kaneohe Marine Corps Base on a historical housing redevelopment project. In addition to the field-engineer role, he will be taking on a more active role in the on-site client service, scope clarification for the structural renovation, refurbishment, and build-out of twenty-two historical homes affected by the Pearl Harbor attack. He attended California Polytechnic State University at San Luis Obispo and graduated in 2008 with a bachelor's degree in construction management. While in college he took an active leadership role in the Construction Management Program's clubs, competitions, and events.

Bart Dickson, LEED AP
Project Manager
Hoffman Construction Company
Portland, Oregon

Bart Dickson has more than ten years' experience in the construction industry and has been with Hoffman for nine years, performing a wide range of preconstruction and construction services. He grew up in construction. Before coming to Hoffman, he gained valuable experience first as a union demolition laborer and asbestos abatement laborer. His experience also includes building custom homes and small commercial buildings, and remodeling churches. After graduating with honors, he joined Hoffman as an engineer for a complex technology facility. Mr. Dickson's strong organizational skills and ability to form effective working relationships with architects have served him well on a variety of complex projects for public and private owners. His project experience includes Eastern Oregon University Student Housing in La Grande, Oregon; the WSU Vancouver Student Services Building in Vancouver, Washington; and the UW IMA Sports Building in Seattle, Washington. Mr. Dickson has a B.S. in construction management from Brigham Young University and a J.D. from the Lewis & Clark Law School.

Dave Engdahl, AIA, DBIA
Senior Vice President—Chief Architect (retired)
The Haskell Company
Jacksonville, Florida

Dave Engdahl recently retired as Senior Vice President and Chief Architect of The Haskell Company, a leading integrated design-build firm headquartered in Jacksonville, Florida. Mr. Engdahl has held licenses as an architect in thirty-four states and has forty-four years of experience in design and construction, of which the last twenty-eight years have been in integrated design-build delivery. He has served as project leader on over 100 commercial, institutional, and industrial design-build projects throughout the United States and has presented and written nationally on design-build project delivery. He is the author of the "Integrated Design-Build Firm" section in the AIA reference book *The Architect's Guide to Design-Build Delivery*.

Bill Flemming
President and CEO
Skanska USA Building, Inc.
Parsippany, New Jersey

Bill Flemming is a President and CEO of Skanska USA Building, Inc., where he is responsible for the company's overall management and operations of offices in Massachusetts, Connecticut, New York, and New Jersey. He also oversees the company's procurement business. Mr. Flemming has piloted Skanska's involvement in high-profile, design-build projects, including New Meadowlands Stadium, the U.S. Census Bureau Headquarters, Carilion Hospital, and the Moffitt Cancer Center. He has also been involved in design-build projects for higher education clients among other market sectors. Through Mr. Flemming's efforts, Skanska has developed and implemented a detailed construction database of system and elemental costs and has provided parametric cost-modeling methods, both of which have proven highly effective in reducing the timeframe and maximizing the accuracy of cost information. This system is employed on a company-wide basis throughout the United States. Mr. Flemming has an M.B.A. from Golden Gate University, a B.S. in construction management, and a B.S. in architectural engineering technology from the University of Cincinnati. He is a member of the International Society of Pharmaceutical Engineers (ISPE).

Joe Flores
Senior Project Manager
Director, Human Resources
The Beck Group Corporate Headquarters
Dallas, Texas

Joe Flores is Director of Human Resources at The Beck Group, a fully integrated design-build firm based in Dallas, Texas. After graduating from Texas A&M in 1992 with a Bachelors of Construction Science, he worked for three years as an estimator before beginning his career at Beck. Once at Beck, Mr. Flores moved through the ranks of Office Engineer, Project Administrator, Senior Project Manager, and Integrated Team Leader (ITL) while completing a number of projects, including the $130 million Texas Motor Speedway, Southlake Town Center, the $77 million Firewheel Town Center, and The Domain, a $105 million project. Mr. Flores's career in human resources began in 2006 when he accepted the position of Director of Human Resources. For a year, he concurrently worked as the Senior Project Manager on The Domain and as Director of Human Resources. He is using his field experience as a Senior Project Manager and ITL to bring a different perspective to the leadership of the organization. His career goals include helping to develop Beck as a more efficient and effective fully integrated design-build company.

Greg Gidez, AIA, DBIA, LEED AP
Corporate Design Manager
Hensel Phelps Construction Co.
Greeley, Colorado

Greg Gidez is Corporate Manager of Design Services for Hensel Phelps Construction Co. Prior to joining Hensel Phelps, Mr. Gidez was a principal with the Denver firm Fentress Bradburn Architects for twenty-six years. He has been practicing as a licensed architect since 1985, and is licensed in Colorado, Texas, Florida, Missouri, and California, as well as NCARB registered. As a principal and senior project manager, he was responsible for the design of over $1.2 billion of constructed projects, totaling over 5.2 million square feet. This included many varied building types: airports, sports facilities, government centers, office buildings, retail, hospitality, courthouses, university, military housing, and research laboratories. Over $1 billion of these projects was delivered utilizing the design-build delivery process. His sensitivity to the need for teamwork and constructive cooperation, and his commitment to technical excellence and the highest levels of performance in design and architectural production have resulted in many award-winning projects. Mr. Gidez was the 2006–2007 president of the Rocky Mountain Chapter of the Design-Build Institute of America (DBIA), where he was active in promoting collaboration and integrated delivery within the design and construction communities. In 2006, he was named to the National DBIA Board of Directors. Responsibilities with the board include chairing the Sustainability Committee and the Research Committee. He is a DBIA-designated design-build professional and a national speaker on the integrated design-build processes. Mr. Gidez has authored numerous articles for industry publications. He was born in Huntington, Long Island, New York in 1956. He received a Bachelor of Business Administration from Rutgers University in 1978 and a Masters of Architecture from the University of Colorado in 1982.

Rebekah G. Gladson, FAIA, AUA
Associate Vice Chancellor and Campus Architect
University of California–Irvine, Design & Construction Services
Irvine, California

Associate Vice Chancellor and Campus Architect for the University of California, Irvine, Rebekah Gladson oversees the design, construction, inspection, and contracting for all major capital projects at both the Irvine campus and UCI Medical Center. Ms. Gladson has spearheaded the broad acceptance of highly successful design-build processes for design and construction on the UCI campus and UCI Medical Center. Through her efforts, UCI has become known in the industry for its alternative delivery of projects and as an active advocate in the use of partnering and other tools for creating spaces conducive to research and learning. She currently oversees a design and construction budget of $1.3 billion, which will support the anticipated doubling in size of the UCI physical plant in square footage over the next decade. Under Ms. Gladson's leadership of Design & Construction Services, UCI has received many national and regional awards for design and building excellence. In 2006 alone, UCI was named Owner of the Year by the American Subcontractors Association (ASA); and Ms. Gladson was personally honored with the Brunelleschi Lifetime Achievement Award from the Design-Build Institute of America (DBIA), as well as the Client Achievement Honor Award from the American Institute of Architects, California Council (AIACC). She is Chair of the National Board for DBIA and serves on the boards of various professional and academic institutions. Ms. Gladson holds a Masters of Architecture from California Polytechnic State University at San Luis Obispo, and certificates from the Institute for Executive Leadership and Management at Stanford University and the University of California Management Institute.

Don Goodrich, DBIA
Vice President
Sundt Construction, Inc.
Tempe, Arizona

Don Goodrich is the Director of Preconstruction for Sundt Construction, Inc. He is responsible for major project reviews and maintaining organizational consistency for the preconstruction efforts of Sundt Construction. He's also responsible for managing the preconstruction efforts for large complicated projects or joint-venture projects, and he oversees the BIM Department. With over thirty years of experience in the construction industry, Mr. Goodrich has been involved in a variety of projects, including: the Apollo Riverpoint Center ($121 million), the Butler Water Reclamation Facility ($106 million), and the Cronkite School of Journalism ($70 million). He is also a part of the Arizona Builders' Alliance and the Leadership Development Forum. He received a B.S. in construction engineering from Arizona State University.

Jim Gray
Chairman and Chief Executive Officer
Gray Construction
Lexington, Kentucky

Jim Gray is chairman and chief executive officer of Gray Construction, a nationally ranked engineering, design, and construction company based in Lexington, Kentucky, with offices in nine U.S. cities and in Tokyo. Gray Construction's main markets are automotive, distribution, and general manufacturing. The company's revenue will top $500 million in 2007. Major private-sector customers include leading companies like Toyota Motors, Hyundai, The Gap, Sears, Boeing, and IKEA. In 2005, Gray Construction was recognized as one of Kentucky's Top 20 Best Places to Work. Mr. Gray has also taken on a number of pro-bono roles, like chairing the Kentucky Governor's Commission on Quality and Efficiency, which conducted a management audit of Kentucky's executive branch, generating budget-reduction recommendations of almost $1 billion, and chairing the capital campaign for the $22 million Kentucky History Center located in Frankfort. He was elected Lexington's Vice Mayor in 2006. Gray is a trustee of Berea College and the restored Shaker Village at Pleasant Hill. He is a graduate of Vanderbilt University and, in 1996, was appointed a Loeb Fellow at Harvard University.

Julie Halvorson, LEED AP
Project Manager II
Ryan Companies US, Inc.
Minneapolis, Minnesota

As a project manager for Ryan Companies, Julie Halvorson works with the project design team to drive as much value into her customers' projects as possible, achieving the best price and schedule while meeting design priorities. She oversees the entire process, from design development, cost estimates, budget preparation, plan and permit reviews and approvals to subcontractor selection and negotiation and all preconstruction and construction activities. Ms. Halvorson excels at working with designers with the objective of developing even better methods for achieving results. She can work with many different kinds of people, taking the time to get to know them and their motivations. The projects Ms. Halvorson has been involved with include: Two MarketPointe, Lexus of Maplewood Service, and Lexus of Wayzata. She received a BS in construction engineering from Iowa State University.

Robert J. Hartung, DBIA
Founder/President
Alternative Delivery Solutions LLC
Laguna Niguel, California

Robert J. Hartung has more than thirty-fives years' construction-management experience in the public and private sectors. He is a former design-build contractor practitioner and founder and president of Alternative Delivery Solutions, LLC. The firm's clients include education, local government, and other public owners. He is a recognized design-build professional by the Design-Build Institute of America (DBIA) and is known as an expert and leader in design-build and alternative-project-delivery methods. Mr. Hartung is a senior instructor for the DBIA's professional designation and owner's series courses. He is an industry representative to the AIA's ongoing work regarding Integrated Project Delivery (IPD) and was a contributing author to its IPD publications. He is a frequent speaker at conferences, workshops, and university classrooms and has authored several articles on the subject of design-build and IPD. He currently serves on the board of directors of the DBIA, Western Pacific Region, where he also chairs the Legislative Committee.

Preston Haskell, PE, DBIA
Chairman
The Haskell Company
Founder, Design-Build Institute of America
Jacksonville, Florida

Preston Haskell is founder and chairman of The Haskell Company, the nation's largest integrated design-build firm in the general buildings sector. Headquartered in Jacksonville, Florida, the company has taken on projects that span a broad range of construction types, including office buildings, data processing centers, institutional buildings, hospitals and medical facilities, continuing-care retirement centers, manufacturing plants, distribution centers, school and university buildings, shopping centers, and parking structures. The firm is a frequent recipient of awards in both design and construction from the AIA, DBIA, ABC, and others. Mr. Haskell graduated with honors from Princeton University with a degree in civil engineering, received an M.B.A. with distinction from Harvard University, and attended Massachusetts Institute of Technology for graduate study in building engineering and construction. He is active in civic and industry leadership at the national, state, and local level. He was founding chairman of the Design-Build Institute of America and served on its board of directors for six years, and was the first recipient of its foremost award, The Brunelleschi Medal. Mr. Haskell has served as a director of the Civil Engineering Research Foundation and is currently a director of the Civil Engineering Forum for Innovation and a member of the National Academy of Construction. He recently completed two terms as a trustee of Princeton University. He is a frequent speaker and panelist on design-build delivery in the United States and abroad, and has authored design-build legislation at the state and federal level. His articles are widely published in industry publications, his most notable contribution in the area of industry research being the landmark paper "Construction Industry Productivity: Its History and Future Direction" (2004).

Diana Hoag
Senior Consultant
Xcelsi Group, LLC
Dayton, Ohio

Diana Hoag is a senior consultant with the Xcelsi Group in Dayton, Ohio. Her expertise comprises acquisition planning and execution for complex systems and design-build construction projects, including development of effective performance-based solicitations, source-selection approaches, and contract incentives. She has over twenty-eight years' experience and demonstrated expertise in acquisition planning and management. During her career with the Department of Defense, she provided expert business advice on the development and production of chemical/biological warfare, military jet engines, weapon system modernization, and covert programs. She also served as a project manager and the Acquisition and Contracting Leader for the Department of Defense's Pentagon Renovation Program, overseeing construction and information technology contracts valued at over $4 billion. She played a leading role in turning around failing construction business processes and introduced new acquisition concepts to the program that are directly accountable for its overwhelming success, including the negotiation and implementation of the program's first and largest design-build contracts. Ms. Hoag also managed the issuance of contracts valued at $760 million for the emergency repair and recovery of the Pentagon as a result of the September 11, 2001, terrorist attack. Recent clients of her consultancy include the United States Air Force and the United States Army Corps of Engineers' Hurricane Protection Office in New Orleans. She has also developed and delivers specialized training for the Design-Build Institute of America.

Jeff Hooghouse, AIA, DBIA, AVS
Deputy Chief Architect of the Corps
United States Army Corps of Engineers Headquarters
Washington, D.C.

Jeff Hooghouse is the Senior Architect at the Headquarters for the United States Army Corps of Engineers, where he currently serves as the Deputy Chief Architect of the Corps. Mr. Hooghouse is the Team Leader for the Corps' Standards & Criteria Team responsible for developing the guidance and policies for Army's MILCON program, including the transformation implementation. Mr. Hooghouse runs the daily operations of the Architectural Community of Practice and is the technical expert for the Department of the Army (DA) in the areas of Industry Standards and Facilities Standardization Program. He started his United States Army Corps of Engineers (USACE) career in 1992. From 1997 to present, he has served in headquarters as Program Architect for several of the Army's largest facilities initiatives: Whole Barracks Renewal Program ($10 billion), Army Facilities Strategy ($10 billion), Army Force Modernization, and now MILCON Transformation (approximately $40 billion). He has served in various positions within the Headquarters, including Special Assistant to the Commanding General and Deputy Chief Value Engineer for the Corps. He has deployed in support of military operations to Afghanistan, Iraq, Bosnia, and Kosovo. Mr. Hooghouse is a registered Architect, licensed in the District of Columbia. He is a Designated Design-Build Professional with DBIA and is certified as an Associate Value Specialist by SAVE International. He is a member of AIA, SAVE International, DBIA, the Society of American Military Engineers, and Army Engineer Association. Mr. Hooghouse represents the Corps on the AIA's Federal Agency Liaison Group for the AIA and is a member of the DBIA Board of Directors. He holds undergraduate degrees in Engineering Technology and Leadership/Management, and is currently working on an M.B.A. in Contracts and Procurement.

Rex D. Huffman, DBIA
Vice President
Gibbs & Register, Inc.
Winter Garden, Florida

Rex D. Huffman is Vice President of Gibbs & Register, Inc., a civil and design-build general contractor headquartered in Orlando, Florida and operating throughout the state of Florida. The company specializes in horizontal projects, including roadwork, underground utilities, park and recreation facilities, building-site work, and streetscape projects. Mr. Huffman has more than thirty-five years of design and construction experience. His primary focus has been in design-build marketing, sales, and project/operations management across numerous market sectors: industrial, highways and bridges, railroads, water, wastewater, buildings, and hazardous waste remediation. Mr. Huffman received a B.S. in civil engineering from Iowa State University. He is a Designated Design-Build Professional and sits on the national board of the Design-Build Institute of America (DBIA).

Neil Johnson, LEED AP
Team Leader
Ryan Companies US, Inc.
Minneapolis, Minnesota

Neil Johnson is known for his dedicated work ethic and attention to customer satisfaction. With years of hands-on experience in carpentry and electrical work, Johnson brings a unique perspective and broad understanding to each project he manages. He is also noted for his ability to help customers envision the overall goals of the project early in the process. As Team Leader, Mr. Johnson's primary responsibilities include design management, scheduling, estimating, budget and revenue control, and overall project administration. Through collaboration with owners, architects, engineers, and subcontractors, he continually strives to bring the highest potential value to Ryan's customers. Among the many projects that Mr. Johnson has been involved in are Target North Campus, Midtown Exchange, Allina Commons, and River Parkway Place. He received a Bachelor of Science in construction management from Mankato State University.

Nat Killpatrick
Vice President, Sales (Texas Division)
Burcamp Steel Company
Dallas, Texas

Nat Killpatrick's primary focus is developing work leads in both the Design-Bid-Build market and the Design-Build market. In this pursuit, he works with the rest of the team to look for opportunities that best match their abilities to provide "turn-key" structural steel solutions for projects. He also develops and improves "BIM"-type processes within the organization's operations while looking for ways to develop integration with outside firms. Burcamp Steel Company strongly believes in the design-build methodology and continues to promote its use throughout the industry. Mr. Killpatrick received a B.S. in Industrial Technologies from Abilene Christian University in 1997. His professional affiliations include DBIA, the Dallas Chamber of Commerce, the Dallas Chapter of AIA, and the Society of Industry Leaders.

Chuck Kluenker, PE, F.CMAA
Associate
Vanir Construction Management
Sacramento, California

Chuck Kluenker, an Associate at Vanir Construction Management, has managed design and construction on public and private projects since 1973. A fellow of the Construction Management Association of America (CMAA), he has served as CMAA's national president, a member of its board and executive committee, chair of its Committee on Standards of Practice, and a member of the CM Certification Board of Governors. Mr. Kluenker also served on the Dean's Advisory Board of the College of Architecture and Environmental Design at California Polytechnic State University, San Luis Obispo, California. He has written numerous articles and papers on construction management and is a frequent speaker on the subject. Mr. Kluenker earned a civil engineering degree from the University of Wisconsin in 1971.

Wayne Lindholm
Executive Vice President
Southern California
Hensel Phelps Construction Co.
Irvine, California

Wayne Lindholm is Executive Vice President and District Manager for the Southern California office of Hensel Phelps Construction Co. As one of seven nationwide offices it was established in 1989 and has since completed more than $3 billion in construction projects in Southern California, Arizona, Nevada, and Utah. Recently completed projects include the St. Regis in Dana Point, the Anaheim Convention Center expansion, Paradise Pier at California Adventure, and The Lofts in San Diego. Projects currently underway include the UCI Medical Replacement Hospital, Medical Research and Chemistry Labs at the University of Arizona, Natural Science II and Computer Science III Labs at UCI, Palo Verde Housing at UCI, and the Coalinga State Hospital. Mr. Lindholm graduated from Colorado State University in 1975 with a bachelor's degree in construction management. He started as a field engineer with Hensel Phelps thirty years ago and worked through all positions of management, moving regularly and tackling the toughest projects available. He was awarded the Honor Alumnus Award from the College of Applied Human Sciences at Colorado State University on May 7, 2004. Wayne has also been actively involved in the community since settling in one spot with the office in Irvine California. He serves on numerous boards and committees, a few of which include the Irvine Chamber Legislative and PAC Committees, the Orange County Associated General Contractors (AGC), the Design-Build Institute of America (DBIA), and the Executive Board and Treasurer for the State of California AGC.

Michael C. Loulakis
President and CEO
Capital Project Strategies, LLC
Reston, Virginia

Michael C. Loulakis has thirty years' experience in representing parties engaged in all aspects of the construction industry. He is the President/CEO of Capital Project Strategies, LLC, a consulting firm that provides strategic support and advice to owners, contractors, designers, and others involved in the development of construction projects. Before starting Capital Project Strategies, Mr. Loulakis served as chairman and president of the national construction law firm Wickwire Gavin, PC, where he represented clients on matters involving procurement, risk management, contract drafting, and dispute resolution. He is well known for his extensive background in design-build and other alternative project delivery systems, and has been one of the industry's most active writers and speakers on the subjects. Mr. Loulakis has been lead counsel on some of the country's most complex design-build projects and public–private partnership programs, where he has helped public agencies formulate effective procurement and contracting strategies. He has also represented contractors, design professionals, and others on construction-business counseling, teaming, contracting, and conflict resolution, often serving as their outside general counsel for construction matters. Mr. Loulakis holds a civil engineering degree from Tufts University and a law degree from Boston University School of Law. He is widely published and has been ranked in Chambers USA as one of the country's best construction lawyers since 2004.

Kim Lum
Executive Vice President
Charles Pankow Builders, Ltd.
Pasadena, California

Kim Lum, a twenty-eight-year veteran in the field of construction, is Executive Vice President at Charles Pankow Builders, Ltd. (CPBL), one of the country's foremost innovators of concrete construction methodologies and an industry leader in design-build project delivery. Mr. Lum's experience includes high-rise residential, commercial, retail, parking garages, hotel, office, retail renovation, and civic center work. Specific responsibilities as Regional Manager include management of project executives, contract negotiations with owners, and pursuit of new construction opportunities. Mr. Lum started his career with Pankow in 1980 as a field engineer in the Hawaii office and worked his way up as Project Engineer, Project Manager, and Project Sponsor, as well as serving for five years as the Regional Manager. In his current position as Regional Manager, Mr. Lum oversees the management of preconstruction and construction activities for all Northern California projects. Throughout his career, Mr. Lum has actively supported the advancement of construction-industry education. He currently serves on the Board of Directors of the California Center for Construction Education, the Board of Directors of the Design-Build Institute of America (DBIA), Western Pacific Chapter, and is a Designated Design-Build Professional. Mr. Lum has a B.S. in civil engineering from Stanford University and graduated Phi Beta Kappa.

John A. Martin, DBIA
Division President
Flintco Inc.
Austin, Texas

John A. Martin provides oversight and leadership of Flintco's Texas operations office, based in Austin. With a twenty-five year track record of excellence in construction management, Mr. Martin's diverse career includes work in commercial, industrial, retail, educational, athletics, lab, high-tech, and historical renovation in both the public and private sectors. Under his leadership, the Flintco's Austin office has won numerous national and regional awards for construction excellence, design-build excellence, and excellence in safety from the Associated Builders and Contractors, the Design-Build Institute of America, and the Associated General Contractors. Mr. Martin has earned the DBIA Designation from the Design-Build Institute of America. He is on the executive board of DBIA's five-state Southwest chapter and served as president of that chapter from 2004 to 2007. He graduated in 1986 from Oklahoma State University with a B.S. in construction management. He has been an executive board member of the OSU Construction Management Advisory Board (CMAB) since 1996 and served as vice president of that organization from 2002 to 2004. Other affiliations and memberships include Texas A&M University Construction Industry Council, California Center for Construction Education Advisory Board, Real Estate Council of Austin, Austin-San Antonio Corridor Council, and the Austin chapter of ACE Mentoring, where he served as Chair in 2005–2006. Mr. Martin has a strong background in leadership development and partnering and is a sought-after speaker and leadership-development facilitator.

Meloni McDaniel
Coordinator of Federal Business Development
The Beck Group
Dallas, Texas

Meloni McDaniel is a Coordinator of Federal Business Development at The Beck Group, a fully integrated design-build firm based in Dallas, Texas. Originally from Vinita, Oklahoma, she attended the University of Oklahoma and majored in architecture with a minor in construction science management. Upon accepting an internship with Beck during the summer of 2004, Ms. McDaniel was fully introduced to the inner workings and advantages of the design-build delivery method. She returned to Beck for full-time employment following her graduation in 2005. Since beginning her career at Beck, she has worked in multiple disciplines for the firm, switching from architecture work with the firm's theaters group to a stint working on scaffolding 60 feet above the ground on curtain-wall quality control on a church construction project in downtown Dallas, to functioning as a full-time construction project engineer on a retail center job site in Austin, Texas, and back again to architecture work. Her career goals include further cross-training experiences on integrated design-build projects with Beck.

John McGarva
President and CEO
Western Water Constructors, Inc.
Santa Rosa, California

John McGarva is the President and CEO of Western Water Constructors, Inc., which specializes in the construction of wastewater and water treatment plants all over Northern California. Mr. McGarva has more than thirty-five years' experience in the construction industry. Prior to becoming president of Western Water, he held a variety of positions in both the residential and commercial industries, including operator, superintendent, foreman, and general contractor. Mr. McGarva is also the CEO and President of two other companies: California Outdoor Cooling, Inc. (incorporated in 2006) and Clean Green, Inc. (incorporated in 2007). He is currently on the Legislative Committee of the Design-Build Institute of America's Western Pacific chapter.

Josh McGarva
Vice President
Western Water Constructors, Inc.
Santa Rosa, California

Josh McGarva got his start in the water and wastewater industry during the summer starting when he was sixteen. He began working for Kirkwood-Bly, which later became Western Water Constructors, Inc. Mr. McGarva attended California Polytechnic State University at San Luis Obispo, where he attained a B.S. in construction management. After graduation, he went to work for Western Water Constructors full-time as a project manager. Over the years with Western Water, he started to manage larger projects and began estimating in 2001. He is currently the chief estimator and operations officer for the company. Western Water is currently delivering about $60 million a year in construction projects. McGarva feels fortunate to be able to work with friends and family in one of the largest merit shop general engineering-construction firms in California.

Brig. Gen. John R. McMahon, PE
United States Army Corps of Engineers
United States Forces Afghanistan, Director of Engineering
San Francisco, California

Brigadier General John R. McMahon has served as the Director of Engineering for United States Forces Afghanistan in Kabul since January 2009. Prior to that, he was the Division Engineer and Commander of the United States Army Corps of Engineers, South Pacific Division, in San Francisco, California. General McMahon entered the Army in 1977 through the ROTC program following his graduation from Syracuse University with a B.S. in biomedical engineering. He has led and trained soldiers at every level, from platoon to battalion, in combat engineering units in Hawaii, North Carolina, Germany, Kansas, and California. General McMahon earned an M.S. in applied mathematics from the Naval Postgraduate School and taught undergraduate mathematics at the United States Military Academy at West Point, New York. He also served as the Professor of Military Science at Rose-Hulman Institute of Technology in Terre Haute, Indiana. In 1990, he was named Director of Engineering and Housing, in Aschaffenburg, Germany. He subsequently participated in the liberation of Kuwait as the Brigade Engineer, 3d Brigade Combat Team,

3d Infantry Division (Mechanized), during Operations DESERT SHIELD/ STORM in Iraq. General McMahon served a joint assignment from 1998 to 2001 as Joint Plans and Requirements Officer, Pacific Command, and later as Division Chief, European Command Customer Operations, National Imagery and Mapping Agency, in Reston, Virginia. In 2001, he served as Commander, United States Army Engineer District, Japan, Camp Zama, Japan. And in 2004, he became the Chief of Staff at Headquarters, United States Army Corps of Engineers, Washington, D.C.

Jordan Moffett, AIA, DBIA, LEED AP
Assistant Project Manager, Architect
The Haskell Company
Jacksonville, Florida

Jordan Moffett joined The Haskell Company in August 2005 as an Architect intern. He worked on several educational facility and government projects before transferring to the construction group in January 2008. As an Assistant Project Manager, he has taken the opportunity to follow his most recent design project, an $80 million student housing facility, into the field. Mr. Moffett has supplemented his education by successfully completing all nine architectural registration examinations on his way to becoming a registered architect with the state of Florida. In addition, he has also become a LEED Accredited Professional and a DBIA Designated Design-Build Professional. As Assistant Project Manager, Mr. Moffett utilizes his comprehensive knowledge of design and building codes to supplement the construction team while taking on the role of Field Architect. His knowledge of sustainability and the LEED program have been invaluable in pursuing LEED certification on his current project; he has also been a valuable member of several proposal teams requiring LEED Accredited Professionals. He received a bachelor's degree in architecture from California Polytechnic State University at San Luis Obispo with a minor in construction management. He is a member of the United States Green Building Council (USGBC), the North Florida Emerging Green Builders, the American Institute of Architects, and the Design-Build Institute of America (DBIA).

Andrea Murray, AIA, NCARB, LEED AP
Architect
Bread Loaf Corporation
Middlebury, Vermont

Andrea Murray is an architect, planner, and LEED Accredited Professional who currently practices Integrated Design-Build at Bread Loaf Architects, Planners, Builders in Middlebury, Vermont. In addition, Ms. Murray teaches 'Architecture and the Environment' and 'Introduction to Architectural Design' at Middlebury College. She focuses her professional energies on integrating sustainable and healthy design elements and strategies into all her buildings and their communities. Having grown up in Vermont, Ms. Murray moved back home in 2003 from New York City, where she had worked on various projects with William McDonough + Partners, Flynn-Stott Architects, and the City of New York. She has a bachelor's degree in English from Syracuse University and a bachelor's degree in architecture from the Pratt Institute. Ms. Murray was on the Board of Directors of AIA Vermont and edited their monthly newsletter from 2003 to 2007. She has published articles on design and building and has received numerous awards for her work.

Michael Ng
Estimator
CDM Constructors Inc.
Rancho Cucamonga, California

Michael Ng has been an estimator for CDM Constructors Inc. since February 2008. Before joining CDM, he attended California Polytechnic State University at San Luis Obispo, and graduated with a B.S. in construction management in 2007. As a student, Mr. Ng gained invaluable experience interning for a variety of commercial contractors, including subcontracting, general contracting, and design-building. While at Cal Poly, he was a member of the school's commercial team at the 2006 ASC-AGC CM competition. They earned first at the regional level and placed second at the national competition. As an estimator for an integrated design-build firm, he consistently coordinates with other in-house engineers to complete his work. He has worked on a wide array of estimates in his short time at CDM, ranging from proposals, engineer's estimates, and opinion of probable cost. To stay connected to the industry, Mr. Ng is also a member of the Young Constructors Forum, a branch of the AGC.

Kimon Onuma, FAIA
President and Founder
Onuma, Inc.
Pasadena, California

Kimon Onuma pioneers fundamental change in problem-solving, knowledge capture, and integrated practice. His award-winning, open-standards design software for real-time, global collaboration represents the future of our profession as demonstrated in his practice. Coming from a family of architects who worked on projects worldwide, Mr. Onuma experienced first-hand the challenges of travel and communication. In 1988, he started his California office and needed to communicate with his father's Tokyo office. This sent him on a nineteen-year journey to define new methods of practicing architecture, which has significantly advanced the profession. In 2008, Mr. Onuma became a fellow in the American Institute of Architects for his work in changing the industry. A global perspective on how architecture impacts the environment became possible when Onuma linked Geographic Information Systems (GIS) to BIM with Onuma Web Services. This occurred for the first time on an Open Geospatial Consortium (OGC) project cosponsored by the GSA, proving that architectural knowledge can be web enabled. Mr. Onuma enabled intelligent building models and intelligent maps to share information that was globally accessible in real time by twenty companies, such as the New York Port Authority, Bentley Systems, and the Hasso Plattner Institute in Germany. Mr. Onuma is a nationally and internationally recognized speaker on topics such as planning, programming, design, sustainability, and integrated practice. He is widely published in professional journals and is a member of BuildSMART alliance, and OGC, two organizations focused on defining open standards.

Lou Palandrani, LEED AP
Senior Vice President
Clark Construction Group
Bethesda, Maryland

Lou Palandrani is a Senior Vice President with the Clark Construction Group, headquartered in Bethesda, Maryland. Mr. Palandrani has been in the construction industry for twenty-eight years, the last twenty-three with Clark. A graduate of the University of Delaware with a civil engineering degree, he maintains a California contractor's license and is a LEED Accredited Professional. He has been involved in the management and completion of significant public and private projects throughout the country. Working with nationally recognized high-profile architectural firms such as Michael Graves Associates, KMD Architects, and SOM, Mr. Palandrani has gained an appreciation for the value of early collaboration between designer and contractor. Over the last 15 years, he has been responsible for the completion of design-build delivery projects ranging in value from $5 million to $250 million. Overall, he has successfully completed over $1 billion in design-build and collaborative delivery projects.

Ed Palmer, AIA, LEED AP
Principal, Senior Associate
Niles Bolton Associates
San Jose, California

Ed Palmer has more than thirty-three years of extensive architectural experience in a wide variety of project types, including housing and mixed-use developments, academic facilities, hotels, office buildings, restaurants, and other public and commercial building types. Mr. Palmer has an extensive background in working with both private and public organizations and has managed projects from concept through design and construction. Currently, he manages the San Jose, California office of Niles Bolton Associates. Mr. Palmer's design-build experience includes a 1,000-bed Student Village at Embry-Riddle Aeronautical University in Daytona Beach, Florida. He led the team in designing the four-building complex, which integrated distinct aeronautical and technological themes into a nontraditional campus plan on 8 acres. At California Polytechnic State University, Mr. Palmer recently managed the design-build project Poly Canyon Village, a 2,700-bed, 992,000-square-foot student-residence development. The project will be the first LEED-certified project on campus and the largest project in the California State University system. At a national level, Mr. Palmer has been an active participant in the American Institute of Architects, and served as the President and Director of the Chattanooga, Tennessee chapter. He has been active in business and civic associations, including the Tennessee Society of Architects and the Sandy Springs Optimist Club. Mr. Palmer is a registered architect in the state of Tennessee and is a frequent speaker at Niles Bolton Associates Continuing Education Seminars. Recently, he has been a guest speaker at Cal Poly's Design-Build "Boot Camp" seminars, and a panelist at the Cal Poly 2006 Design-Build Educator Workshop. Mr. Palmer received a bachelor's degree in architecture from the University of Tennessee in 1972.

John Parnell, LEED AP
Project Manager
Ryan Companies US, Inc.
Phoenix, Arizona

John Parnell is currently a Project Manager working for Ryan Companies US, Inc. in its Phoenix office. He received his B.S. in construction management from California Polytechnic State University at San Luis Obispo in the spring of 2006. While completing his degree, he worked with companies such as Pulte Homes, Rudolph and Sletten, and Walt Disney Imagineering. In the fall of 2006, he worked in Africa for four months with Engineering Ministries International, serving in project management and civil engineering roles. He pursued his passion for design-build project delivery by beginning his long-term career with Ryan Companies US, Inc. in January 2007. Currently, he is working on his LEED AP certification and plans on pursuing a master's degree in environmental engineering.

Bill Quatman, Esq., FAIA, DBIA
Attorney at Law
Shughart, Thomson & Kilroy, PC
Kansas City, Missouri

G. William Quatman is a licensed architect and practicing attorney with the law firm of Shughart, Thomson & Kilroy, PC in Kansas City, Missouri. Quatman is a nationally recognized author and speaker on design-build. He is the past president of AIA Missouri and is the current national chairman of the DBIA's Laws and Regulations Committee. He has spoken at multiple AIA and DBIA national conventions on design-build, and he has given courses on design-build nationwide, as well as in Puerto Rico and Canada. He teaches courses on design-build at the University of Kansas and is the author of *Design-Build for the Design Professional*, as well as the chief editor of AIA's *The Architect's Guide to Design-Build Services*. He serves on AIA's Knowledge Community Advisory Group and was the first architect to become both a fellow of the AIA (FAIA) and a Designated Design-Build Professional of the Design-Build Institute of America. In 2004, Mr. Quatman was named to the Best Lawyers in America for construction law, and he was the 2004 Distinguished Alumnus for the University of Kansas School of Architecture. In March 2005, Mr. Quatman received AIA Missouri's Distinguished Service Award.

Victor Sanvido
Senior Vice President
Southland Industries
Garden Grove, California

Victor Sanvido is responsible for developing and delivering custom design-build-maintain services to targeted end users in Southland's California markets. These solutions include the practical application of lean principles, BIM, project-incentive programs, sustainability, and integrated project delivery and contracting, particularly design-build. Mr. Sanvido received his Ph.D. in civil engineering/construction engineering and management from Stanford University. Formerly a professor of architectural engineering at Penn State University, he has written over 100 publications and is a leader in several industry organizations, including DBIA and LCI. He is co-author of the book *Selecting Project Delivery Systems* (PDI, 1999). He speaks frequently on the implementation of integrated lean project delivery at a national level. He has organized several project delivery conferences for health care, education, community colleges, public owners, and specialty contractors.

Michael A. Schmieder, AIA
Project Design Principal
The Haskell Company
Jacksonville, Florida

Michael A. Schmieder is a registered architect and Project Design Principal with Haskell, a design-build firm headquartered in Jacksonville, Florida. He has over thirty years of design/design-management experience serving large national A/E/C firms, with a specialty in university R&D laboratories, classroom buildings, conferencing facilities, performing-arts facilities, student housing and numerous other public-sector project/client types. Mr. Schmieder's most recent projects include the University of Florida Osprey Fountains Student Housing ($80 million), the Jacksonville University Oak Hall Student Housing and Parking Structure ($30 million), and the University of West Florida Heritage Hall Student Housing ($13.8 million). He is a member of the American Institute of Architects (AIA) and the South Eastern Association of Housing Officers (SEAHO). He is a registered architect in Florida, Georgia, North Carolina, Virginia, Tennessee, Kentucky, Mississippi, and Texas. He is the recipient of a number of awards, including an AIA Merit Award for the Jacksonville University Swisher Auditorium project, the Ahead of its Time Award from the Northeast Florida Regional Planning Council for the Northbank Riverwalk project in Jacksonville, and two Significant Concrete Structures Awards from the American Concrete Institute (one for the State of Florida Department of Health and Rehabilitative Services, and one for the City of Jacksonville Pre-Trial Detention Facility. He earned a Bachelor of Architecture from the University of Tennessee in 1976, graduating with honors.

Alan Schorfheide, DBIA
Senior Director of Integrated Services
The Korte Company
Highland, Illinois

Alan Schorfheide has more than thirty years of construction experience. He is currently part of the Business Development Team at The Korte Company and serves as the Senior Director of Integrated Services. Mr. Schorfheide is responsible for assembling teams of design and construction professionals to prepare design-build project solutions for owners. He leads each team to develop preliminary cost estimates, design and construction schedules, and proposal documentation in response to owner requests. Prior to 2002, he was the Executive Vice President and Chief Operating Officer at Korte. Prior to that, he worked in the Preconstruction Services Department, where he had overall responsibility for reviewing and approving all of Korte's estimates prior to submission. He pioneered the Korte Team approach toward design-build contracts for the United States Postal Service and United States government agencies. The approach focused on common-sense design solutions for initial and life-cycle cost savings. Mr. Schorfheide has worked on projects around the country in a wide range of areas, including military, distribution centers, warehouse/manufacturing facilities, banks, educational facilities, housing, retail, offices, and recreation. He received a B.S. in civil engineering from the University of Illinois.

Paul Shea, PE, BCEE, DBIA
President
CDM Constructors Inc.
Camp, Dresser, & McKee Inc. (CDM)
Denver, Colorado

Paul Shea has more than thirty-seven years' experience managing major environmental and general infrastructure design and construction projects throughout the United States. His experience includes construction scheduling, cost estimating, materials and equipment procurement, contract administration, subcontractor coordination, quality control, and health and safety management on projects valued up to $80 million. Mr. Shea is currently leading CDM's design-build and general construction activities globally. In this capacity, he is responsible for managing quality, interfacing with clients, allocating resources, project pricing, managing business operations, and developing and implementing strategic directives. In his five years with CDM as president of CDM Constructors Inc., he has championed the growth of the firm into a major design-build company doing over $0.25 billion in annual revenues globally. Recent projects include the Newport News Waterworks Facility ($70 million), the Damascus Wastewater Treatment Plant ($13 million), and the Potomac Water Treatment Plant Improvements ($18 million). Mr. Shea is a member of the National Society of Professional Engineers, the Water Environment Federation, and the Design-Build Institute of America.

Philip J. Sheridan, PE, DBIA
Project Executive
Clark Civil, LLC
Bethesda, Maryland

Philip J. Sheridan is a project executive for Clark Civil. With over twenty-four years in the construction industry, he has been involved with a number of design-build projects on the East Coast. These include an ICC Contract C–MdSHA highway project for Montgomery and Prince George's counties in Maryland ($513 million), the Navy Yard Station Upgrade of the existing Metro entrance in Washington, D.C. ($18 million), and the Brentwood Shop renovation and upgrade of its heavy-rail maintenance facility in Washington, D.C. ($43 million). Mr. Sheridan received a B.S. in civil engineering from Worcester Polytechnic Institute. He is a member of the Design-Build Institute of America (DBIA) and president of his DBIA region. He is also a member of the American Society of Civil Engineers (ASCE). Several of Mr. Sheridan's projects have received both regional and national awards for design and construction excellence.

Gregory L. Sizemore, Esq.
Executive Vice President
The Construction Users Roundtable
Cincinnati, Ohio

Gregory L. Sizemore is the Executive Vice President of The Construction Users Roundtable. Mr. Sizemore has more than twenty years of construction industry and association management experience. Immediately out of graduate school, he served for three years as the Director for Labor Relations at the Associated General Contractor's and Allied Construction Industries in Cincinnati, Ohio. For eighteen years, he was the Executive Director of the Construction Owners Association of the Tri-State (COATS) in Cincinnati, OH. In 1989, he started Sizemore & Company, LLC, a construction-industry association management and consulting firm. Some of Sizemore & Company's other client organizations include COATS, the South Georgia Construction & Maintenance Council, and the Appalachian Construction Users Council. Additionally, Mr. Sizemore serves on various industry and civic boards, including the University of Cincinnati's Industrial Advisory Council, the Advisory Committee to Albany Technical College, the West Virginia Construction Coalition Conference, and the Design-Build Institute of America's (DBIA) Board of Directors. Mr. Sizemore also teaches construction law and leadership and decision-making classes as an adjunct professor at the University of Cincinnati. He holds a B.S. in economics and an M.S. in industrial and labor relations from the University of Cincinnati. He earned his J.D. from Northern Kentucky University's Salmon P. Chase College of Law and is a licensed attorney in the state of Ohio.

Tom Sorley
Chairman and CEO
Rosendin Electric, Inc.
San Jose, California

Tom Sorley is responsible for the overall management, vision, and direction of Rosendin Electric, a 100% employee-owned firm. With a focus on customer service and relationship-based business, the company has grown over 400% under his leadership. Combining strength of management with a highly skilled workforce, Rosendin Electric is able to seamlessly deliver engineering, construction, maintenance, and service work across a broad spectrum of industries. Established in 1919, the company is consistently ranked as one of the top ten electrical contractors in the United States. Active in industry affairs, Mr. Sorley serves on several joint IBEW/NECA committees. He also serves on the board of the San Francisco Bay Area Boy Scouts of America, the national board of the Design-Build Institute of America, and the national board of the Building Futures Council. Mr. Sorley is a graduate of the University of Nebraska.

Julie M. Stites, AIA, LEED AP
Architect
The Haskell Company
Jacksonville, Florida

Julie M. Stites is an architect and a LEED Accredited Professional working in a design-build firm based in Jacksonville, Florida. She graduated from California Polytechnic State University at San Luis Obispo in 2005 with a Bachelor of Architecture and a minor in integrated project delivery systems. Since graduation, she has been employed by The Haskell Company, where she works with interdisciplinary teams on projects varying from health care to government to general commercial facilities. Early on in her career, Ms. Stites became one of Haskell's BIM pilot-team members and became proficient with the Revit software. She is currently a Revit resource for her co-workers as Haskell works to transition from AutoCAD to the Revit platform.

Robynne Thaxton Parkinson
Attorney
Robynne Thaxton Parkinson PLLC
Mercer Island, Washington

Robynne Thaxton Parkinson is a Seattle-based lawyer and the principal of the law offices of Robynne Thaxton Parkinson PLLC. She is also of counsel with Groff Murphy PLLC, one of the premier construction law firms in the Northwest. Ms. Parkinson provides owners, design-builders, general contractors, and designers with practical and proven advice. With more than nineteen years of experience, she is one of the leading experts in construction law and alternative procurement, both in the state of Washington and on a national basis. She participates on the legislative subcommittees of the Washington Capital Projects Advisory Review Board, which reviews all legislation involving capital projects in the state. She is the president of the Northwest Region of the Design Build-Institute of America. She also serves on the National DBIA task force to revise its contract documents, the DBIA Education Committee, the DBIA National Convention Planning Committee and is one of DBIA's highest-rated speakers on the subject of design-build contracts and risk management. She was an adjunct professor for Seattle University's Albers School of Business and has been a speaker for many groups, such as the Washington State Bar Association, Associated General Contractors, the National Business Institute and Lorman Business Seminars. She also serves as an advisor to New Construction Strategies. Ms. Parkinson received her J.D. from the University of Colorado School of Law and her undergraduate degree from the University of Texas at Austin.

Peter Tunnicliffe, PE, BCEE, DBIA, CIRM
Corporate Director of Project Development
Chair of the Risk Management Committee
CDM
Cambridge, Massachusetts

Peter Tunnicliffe is currently leading CDM's design-build and design-build-operate project development and serving as CDM's corporate account manager for two national private water companies with regulated and non-regulated water and wastewater projects throughout the United States. In this role, he is responsible for coordination of all CDM work with the water companies, the evaluation of major project opportunities, teaming arrangements, development of capture strategies, and pricing input. He is also responsible for CDM's service delivery on an overall basis, for conducting periodic project reviews, for participating in project execution, and for reviewing with clients how CDM can improve its service delivery. From 1992 into 2002, he was the division manager for the CDM organization's design-build and construction subsidiary, CDM Engineers & Constructors. Mr. Tunnicliffe was responsible for developing and implementing the construction and design-build strategic directives for the organization that advanced the firm to one of the top 400 general contractors in the United States (Engineering News Record, 2001) and one of the top 10 design/build firms in the water/wastewater industry (Engineering News Record, 2001). In this role, his duties included market analysis, client interaction, contract development, major project evaluations, P&L responsibility, and inter-unit integration. He was also responsible for leadership development for the key staff within the CDM E&C unit. He has more than thirty years' experience in water and wastewater treatment facilities and conveyance system design and construction management. He has served as an expert witness in private water company rate cases, litigation, arbitration services, and he has performed post-construction facility analysis as part of claims analysis work. Mr. Tunnicliffe received his BS in civil engineering from Worcester Polytechnic Institute and his JD from the Brooklyn Law School.

Craig Unger, DBIA
CEO and President
Unger Security Solutions, LLC
Knoxville, Maryland

Craig Unger is CEO and President of Unger Security Solutions, LLC, a private business providing consulting services for acquisition management, project delivery, and the security and detention environment. Mr. Unger founded the business upon his retirement from the federal government in 2003. Additionally, Mr. Unger served as President of the Design-Build Institute of America (2003–2004), a national nonprofit organization dedicated to the successful use of integrated design and construction project delivery services, serving as the organization's lead spokesman and advocate for legislative, regulatory, and government actions in design-build at the national, state, and local levels. Mr. Unger is regarded as one of the leaders of the design-build movement in the public sector. Before joining DBIA, he served for twenty-six years at the United States Department of Justice. Mr. Unger was the procurement executive for the Federal Bureau of Prisons, where he led that agency's design and construction program from a one-dimensional "low bid" approach to one that embraced alternate delivery systems. Additionally, Mr. Unger has been a senior consultant to the Headquarters of the US Army Corps of Engineers for the Army MILCON Transformation Initiative, which includes: development of performance specifications; acquisition planning; value engineering of the Brigade Combat Team requirements; moderator for the Association Industry Workshop (AIA, ACEC, AGC, DBIA); facilitator in the RFP risk-analysis exercise, and development of evaluation criteria for best value source selection. Mr. Unger has an extensive background in various alternative disputes resolution (ADR) techniques, including Partnering Workshops and Mediation. He is currently a member of the Washington Metropolitan Area Transportation Authority (WMATA) Disputes Resolution Board (DRB) for several design-build multimillion dollar infrastructure projects. Mr. Unger is a nationally recognized lecturer throughout the industry and has authored numerous articles and publications, including congressional reports and testimony. He received a B.A. in accounting from Bloomsburg University, an M.B.A. from West Virginia University, and attended Harvard University for postgraduate work. He is also a veteran of the United States Marine Corps.

Carter Vecera
Director of Preconstruction
The Beck Group
Dallas, Texas

Carter Vecera brings a variety of attributes to his role as Director of Preconstruction. He has thirteen years of preconstruction and project management experience in the construction industry and is well trained in the use of Timberline and cost database management. His recent experience includes preconstruction services for five large hospital and hospital renovation projects. He also has extensive experience with MEP systems and $100 million projects, both during preconstruction and project life cycle estimating and cost management. As Director of Preconstruction Services, Mr. Vecera is responsible for review of estimates for completeness and correctness, assignment of preconstruction resources, assuring compliance with Beck's estimating philosophy accustomed to the client's needs, and operation responsibility for the preconstruction group. Mr. Vecera's projects are in a variety of sectors, including office, health care, data centers, retail, hospitality, and specialty. He received a B.S. in construction science from Texas A&M University.

Stephen Vrabel, DBIA, LEED AP
Preconstruction Manager—Major Projects
PCL Construction Enterprises, Inc.
Denver, Colorado

Stephen Vrabel started his career with PCL Construction (Poole back then) in 1976 in Edmonton, Alberta, Canada in the costing and accounting area. He then transferred to the construction side of the market when an opening came up in 1979 as a field engineer, project engineer, project coordinator, and project manager. He also worked in estimating as senior estimator, chief estimator, preconstruction manager, and director of preconstruction services. He is currently Preconstruction Manager of major projects working in PCL's U.S. corporate office in Denver, Colorado. His experience has been quite varied throughout his career, offering exposure to the building side as well as light civil, infrastructure, and industrial projects. More recently, he has returned to major building projects. He works with PCL's U.S. Districts assisting with pursuits during the RFQ/RFP and preconstruction phase of the projects. In 1992, he was transferred from Edmonton, Alberta to PCL's Irvine, California district. In 1999, he was transferred to PCL's U.S. corporate office, where he is at present.

Barbara Wagner
Senior Vice President
Clark Construction Group, LLC
Costa Mesa, California

Barbara Wagner is a Senior Vice President for Clark Construction Group's Western Region. Clark Construction is one of the largest general contractors in the country, with over $3.5 billion of annual revenue. Ms. Wagner has more than twenty-five years' experience in the construction industry. She has extensive experience in both the design-build and sustainable markets. In addition to heading Business Development for Clark in the Western Region, she is Officer-in-Charge for projects at Pasadena Convention Center, Dominican Hospital, and Northridge Hospital. Her extensive experience in the design-build market started in the 1980s with the United States Postal Service (USPS). Ms. Wagner currently serves as president of DBIA's (Design Build-Institute of America) Western Pacific Region. Ms. Wagner also sits on the board of the California Center for Construction Education, the executive board of the School of Engineering at Catholic University of America, and as an appointed member of the Facilities Program Advisory Committee for the State of California Department of Corrections. She is an active member of the USPS' Industry Advisory Council and the American Society of Healthcare Engineers. She has a bachelor's degree in architecture from Catholic University and a master's in civil engineering and construction management from the University of Maryland. Her professional licenses and accreditations include a contractor and engineer's license in the state of California and LEED accreditation.

Coleman Walker
Director of Talent Acquisition
The Haskell Company
Jacksonville, Florida

Coleman Walker has more than twenty years of combined experience in construction, executive staffing, and, most recently, human resources as the Director of Talent Acquisition for an integrated design-build firm. During a diverse career path, Mr. Walker has always maintained his focus on two passions: leading and developing people/teams and improving the building/construction industry. His experience in directing on-site teams to success on many complex projects for exceptionally demanding clients in the entertainment industry reinforced his belief that by providing the right tools and empowering people you will create successful results. He developed a reputation for turning bad projects/client experiences into positive results through his extensive experience and capabilities as a manager. In his current role, Mr. Walker's ability to identify and acquire exceptional talent with emphasis on recent college graduates has sparked a cultural shift, which will be the future of his current company and the AEC industry. Mr. Walker obtained his bachelor's degree in business administration, with a minor in psychology, from the University of Central Florida.

Nick Watry, PE, Licensed Architect
Founder (retired)
Watry Design Group
Redwood City, California

As the 1975 founder of the Watry Design Group, now known as Watry Design, Inc., Nick Watry gained a reputation as a premier designer of post-tensioned concrete framed structures in the United States, most notably parking structures. Ten years prior to 1975, Mr. Watry was employed by the largest post-tensioning sub-contractor in America and was the Hawaii Division Manager. Since the fall of 2002, he has taught in the Construction Management Department of the College of Architecture and Environmental Design at California Polytechnic State University, San Luis Obispo, California. All through Mr. Watry's professional career he has passionately participated in community, professional, and educational endeavors and "has made a difference." He now offers his over forty years' experience as a consultant to the design and construction industry, particularly to provide the integration of architecture, structural engineering, and parking principals to the design of concrete framed structures, constructability analysis of concrete framed structures, code analysis and interpretation of concrete framed buildings, concrete building frame quantities and costs, consensus of design approach with building officials, and consensus of design approach with "Stakeholders," particularly project neighbors. Mr. Watry received a B.S. in architectural engineering and an M.S. in architecture from Cal Poly. His many awards include the George Hasslein Interdisciplinary Studies Endowed Chair, the Community Involvement Award from the Structural Engineers Association of Northern California (2003), the Legend of Post-Tensioning Award from the Post-Tensioning Institute (2004), and the "Distinguished Teacher of the Year" from the Design Build-Institute of America (2004). Mr. Watry is a member of many organizations, including the Structural Engineers Association of Northern California, the Prestressed Concrete Institute, the Post-Tensioning Institute (former director), the College of Architecture and Environmental Design Alumni Association (former president), and the American Concrete Institute (former president).

Jay Whisenant, AIA, CCS
Principal, COO
NTD Architecture
San Diego, California

Jay Whisenant has more than thirty-five years' experience in project management, firm operations, and documentation methods for a wide range of institutional, commercial, and mixed-use projects. In his role as Principal and COO, he has led the firm's development of Building Information Modeling standards, as well as procedures to address contract development, project systems selection criteria, and project delivery methods. In addition, Mr. Whisenant is responsible for NTD's comprehensive quality-control program, including establishing firm-wide standards. Mr. Whisenant is a registered architect in California and Arizona, and is a Certified Construction Specifier. Additionally, he is past chairman of the American Institute of Architects California Council Codes Committee, has represented the national American Institute of Architects at the ICC and ICBO code development conference, and is the past chairman as professional member on the California Building Standards Commission Technical Advisory Committee for Accessibility. Mr. Whisenant's background includes lecturing for eight years at California Polytechnic State University at San Luis Obispo, and conducting lectures and workshops for clients such as California Polytechnic State University's School of Architecture and Environmental Design; San Diego State University; University of California, San Diego; the International Conference of Building Officials, and the American Institute of Architects. He was named Outstanding Instructor for the UCSD Extension Program in 1993. He has published articles in Facility Management Journal and in Business Excellence on professional practice issues.

Chuck Williams, PE, DBIA, PMP
President
ODC Synergy
Overland Park, Kansas

For more than thirty years, Chuck Williams has provided vision and leadership for alternative project delivery strategies, including CM-at-risk, Design-Build, and additional services such as DB Operate (DBO) and DB Finance. He previously served as Senior Vice President at Earth Tech for the firm's design-build program for North America, and held leadership and senior management positions in other national firms, where he significantly expanded their design-build activities and capabilities. Mr. Williams has planning, design, and construction experience in industrial plants, water and wastewater facilities, transportation, and general building projects. He is currently President of ODC Synergy, an independent consultant assisting owners in their planning and procurement of design-build projects, and advising design-build teams to compete for and execute projects. He has long advocated alternative delivery strategies to provide more value and tools for owners to drive greater project success. Mr. Williams is a Designated Design-Build Professional of the Design-Build Institute of America (DBIA) and founding president of the Mid-America Chapter. He serves on the DBIA National Water and Transportation committees, and teaches several DBIA courses. Mr. Williams regularly presents at national conferences on design-build topics and projects, and serves on the Water Design-Build Council, a trade-industry group. He has written several articles in national publications regarding design-build best practices and the means for owners to obtain best value. He has a bachelor's degree in civil engineering from Villanova University, a master's degree in industrial engineering from Stanford University, and is a registered professional engineer in California. Mr. Williams is also a certified Project Management Professional (PMI), and chaired an ASCE Alternative Project Delivery Task Force.

Alan Wilson, AIA
Chief Architect
The Haskell Company
Jacksonville, Florida

Alan Wilson is Chief Architect at The Haskell Company, one of the country's largest integrated design-build organizations. He received his B.A. from Columbia College and his Master of Architecture from the Harvard University Graduate School of Design. Mr. Wilson is active in leadership and service to both the profession and community. He has served in many capacities for the Jacksonville chapter of the American Institute of Architects (AIA), including chapter president. His charitable work includes CANstruction, an annual construction of food cans that benefits the Second Harvest Food Bank, and The Greeley Foundation, a charitable and educational arm of AIA Jacksonville. Currently, he serves as president of the Tilt-Up Concrete Association (TCA), an international trade association. He was the first recipient of the TCA Architectural Achievement Award, given annually for outstanding design contributions in Tilt-Up. Educating and mentoring children is of primary importance to Mr. Wilson. He has educated students at all levels about architecture, including classroom visits to many area schools. Wilson has also served as a soccer coach and is currently scoutmaster for a Boy Scout troop. His passion for architecture is expressed in many ways: through award-winning work, serving on student design juries, acting as a mentor for architectural interns, speaking engagements, and community design workshops. In 2006, Mr. Wilson was presented with the John Dyal Award from AIA Jacksonville, the purpose of the award "to recognize an architect whose leadership and service has been a direct benefit to the profession or community" over an extended period of time.

Jill Wilson
Vice President, Marketing
Gray Construction
Lexington, Kentucky

As Gray Construction's top marketing officer, Jill Wilson skillfully and creatively markets design and construction services to customers, communities, and allied organizations around the world. Through her networking efforts, she assists in generating project leads, developing customer relationships, and coordinating proposal and presentation efforts. Her overall marketing efforts also include customer events, marketing materials, public relations efforts, multimedia presentations, the company corporate database, and historical archives. Ms. Wilson also assists Gray Construction's sister companies—Gray-ICE Builders, Anaheim, California; Operations Associates, Greenville, South Carolina; and WS Construction, Versailles, Kentucky—with various marketing efforts. She joined Gray in 1988 as Marketing Specialist. During her twenty years at Gray, she has also held the positions of Marketing Manager and Director of Marketing. Realizing networking is vital to a successful marketing program, she is actively involved in numerous organizations. She serves on the Kentucky Chamber of Commerce board, is a board member of the Commerce Lexington Workforce Development Partnership, serves on the Associate Advisory Committee for the American Frozen Food Institute, serves on the Junior Achievement of Lexington board; chaired the National Design-Build Awards Committee for the Design-Build Institute of America; has served on the board of the Ohio Valley chapter of the Design-Build Institute of America; is past president of the Kentucky Association for Economic Development; and is past president of The Society for Professional Marketing Services, Kentucky chapter. Ms. Wilson is a 1984 Georgetown College graduate with a B.S. in business administration.

Edward C. Wundram, AIA Emeritus, CSI, DBIA
Senior Project/Construction Manager
Cumming
LACCD—Los Angeles Southwest College
Los Angeles, California

Edward C. Wundram has forty years' experience in the design and procurement of major public facilities, including thirty years of developing and utilizing performance-based specifications for the acquisition of building systems and complete facilities. He has managed his own professional services firm since 1980, specializing in the organization and administration of design-build procurement programs for significant public facilities. Mr. Wundram has considerable experience managing large public projects as the design coordinator of an international architectural, engineering, and project management firm. He received a bachelor's degree in architecture from Georgia Tech. Affiliations include the American Institute of Architects, the Construction Specifications Institute, and the Design-Build Institute of America. Mr. Wundram has published a number of materials about design-build, including several articles for The Construction Specifier and a textbook entitled *Design-Build: Planning Through Development* with Michael Loulakis and Jeffrey Beard.

Drew Yaggy, PE, DBIA, LEED AP
Chief Engineer and Preconstruction Manager
TDIndustries, Inc.
Dallas, Texas

Drew Yaggy serves as Chief Engineer and Preconstruction Manager for TDIndustries in Dallas, Texas, a large facilities service and specialty construction company that primarily serves the southwest region of the United States. Mr. Yaggy is responsible for procuring and developing new mechanical and plumbing construction work, with a major focus on design-build projects. TDIndustries is an employee-owned company with over 1600 partners operating under the management style known as "Servant Leadership." Having been named to Fortune magazine's "100 Best Companies to Work For" list for the last twelve years, TDIndustries has become a Fortune "All Star" for being one of only thirteen companies nationwide to attain that distinction since the list's inception in 1998. Mr. Yaggy graduated from Texas Tech University in 1970 with a B.S. in engineering. He has over forty years' experience in the construction industry, working first in his father's cabinet and millwork shop and then with consulting engineers, equipment manufacturers, and mechanical contractors. He has worked as a designer, engineer, estimator, project manager, and construction division manager on a wide variety of commercial, industrial, institutional, and sports-facility projects. Mr. Yaggy joined TDIndustries in 1993 as Chief Engineer. He is a registered professional engineer in Texas. He is a Designated Design-Build Professional by the Design-Build Institute of America (DBIA) and currently serves as secretary of the Southwest Region of DBIA. He is an active member of ASHRAE and serves on the Regional Codes Coordinating Committee for the North Central Texas Council of Governments, which covers a sixteen-county region of Texas.

Jim Zahn, LEED AP
Project Director
Ellerbe Becket
Minneapolis, Minnesota

Before joining Ellerbe Becket in 2009, Jim Zahn used his twenty-five years of design-build experience to lead the preconstruction activities at Ryan Companies US. As a part of Ryan's Lean Initiative, he was integral to formulating, refining, and sustaining process improvements throughout the entire Ryan organization. His project experience is deep and broad, having managed high-rise residential buildings, downtown skyscrapers, suburban corporate centers, mission critical facilities, and industrial buildings. These projects include the Mississippi Plaza in Davenport, Iowa, the River Parkway Place in Minneapolis, Minnesota, and the Children's Hospital and Clinic in Roseville, Minnesota. Mr. Zahn is well respected by his customers, team members, and the design community, and strives for the highest level of professional service in all his endeavors. He received a B.S. in civil engineering from Marquette University. He is a member of the Design-Build Institute of America (DBIA) and the American Society of Civil Engineers (ASCE).

INDEX

A

acquisition strategies, 107–112
Adams, Marilee, 308
ADePT, 257–258, 258f
ADePT Design Builder™, 258–259, 270
ADePT Design Manager™, 269–270
Adept Management Ltd., 238, 258
adjusted low-bid method, 127, 127f
AEC. *See* architecture, engineering, and construction (AEC)
AIA. *See* American Institute of Architects (AIA)
AIArchitect, 341
Alberti, Leon Battista, 4–5
alignment, 246–247
Alley, John, 325
Alliance Contract, 332
alliance contracting, 35–36, 276, 333f
Alliance Contracting IQ, 35
Alternative Delivery Solution, LLC, 141
alternative project delivery methods, 11–13
AMEC Construction Inc., 22, 23
American Council of Engineering Companies (ACEC), 21, 350
American Institute of Architects (AIA), 5, 12, 19, 32, 332, 334
American Institute of Architects California Council (AIACC), 33–34
American Institute of Constructors (AIC), 5, 13–15
American Society of Civil Engineers (ASCE), 5, 18, 275, 351
American Society of Heating, Refrigerating and Air Conditioning Engineers (ASHRAE), 91
American Subcontractors Association (ASA), 5
Analytical Design Planning Technique (ADePT). *See* ADePT
annotated drawings, 255
Apple, 8
ArchiCAD®, 339, 340, 342
architect's estimate, 194
architecture, engineering, and construction (AEC), 4, 12, 323, 326, 342, 352, 354
Training Technologies, 68
Arizona State University, 55
ASCE. *See* American Society of Civil Engineers (ASCE)
ASCE's Design-Build Task Force, 351
Associate Constructors (ACs), 14
Associated Builders and Contractors (ABC), 5
Associated General Contractor (AGC), 5, 12, 19, 342, 350
Associated Schools of Construction (ASC), 15
Atwell Hicks, LLC, 312
Austin Company, The, 16–17
Austin Method®, The, 17
authentic trust, 304–305
Autodesk, 342
award celebration, 178
award fees, 144–147
awards, 74–75

B

background of relatedness, 247
Ballard, Glenn, 35
Base Realignment and Closure Commission (BRAC), 79–80
Beamer, Todd, 25
Beard, Jeffrey L., 5, 15, 18, 19, 20, 350
Beck, Peter, 162, 322
Beck Group, The, 109, 111, 162, 208, 278, 312, 314, 317
Bedner, Nirsch, 197
Beiswenger, Hoch and Associates, Inc. (BHA), 219–220
Berard, John, 350
best and final offer (BAFO), 128
best-in-class technical knowledge, 291
best practice capture log, 272–273
best value phase, 55, 116–119
best value procurement method, 116–117
Betamax, 336
BIM Bombs™, 340
BIMstorm™, 337–343, 348
BIMstorm™ LAX, 337, 341–342
Black & Veatch, 60–63
Bloch, Arthur, 40
Block, Peter, 324
Bohm, David, 320
Boston University School of Law, 274
Bradberry, Travis, 310
Bradfish, Earl, 122
Brafman, Ori, 289
Brafman, Rom, 289
brand diffusion, 313
Bread Loaf Corporation, 344
bridging, defined, 113, 139
bridging design, 139, 139f
Brooks, Herb, 157–158
Brooks Act, 7, 114
Brozovich, Erin, 66
Brunelleschi, Filippo, 4
budget for design-build services, 192–210
  concept estimates phase of, 206
  conceptual design stage of, 193
  construction document stage of, 194–195
  cost model for, 203–204
  design development estimates phase of, 207
  design development stage of, 194
  detailed estimates phase of, 207–208
  developing, 195, 202–204
  estimating churn phase of, 208
  feasibility estimates phase of, 205–206
  programming stage of, 193
  progressive estimating phases of, 204–210
  schematic design stage of, 193
  schematic estimates phase of, 206–207
  stages of, 192–195
building information modeling (BIM), 3, 239, 332
  architectural implementation of, 32
  defined, 31, 334
  examples of, 32f
  future of, 32
  goal of, 335
  Kimon Onuma's role in, 336–337
  purpose of, 31–32, 336
buildingSMART™, 336
Building Trust in Business, Politics, Relationships, and Life (Flores/Solomon), 304
Built to Last (Collins/Porras), 309
Burcamp Steel Company, 242
Burns, Pat, 129
business enabler, 153
buying design-build services, 107–147
  acquisition strategies for, 107–112
  adjusted low-bid method for, 127
  competitive factors for, 119–124
  contract types for, selecting, 141–147

equivalent design/low bid method for, 128–129
evaluation methods for, 126–133
fixed price/best design method for, 127
meets criteria/low bid method for, 128
negotiating, 132–133
project enhancement approach to, 129–132
sources for, 112–117

C

California American Institute of Architects (AIA), 332
California Center for Construction Education (CCCE), 354
California Code of Regulations, 201
California Polytechnic State University, 211, 332, 334, 344
Calloway, Joe, 156
Caltrans (California Department of Transportation), 164–169
Camp, Dresser, & McKee Inc. (CDM), 85
Campbell Shipyard, 196
Capital Project Strategies, LLC, 274
Carroll, Jim, 106
cascading communication, 320
CDM Constructors Inc., 66, 211, 310
Cebulla, Michael, 280
Certified Design-Build Professional™, 20, 354–355, 355f
Certified Professional Constructor (CPC), 13, 14
Charles Pankow Builders, 14, 17, 19, 20, 76
Charles Pankow Foundation, 246, 354
Chicago Public Library, 33
Childress, Blake, 62
churn phase, 208
Citi Field, 234
Clampitt, Phillip G., 223, 308
Clark Civil, 266

Clark Construction Group, LLC, 164, 165–169, 253, 305
CM-at-Risk project delivery, 13
advantages of, 49–50
choosing, primary reasons for, 51
components of, 48
configurations for, 47f, 49f
disadvantages of, 50
importance of, 48
purpose of, 47–48
cognitive intelligence (IQ), 310
Cohen, Alan, 72
co-leadership, 324
collaboration, 28–31, 321–323
collaboration economy, 348–252
Collaboration in the Building Process, 28
collaborative process, 290
integration vs., 322f
overview of, 29f
Collaborative Process Institute (CPI), 158
concepts of, 29
maxims of, 28
organization of, 28
overview of, 29f
College of Liberal Arts at Tufts University, 274
Collins, Jim, 306, 309
co-location, 120, 124, 261
commitment management systems, 248
committed listening, 319
committed openness, 305
committed speaking, 319
communicating performance, 90
communication
approaches to, practicing, 321
assumptions and, 320
cascading, 320–321
design-build services and, 152–153
design-build team and, 308–321
methods of, 177–178
model for, 318f
performance requirements for, 90
techniques/tools for, 319
competing for design-build services, 151–179
award celebration and, 178

communication and, importance of, 152–153
marketing and, 153–155
procurement workshops for, 155–156
rules for, 152
team for, selection of, 157–163
Computer Aided Design (CAD), 31, 335, 338, 339
Concept Design Package, 197
concept estimates phase, 206
concept estimating process, 189f
conceptual design, 120, 124
conceptual design-build services, 186–192
conceptual design stage, 193
conceptual estimating, 96, 186
Conrad Hotels, 196
constructability reviews, 241
construction associations, 5f
construction document stage, 194–195
Construction Exchange, 349f, 352
Construction Industry Presidents Forum (CIPF), 21
construction management (CM), 345
alternative project delivery and, 11–13
college degrees in, 15
Construction Management Association of America (CMAA), 5, 12
construction management-at-risk, 47–51
Construction Manager, 277
Construction Operations, 222
Construction Specifications Institute (CSI), 5, 187
Construction Users Round Table (CURT), 5, 334
concepts of, 30–31
mission of, 29
organization of, 29
task force for, 30
constructor. *See* contractor
Constructor Certification Commission, 13

contingencies, 211, 215–218
contingency development, 212
contingency round table, 215
continuous team debriefs, 264–265
contract cost-risk, 144f
Contract Documents, 100–101
contractor
concept of, 14
design-build project delivery and, role of, 75–76
importance of, 15
responsibilities of, 14–15
transformation of industry and, 13–15
contractor-led design-build project delivery, 78, 84
contract performance phase, 296
contracts
approaches to, 98–103
award fees for, 144–147
design-build, 100–101
governing documents in, 98–100
high performance, 102–103
relational, 102
target cost, 143–144
controlling party, 86
Corona Maintenance Shop and Car Washer Facility, 232–236
Corona Plaza, 234
cost model, 174, 203–204
design-build project delivery vs., 96–97
Cost Model Briefing Session, 204–205
cost-plus, 141
Cramer, James P., 37
Crawford, David S., 74, 178
Crockett, Jim, 186, 221
cultural fit, 160, 291
Cumming Corporation, 121
Cupka, Jeffrey, 120
Cushman, Frank, 111

D

Dahlberg, Matt, 16
data-centric activity, 322
Davidson, Charlie, 20
Day, Brian, 169
DBIA. *See* Design-Build Institute of America (DBIA)

"Declaration of Information Independence," 337, 337f
DeKoch, Robert J., 223, 308
DelaRiva, Ryan, 290
Deming, W. Edwards, 228, 284
Denver's Downtown Aquarium, 33
Dependency Structure Matrix (DSM), 257
design aesthetics, 213–214
design associations, 5f
design-bid-build-bust, 53, 54f
design-bid-build project delivery
  advantages of, 43–44
  choosing, primary reasons for, 46–47
  design-build services vs., 92–95
  disadvantages of, 44–46
  process of, 43
  sequence for, 43f
design-build charrette, 172
Design-Build Consulting Group, 20
design-build contract, 51f, 100–101
Design-Build Contracting Handbook, 111, 142
Design-Build Dateline, 20
Design-Build Educator Workshop, 354
design-build entity, 51–52, 51f, 84f
design-builder
  characteristics of, 306
  ego of, 306–307
  emotional intelligence of, 310–311
Design-Build Institute of America (DBIA)
  Annual Conference, 354
  Board of Directors, 275
  Certification Board, 354
  certification programs for, 20
  Certified Design-Build Professional, 354
  Code of Ethics, 20
  Contracting Guide, 101–102
  DBIA Task Force, 275
  DBIA University Program, 354
  Design-Build Contracting Guidelines, 275
  founders of, 18–19
  legislation map for, 55f
  legislative initiatives by, 20–21
  Manual of Practice, 20, 73, 275, 351

members of, elected, 20
  official meetings of, first, 20
  regulations for, modification of, 21, 28
  steering committee for, 19–20
Design-Build in the Public Sector (Beard), 18
Design-Build Knowledge Community, 33
Design-Build Lessons Learned (Loulakis), 102, 140
design-build management team
  organizational structure of, 278f
  project managers, 278–279
  roles/responsibilities of, 276–277
  superintendents, 279–280
design-build-operate-maintain (DBOM), 123
Design-Build-Planning through Development (Beard/Loulakis/Wundram), 5, 15
design-build plus, 64, 138
Design-Build Professional Group, 124, 353
design-build project delivery, 51–57
  advantages of, 52–53
  characteristics of, 73–104
  choosing, primary reasons for, 54–57
  concept of, 51
  contracting approach, 98–103
  contract issues concerning, 100–102
  contractor in, role of, 75–76
  contractor-led, 78, 84
  controlling, concept of, 85–86
  cost vs., 96–97
  design-bid-build vs., 92–95
  designer-led, 84
  disadvantages of, 54
  formats in, 97
  future of, 103–104
  innovative solutions using, 59, 64
  integrated, 85
  management of, 65–66
  performance requirements, 89–92

proactive vs. reactive response to, 76–78
  purpose of, 51–52
  risks, 87–88
  schedule, 97–98
  as single entity, 78, 84–86
  solicitation/awards, 74–75
  team approach, 75–78
  total-facility-solution types of, 64–65
Design-Build Project Manager (DB-PM), 276, 279f
design-build project managers, 278–279
design-build schedule, 99f
design-build sequence, 52f
design-build services
  concept of, 15
  curriculum for, 354–355
  definition of, 3
  design-bid-build project delivery vs., 92–95
  education on, 354
  future of, 331–357
  practitioners of, 16–17
  public funds for, use of, 15
  reemergence of, 15–21, 28
design-build superintendent, 279–280
design-build team, 285–326
  advocating project to, 311, 318
  collaboration and, 321–323
  communication and, 308–321
  integration and, 321–323
  leading, 323–325
  members of, 292–294
  as "No Gaps Game," 286
  results for, 290–292
  rules for, 294–300
  success to building, 306–311
  transitioning of, 286–290
  trust and, 300–305
design commitment, 260, 267
Design-Construct Coordinator, 277
design-construct interface, 241, 242
design-construct stage, 267–270
  for design-build services, 267–270
  management task for, 268, 268f

teaming strategies for, 268–269
  techniques for, 269
  tools for, 269–270
design coordinator, 277
design-cost interface, 241, 242
design-cost reconciliation, 208, 259–261
Design Criteria Design-Build approach, 139
design delivery schedule alignment, 268, 270f
design development estimates phase, 207
design-development stage, 194, 259–267
  for design-build services, 259–267
  management task for, 259–261, 260f, 262–264
  teaming strategies for, 261, 264–265
  techniques for, 261, 265–266
  tools for, 262, 266–267
designer-led design-build project delivery, 84
design evolution log, 266–267
design impact, 213
design impact risks, 214, 215f
DesignIntelligence, 37
design management, 231
Design Manager, 277
design narrative, 210
Design-of-Record (DOR), 277
design parameters, 187
design-performance interface, 243
design-performance reconciliation, 260f, 262–264
design process, 237
design process planning, 255–256
design-risk shifting, 140f
design services, 213
design specifications, 89, 90
design stability, 214
design validation, 252–255, 260f
design work breakdown structure (WBS), 257
detailed estimates phase, 207–208
detailed estimating process, 186f

diagnosis bias, 289
dialogue, 320
Dickson, Bart, 50
Dielschneider Associates, 122
Direct design-build approach, 138–139
discussions, 128
Dodger Stadium, 337
Dome of Santa Maria del Fiore, 4
Dome-Tech, Inc., 233
Dondiego, Robert L., 236
Donnellon, Anne, 311, 319
draw-build, 113

E

early release packages, 269, 269f
Ecotect, 341, 342
Edison, Thomas, 150
Educational Facilities Laboratory (EFL), 122
Elite CAD, 342
Elite Software, 342
El Taller Colaborativo, PC, 232
Elton, Chester, 247
Embracing Uncertainty: The Essence of Leadership (Clampitt/DeKoch), 223–224, 308
EMBT Architects, 165
emergency award, 112
Emery Roth & Sons, 123
emotional intelligence (EQ), 310
Emotional Intelligence Quick Book, The (Greaves/Bradberry), 310
Engdahl, Dave, 31
Engineering News Record (ENR), 351
Engineer-Procure-Construct (EPC), 274
engineer's estimate, 194
Engineer's Joint Contract Documents Committee (EJCDC), 101
enhancement criterion, 130
environment of trust, 300
equivalent design/low bid method, 128–129
estimates on developing design-build services, 183–224
  accounting for, 211–218
  conceptual, 186–192
  contingencies and, 215–218
  estimators for, role of, 221–222
  pricing for, 184–186
  qualifications for, 210–211
  risks and, assessment of, 213–215
  uncertainties to consider, 222–224
  value engineering and, 218–221
  value/quality vs., 185–186
estimating format, 188f
estimating phases, 205f
estimating progression, 209f
Eugene IV, Pope, 5
Evey, Walker Lee, 21, 22–24, 27, 92
Expedia.com, 340
external trend management, 265f

F

Facebook, 348
feasibility estimates phase, 205–206
Federal Acquisition Regulations (FAR), 21, 28
Fifth Discipline Fieldbook, The (Senge), 287
Final Revised Proposals, 128
Five Dysfunctions of a Team, The (Lencioni), 300
Five Star Electric, 233
fixed price/best design method, 127
Flemming, Bill, 96
Flickr, 348
Flintco, Inc., 279
Flores, Joe, 109, 278, 304
Flores Lund Consultants, 197
From Good to Great (Collins), 306
full disclosure, 212
functional requirements, 90
function-oriented focus, 219f
Future of Management, The (Hamel), 230

G

gainshare, 333
Gavin, Don, 274
Genius of AND, The, 309
Gibbs and Register, Inc., 112
Gidez, Greg, 45, 345, 355
Gladson, Rebekah G., 59

Goldsmith, Neil, 123
Goodrich, Don, 203
Google Earth, 342
go-or-no-go decision, 169
Gostick, Adrian, 247
governing documents, 98–100, 100f
Graphisoft, 342
Graves, Michael, 123
Gray, Jim, 19, 347
Gray Construction, 17, 19, 154, 347
Greaves, Jean, 310
Greenberg, Bennett, 20
Green Building Studio, 342
Greenburg, Bennett, 275
green elements, 315
Greenleaf, Robert K., 323
guaranteed maximum price (GMP), 136, 142, 199–201

H

Haddon, Bruce, 330
Halvorson, Julie, 238
Hamel, Gary, 3, 230, 244
Hammond, David, 343
Harms Associates Engineers, 350
Hartung, Robert J., 141
Harvard Business Review, 12
Haskell, Preston, 16–21, 123–124, 275, 351
Haskell Company, The, 16, 17, 64–65, 134–137, 159, 174, 239, 334, 351, 353
HDR Architecture, 22, 24, 25
Heery, George, 121–122
Heery-Farrow, Ltd., 122
Hensel Phelps Construction Co., 24, 25, 45, 196–197, 201, 345, 355
H.H. Robertson, 350
high performance contracts, 102–103
Hilton Hotels Corporation, 196–197
Hilton San Diego Bayfront Hotel, 184, 196–201
Hoag, Diana, 102–103
Hoffman Construction Company, 50, 123
Holmes Hepner & Associates Architects, 312
honorarium, 118
Hooghouse, Jeff, 56
Howe, Neil, 352, 353
Howell, Greg, 35
Huffman, Rex D., 112
Hyatt Regency, 10

I

incentive award fee evaluation form, 145f
information inputs, 240f
information interviews, 154–155
information outputs, 240, 240f
integrated design-build project delivery, 85
integrated design-build services, 94f
integrated design-build team, 286
Integrated Design Review (IDR), 239
Integrated Form of Agreement (IFOA), 35–36, 102
integrated partnering program, 247–248
Integrated Project Delivery (IPD), 32–33, 276, 332–333
  model for, 34f
Integrated Project Leader (IPL), 276
integration, 28–31, 321–323, 322f
integration manager, 248, 276–277
interdisciplinary fluency, 286
internal trend management, 265f
International Express, 233
International Parking Design, Inc., 197
International Roughness Index (IRI), 243
Invisible Employee, The (Elton/Gostick), 247
Invitation for Bids (IFB), 74–75
Iowa State University, 238
Irresistible Pull of Irrational Behavior, The (Brafman/Brafman), 289

J

J.A. Jones Construction, 20
JBA Consulting Engineers, 197
John Portman and Associates, 197
Johnson, Neil, 160
Johnson, Philip, 123
joint risk assessment, 172
Jones, Dan, 346
Joseph Wong Design Associates, 197

J.S. Alberici Constructors, 60–63

K

kaizen, 347–348
Kajima Corporation, 17
Kashiwagi, Dean, 55
Kelly Masonry Corporation, 233
Kersey, Marc, 165
Killpatrick, Nat, 242
Kiss + Cathcart, 233
Kleinfelder, Inc., 197
KlingStubbins, 79–83
Kluenker, Chuck, 11
knowledge-centric activity, 322
knowledge-in-action, 308
Konchar, Mark, 57
Koolhaas, Rem, 165
Korte Company, 190
Korte Construction, 17
Kose, Dane, 345
Kreikemeier, Kraig, 19, 275
Krispy Kreme Doughnut Corporation, 134–137
Kuhn, Thomas, 231
Kunnath, Rik, 19, 20, 275, 351

L

Last Planner™, 35
Layton Construction Inc., 325
Leadership in Energy and Environmental Design (LEED), 36–37, 159, 169, 314, 316, 344, 346
Leading the Revolution (Hamel), 3
Lean Construction Institute (LCI), 5, 34–35, 102, 332, 346–348
lean manufacturing, 346
Lean Project Delivery, 102
LEED. *See* Leadership in Energy and Environmental Design (LEED)
LEED Accredited, 353
Lencioni, Patrick, 300, 320
Lichtig, Will, 35
Lindholm, Wayne, 185
Linscott, Law Y Greenspan Engineers, 197
liquidated damages, 356
L.M. Drown, 196
Lockheed Corporation, 121
Loulakis, Michael C., 5, 15, 20, 102, 111, 274
low-bid design-build, 113

low-bid mentality, 6–8
low-bid procurement method, 113–114
Lum, Kim, 14, 76
lump sum price, 141

M

M.A. Mortenson Company, 79–83, 129
Machine That Changed the World, The (Roos/Jones/Womack), 346
Macintosh®, 8
MacLeamy Curve, 237, 237f
Main and First Design/Build Associates Inc., 164
managing design-build services, 229–281
design-construct stage for, 267–270
design-development stage for, 259–267
design in, 231, 237–244
management team for, 273–280
postaward stage for, 252–259
postconstruction stage for, 270–273
process for, 230–231
proposal stage for, 246–252
stages of, 245–273
standard practices for, establishing, 244–245
team for, 245
techniques for, 245
tools for, 245
marketing, 153–155
MARS Facility Cost Forecast System, 342
Martin, John A., 279
Martin and Peltyn, 197
Massachusetts Institute of Technology, 166–167
Master Builder, 4–6, 37–38
MasterFormat™, 97, 187, 204
Mayne, Thom, 165–167
McCullough, Grant, 275, 351
McDaniel, Meloni, 111
McGarva, John, 46, 212
McGarva, Josh, 349, 352
McGraw-Hill, 20
McMahon, BG John R., 301
McNeil, Al, 20
Mechanical Contractors Association of America (MCAA), 5

meets criteria/low bid method, 128
mental model, 287
Millennial Generation, 352–353, 354
Millennials Rising: The Next Generation (Howe/Strauss), 352–353
Miller Act, 6
Miracle (movie), 157
Miralles, Enric, 165
Mitchell, Fred W., 137
Moffett, Jordan, 159
Molenaar, Keith, 67
Momsen, Charles B., 79
Moraga, Cherrie, 2
Morphosis Architects, 164, 165, 168
Morris Forman Wastewater Treatment Plant, 60–63
Mortenson Construction, 345
Mötley Crüe, 8
Murray, Andrea, 344
MySpace, 348

N

National Design-Build Conference, 351
National Guard, 21
National Institute of Building Sciences (NIBS), 336
National Society for Professional Engineers, 21, 275
Navis Works, 342
Nemetschek North America, 342
New London Naval Submarine Base, 79–80
New York City Transit Authority (NYCT), 232–236, 235
New York Mets, 234
Next Architect: A New Twist on the Future of Design, The (Cramer), 37
Ng, Michael, 211
Nigro, William T., 239
Niles Bolton Associates, 44
"No Gaps Game," 286
non-linear innovation, 3
notice of trend form, 262f
NTD Architecture, 335

O

ODC Synergy, 163
Office of Metropolitan Architecture, 165
O'Hare Airport, 20

Ohio State University, 66
one-on-one briefings, 247
Onuma, Inc., 336, 337
Onuma, Kimon, 336–337, 338–343
ONUMA Planning System (OPSTM), 337, 339, 340–343
open-book policy, 212
open standards, 337
Opus Group, 17
outline specifications, 210
Overcoming the Five Dysfunctions of a Team: A Field Guide for Leaders and Facilitators (Lencioni), 320

P

painshare, 333
Palandrani, Lou, 253
Palmer, Ed, 44
Pankow Builders, 290
parametric formulas, 187
Parkinson, Robynne Thaxton, 88, 241
Parkinson PLLC, 88
Parnell, John, 77
Parsons Brinckerhoff, 231, 232, 236
partnering program, 178
partnering rating form, 248–252, 249–251f
pass or play decision, 169
Patuxent Naval Air Station, 350
Pavarini Construction Co., 123
PCL Civil Constructors, Inc., 219–220
PCL Construction Enterprises, Inc., 95, 220
Pentagon Renovation Project (PENREN), 21, 22–27, 103
specifications for, 92f
"perfect storm," 331–332
performance criteria, 91–92, 92f
performance requirements
communication, 90
defined, 89
design specification, 89, 90
incentivizing for, 143
perspective specifications *vs.*, 89f
verification of, 91
waste reduction through, 92
performance standards, 243

Permadur Industries Inc., 233
perspective specifications, 89f
Phelps Portman San Diego LLC, 196–197
Phoenix Project, 22–27
placemaking, 315
platform skills, 176
Platinum LEED, 346
Polytechnic State University, 290
Porras, Jerry, 309
Portland Building, 122
postaward stage, 252–259
 for design-build services, 252–259
 management task for, 252–253, 253f, 255–256
 team strategies for, 253–254, 256–257
 techniques for, 254–255, 257–258
 tools for, 255, 258–259
postconstruction evaluation, 271
postconstruction stage, 270–273
 for design-build services, 270–273
 management task for, 271, 271f
 teaming strategies for, 271–272
 techniques for, 272
 tools for, 272–273
Practice Management Digest (AIA), 334
preconstruction services, 12, 47, 222
preemptive marketing, 154–155
preliminary design design-build approach, 139–140
prescriptive specifications, 89
price proposal, 99, 100
PricewaterhouseCoopers (PWC), 345
procurement workshops, 155–156
production control process, 35
program estimate, 206
programming stage, 193
progressive design-build, 115–116, 115f
progressive estimating, 204–210
Project Advocate, 158
Project Alliance Agreement, 332–333
project characteristics workshop, 256–257

project debriefs, 272
project delivery, defined, 41
project delivery methods, 42–57. *See also* design-build project delivery
 CM-at-Risk, 47–51
 comparing, 57–59
 design-bid-build, 42–47
 selection of, fundamental factors for, 67–69
 trend lines for, 57f
project delivery selection matrix, 68, 68f
Project Director, 158
project enhancement approach, 129–132n, 131f
project enhancement evaluation, 129–132, 131f
project enhancement score sheet, 131f
Project Executive, 158
project-oriented focus, 219f
proposal response team, 158
proposal stage, 246–252
 for design-build services, 246–252
 management task for, 246–247, 247f
 team strategies for, 247
 techniques for, 247–248
 tools for, 248–252

Q

qualification phase, 118
Qualifications Based Selection (QBS), 7, 21, 114–116
qualifications-based selection method, 114–116
qualifications package, 114
quality points, 130, 185–186
Quatman, Bill, 33
question thinking, 308

R

recognition luncheon, 271–272
RediCheck Interdisciplinary Coordination (Nigro), 239
redundant contingencies, 215
Rees, William, 344
reflecting-in-action, 308
relational contracting, 35–36, 102
relational contracts, 102
release for construction (RFC), 194

Request for Information (RFI), 242
Request for Proposal (RFP)
 analyzing, 172
 delivering, 175, 176–178
 design-build services and, competing for, 163, 169–170
 design strategies for, 138f
 documents for, producing, 174–175
 Invitation for Bids *vs.*, 74–75
 oral presentation of, 175–178
  communication, methods of, 177–178
  focused, staying, 176
  justification, 177
  preparing for, 175–176
  rules for, 176–178
  talking heads syndrome, 176–177
 process for
  best value phase of, 118–119
  design approaches for, 138–141
  qualification phase of, 118
  scoring of, 124–126
  two-phase procurement for, 117–118
Request for Qualifications (RFQ), 114, 118
 analyzing, 170–171
 responding to, 171
 short list and process of, 171
responsibility matrix, 296, 297–298f
Revit Architecture, 342
RFP. *See* Request for Proposal (RFP)
RFQ. *See* Request for Qualifications (RFQ)
Right Stuff, The (Wolfe), 350
Riley, David R., 53
risk assessment, 212
risk identification, 173f
Risk Management Committee, 310
risk mitigation plan, 172
risk registers, 240
Ritz Carlton, 19
Robynne Thaxton Parkinson PLLC, 88
Ronald Reagan Washington National Airport, 19

Roos, Dan, 346
Rosendin Electric, 293
Rough Order of Magnitude (ROM), 205–206
Rudolph, Vernon, 134
Rutgers College, 350
Ryan Companies USA, Inc., 17, 77, 120, 160, 238, 280

S

San Diego Bayfront Hilton Hotel, 201
Santa Maria Novella, 5
Sanvido, Victor, 57, 347
Sarnafil, 167
schedule, 97–98
schematic design stage, 193
schematic estimates phase, 206–207
Schmieder, Michael, 239
Schön, Donald, 308
Schorfheide, Alan, 190, 221
seamless team, 176–177
Second Life, 348
Sector Command Centers, 340
segregated services concept, 4–6
Selecting Project Delivery Systems, 57
Senge, Peter, 287
Serco Institute, 35
Servant as Leader, The (Greenleaf), 323
servant leadership, 323
Servant Leadership: A Journey into the Nature of Legitimate Power and Greatness (Greenleaf), 323
Seyfarth Shaw LLP, 20
Shalom Baranes Associates & Studios Architecture, 25
Shea, Paul, 85
Shea Stadium, 234
Sheridan, Philip J., 266
shortlisting, 118
Shughart, Thomson, and Kilroy, PC, 33
signature architect, 158
Simmons Machine Tool Corporation, 233
simple trust, 304
Simpson, Scott, 37
single entity, 78, 84–86
single source responsibility, 51–52
Sisel, Dan, 345
Sizemore, Gregory L., 30
Skanska USA Building Inc., 96

Skanska USA Civil Northeast, 231, 232, 236
Sketch Up, 342
Smith, Stan, 275
Solibri, 342
solicitation, 74–75
Solomon, Robert, 304
Sonnier, Keith, 166
Sorley, Tom, 293
source selection, 112–117
  best value procurement method for, 116–117
  defined, 112
  low-bid procurement method for, 113–114
  qualifications-based selection method for, 114–116
Southland Industries, 347
Spearin Doctrine, 87, 88
Special Theory of Relativity, The (Bohm), 320
specialty contractors, 84
specifications, 89, 91f
Squalus, 79
Square One Research, 342
Steele, John, 238
stereotypes, 288
stipend, 118
Stites, Julie, 334
Stone, 350
strategic branding programs, 312–313
Strategic DB-Teaming Initiative, 291
strategic partners, 293–294
Strauss, William, 352, 353
structural insulated panel (SIP), 219
Structure of Scientific Revolutions, The (Kuhn), 231
Structures One PL, 312
Stubbins Associates, 37
subcontractor-vendor workshops, 268–269
Suitt Construction, 17, 19
Sundt Construction, Inc., 74, 178
Support Services, 222
surrogate owner, 156, 157
Sustainable Design, 332, 344–346
Sutter Health, 35, 102
Sverdrup Engineering, 19
systems thinker, 279

T

Tagliabue, Benedetta, 165
talking heads syndrome, 176–177
Tapscott, Don, 348
Tardif, Michael, 341
target cost contract, 143–144
target design schedule, 269
target numbers, 197
target price, 333
TD Industries, 323
team alignment, 160–161, 161f
team approach, 75–78
team chatter, 321
teaming agreements, 161, 294
teaming profile, 161
teaming session, 178
team integration, 273
Team Talk: The Power of Language in Team Dynamics (Donnellon), 311
Technical Evaluation Criteria, 131
technical expertise, 291–292
technical leveling, 128
technical proposal, 98–99
technical review team, 265–266
technical transfusion, 128, 129
Tekla, 342
Times Square, 315
TLC Engineering for Architecture, 312
total-facility-solutions, 64–65
Toyota Production System (TPS), 346
traditional project delivery method, 42
traditional service models, 94f
transformation, defined, 3
transformation of industry, 3–38
  alternative project delivery methods and, 11–13
  collaboration/integration and, increased, 28–31
  construction management and, 13–15
  design-build and, reemergence of, 15–21, 28
  economic conditions/outside influences on, 8–11
  future of, 37–38
  low-bid mentality and, 6–8
  Master Builder and, 4–6
  new initiatives/strategies for, 33–37
  technology in, role of, 31–32
Transit Technologies, 233
trend estimate report, 263f
trend log summary, 264f
trend management, 209, 261, 265f
trend racking documents, 262
trend register logs, 262
Trevi Fountain, 315
trust
  building, 300–305
  character traits and establishing, 303–304
  competency and ability to, 303
  design-build team and, 300–305
  distinguishing, 301–302
  past and gaining, knowledge of ones, 302–303
  relative power/authority and, 304–305
  similarities *vs.* differences and, 303
  spectrum for, 302f
Tufts University, 274
Tunnicliffe, Peter, 310
Turner, Michael, 322
Turner Construction, 20
two-envelope proposal, 126
two-phase procurement, 117–118, 119f
2008 R.S. Means Square Foot Costs, 187

U

Unger, Craig, 142
Unger Security Solutions, LLC, 142
UniFormat™, 97, 187, 204, 205f
unifying mechanisms, 253–254
Union Switch & Signal, 233
United Nations Brundtland Commission, 344
United States Army Corp of Engineers, 21, 56, 301
United States Coast Guard, 340, 343
United States Green Building Council (USGBC), 5, 36, 169, 314, 344
United States Navy, 21, 79–83
United States Open tennis tournament, 234
unit price, 142
University of British Columbia, 344
University of California – Irvine, 59
University of Colorado, 67
University of Oklahoma, 111
University of Southern California, 342

V

value engineering, 195, 218–221
values, 185–186
Vanir Construction Management, 11
Vecera, Carter, 209
VectorWorks, 342
Verrastro, Mike, 197–198, 201
virtual building models, 338
Vrabel, Stephen, 95

W

Wagner, Barbara, 305
Walker, Coleman, 353
Walker, Nancy, 314, 316, 317
Walker Brands, 312–317
Warren, Don, 19, 275, 351
Washington Post, 24
waste reduction, 92
Watry, Nick, 7
Watry Design Group, 7
Webster, 350
weighted criteria matrix, 125, 125f, 126
Welsbach Electric Corp., 233
Western Water Constructors, Inc., 46, 212, 349
Wheatley, Margaret, 224
Wheeler Goodman Masek Architects, 350
Whisenant, Jay, 335
Whitestone Research, 342
Wickwire, Jon, 274
Wickwire Gavin, P.C., 20
Widham, Chet, 21
Wikinomics; How Mass Collaboration Changes Everything (Williams/Tapscott), 348
Williams, Anthony, 348
Williams, Chuck, 163

Wilson, Alan, 174
Wilson, Jill, 154
window system, 210
win strategy, 171
Wizard of Oz, The, 69
Wolfe, Tom, 350

Womack, Jim, 346
Work Like You are Showing Off (Calloway), 156
Wundram, Edward, 5, 15, 20, 121–124, 275

X
Xcelsi Group, LLC, 103

Y
Yaggy, Drew, 323
YouTube, 348

Z
Zahn, Jim, 190
zero dollar change order, 255
ZweigWhite Information Services, 52, 84